GLOBAL POWER KNOWLEDGE

Science and Technology in International Affairs

EDITED BY

John Krige and Kai-Henrik Barth

OSIRIS | 21

A Research Journal Devoted to the
History of Science and Its Cultural Influences

Osiris

Series editor, 2002–2012

KATHRYN OLESKO, *Georgetown University*

Volumes 17 to 27 in this series are designed to dissolve boundaries between history and the history of science. They cast science in the framework of larger issues prominent in the historical discipline but infrequently treated in the history of science, such as the development of civil society, urbanization, and the evolution of international affairs. They aim to open up new categories of analysis, to stimulate fresh areas of investigation, and to explore novel ways of synthesizing major historical problems that demand consideration of the role science has played in them. They are written not only for historians of science, but also for historians and other scholars who wish to integrate issues concerning science into courses on broader themes, as well as for readers interested in viewing science from a general historical perspective. Special attention is paid to the international dimensions of each volume's topic.

17 LYNN K. NYHART & THOMAS H. BROMAN, EDS., *Science and Civil Society*

18 SVEN DIERIG, JENS LACHMUND, & J. ANDREW MENDELSOHN, EDS., *Science and the City*

19 GREGG MITMAN, MICHELLE MURPHY, & CHRISTOPHER SELLERS, EDS., *Landscapes of Exposure: Knowledge and Illness in Modern Environments*

20 CAROLA SACHSE & MARK WALKER, EDS., *Politics and Science in Wartime: Comparative International Perspectives on the Kaiser Wilhelm Institute*

21 JOHN KRIGE & KAI-HENRIK BARTH, EDS., *Global Power Knowledge: Science and Technology in International Affairs*

Cover Illustration:

Geneva Conference of Experts, 1958. Scientists from both sides of the Iron Curtain negotiated technical requirements for the detection of nuclear explosions, laying the technical-political foundation for the Limited Nuclear Test Ban Treaty of 1963. Courtesy of the Hans Bethe Papers, Division of Rare and Manuscript Collections, Carl A. Kroch Library, Cornell University.

GLOBAL POWER KNOWLEDGE: SCIENCE AND TECHNOLOGY IN INTERNATIONAL AFFAIRS

JOHN KRIGE AND KAI-HENRIK BARTH: *Introduction: Science, Technology, and International Affairs: New Perspectives* 1

IN THE SHADOW OF THE SUPERPOWERS

GABRIELLE HECHT: *Negotiating Global Nuclearities: Apartheid, Decolonization, and the Cold War in the Making of the IAEA* 25

ITTY ABRAHAM: *The Ambivalence of Nuclear Histories* 49

RONALD E. DOEL AND KRISTINE C. HARPER: *Prometheus Unleashed: Science as a Diplomatic Weapon in the Lyndon B. Johnson Administration* 66

ALEXIS DE GREIFF: *The Politics of Noncooperation: The Boycott of the International Centre for Theoretical Physics* 86

STUART W. LESLIE AND ROBERT KARGON: *Exporting MIT: Science, Technology, and Nation-Building in India and Iran* 110

SCIENCE AND U.S. FOREIGN POLICY

CLARK A. MILLER: *"An Effective Instrument of Peace": Scientific Cooperation as an Instrument of U.S. Foreign Policy, 1938–1950* 133

JOHN KRIGE: *Atoms for Peace, Scientific Internationalism, and Scientific Intelligence* 161

KAI-HENRIK BARTH: *Catalysts of Change: Scientists as Transnational Arms Control Advocates in the 1980s* 182

SCIENCE, TECHNOLOGY, AND GLOBALIZATION

JACOB DARWIN HAMBLIN: *Hallowed Lords of the Sea: Scientific Authority and Radioactive Waste in the United States, Britain, and France* 209

PAUL N. EDWARDS: *Meteorology as Infrastructural Globalism* 229

JEAN-PAUL GAUDILLIÈRE: *Globalization and Regulation in the Biotech World: The Transatlantic Debates over Cancer Genes and Genetically Modified Crops* 251

SHEILA JASANOFF: *Biotechnology and Empire: The Global Power of Seeds and Science* 273

NOTES ON CONTRIBUTORS 293

INDEX 295

Introduction:
Science, Technology, and International Affairs

By John Krige and Kai-Henrik Barth[*]

Science and technology play a significant role in international affairs.[1] Many global environmental issues, in particular global climate change and ozone depletion, depend on science and technology.[2] Equally, many security concerns focus on the proliferation of weapons technologies and the scientific and technical knowledge associated with uranium enrichment, plutonium reprocessing, radiological sources, viruses, bacteria, and chemical agents.[3] Other international debates focus on genetically-modified organisms, intellectual property rights, the exploitation of oceans and marine resources, and the numerous threats to global health by infectious diseases such as AIDS, tuberculosis, SARS, and the recent H5N1 avian flu virus.[4] Finally, information technologies have also enabled a shift of economic power from states to markets and to international financial bodies such as the World Bank and the International Monetary Fund (IMF). Computerized financial markets, high-tech competence, and competition now also shape economic development and the interactions of states. The relevance of science and technology to the conduct of international affairs, to the relationships between states, to the changing balance of power between the nation-state

[*] John Krige, School of History, Technology and Society, Georgia Institute of Technology, Atlanta, GA 30332; john.krige@hts.gatech.edu. Kai-Henrik Barth, Security Studies Program, Edmund A. Walsh School of Foreign Service, Georgetown University, Washington, D.C. 20057; khb3@georgetown.edu.

We would like to thank Martin Collins for a close and constructive reading of a draft of this chapter.

[1] Even the traditionally science-skeptical State Department is rethinking how science and technology has, can, and possibly should interact with foreign policy and diplomacy. For a summary of the kinds of areas in which American diplomacy requires "good science," see Secretary of State Madeleine Albright's speech to the American Association for the Advancement of Science, Feb. 21, 2000, Washington, D.C., available online at http://secretary.state.gov/www/statements/2000/000221.html (accessed Nov. 21, 2005). See also National Research Council (U.S.) Committee on Science Technology and Health Aspects of the Foreign Policy Agenda of the United States, *The Pervasive Role of Science, Technology, and Health in Foreign Policy: Imperatives for the Department of State* (Washington, D.C., 1999). For a critical review of the State Department's perspective on science and technology in foreign policy, see Daniel S. Greenberg, *Science, Money, and Politics: Political Triumph and Ethical Erosion* (Chicago, 2001), 305–29.

[2] Spencer R. Weart, *The Discovery of Global Warming* (Cambridge, Mass., 2003); Clark A. Miller and Paul N. Edwards, *Changing the Atmosphere: Expert Knowledge and Environmental Governance* (Cambridge, Mass., 2001).

[3] Joseph Cirincione, Jon B. Wolfsthal, and Miriam Rajkumar, *Deadly Arsenals: Nuclear, Biological, and Chemical Threats,* 2nd ed. (Washington, D.C., 2005).

[4] Laurie Garrett, *The Coming Plague: Newly Emerging Diseases in a World Out of Balance* (New York, 1994); idem, *Betrayal of Trust: The Collapse of Global Public Health* (New York, 2000); idem, "The Next Pandemic?" *Foreign Affairs* 84 (July/Aug. 2005): 3–23.

 0369-7827/06/2006-0001$10.00

and regional and global groupings of states, not to say between the state itself and the market, is evident and expanding.[5]

The use of science and technology as an instrument of state power is not new: the rise of the nation-state, the colonial enterprise, and the consolidation of empire went along with the patronage of selected sciences, and the exploitation of local knowledges and techniques on the "periphery" to the benefit of the "metropolis."[6] World War II did not simply consolidate the links between science and the state to achieve specific practical objectives; it irreversibly embedded science at the heart of political processes. The theater of war displayed the immense contribution that science-based technologies, such as the atom bomb, proximity fuses, guided missiles, and radar, could make to national defense and that new drugs, such as penicillin, could make to national health.[7] The conflict also spawned entirely new fields such as operations research, which applied statistical methods to improve the efficiency of resource allocation in both military and industrial systems.[8] By 1945, many people were persuaded that scientists, even those in esoteric fields such as mathematics and theoretical physics, along with engineers, were an essential national and strategic asset to be coveted like any other "raw material" crucial to the health and wealth of nations, as well as to their military might and political stability.[9] The superpowers first, but all industrialized and industrializing nations thereafter, began to treat scientists and engineers as "manpower," an essential pool of skills and of knowledge that had to be quantified, constantly accumulated, and stockpiled. Beginning in the 1960s, and stimulated by the Organization for Economic Co-Operation and Development (OECD), use of the percentage of gross national product (GNP) devoted to research and development as a key indicator of national strength became common, along with the number of trained scientists and engineers available to make that R&D productive and compatible with the states' national and international agenda. The conviction that the governments in the western industrialized countries should implement formal policies that matched science with national priorities, a notion previously anathema to many scientists and laissez-faire economists, emerged at this time as well.[10] In the first fifteen years after the war, then, science and technology became an affair of the state, and reasons of state became central to their patronage and promotion.

It is useful to be reminded of the extent and character of federal support for science in the first decade of the cold war. Before the outbreak of hostilities, the federal government gave little support to scientific research—that was seen as the responsibility

[5] Eugene B. Skolnikoff, *The Elusive Transformation: Science, Technology, and the Evolution of International Politics* (Princeton, 1993).

[6] For a recent survey of the literature, see Roy MacLeod, ed., *Nature and Empire: Science and the Colonial Enterprise, Osiris,* 2nd ser., 15 (2000).

[7] Daniel J. Kevles, *The Physicists: The History of a Scientific Community in Modern America* (New York, 1978), describes some of these key developments.

[8] See, notably, Michael Fortun and Sylvan S. Schweber, "Scientists and the Legacy of World War II: The Case of Operations Research (OR)," *Social Studies of Science* 23 (Nov. 1993): 595–642.

[9] Vannevar Bush was, of course, an enthusiastic proponent of this idea: Bush, *Science, the Endless Frontier: A Report to the President* (Washington, D.C., 1945). On scientists as a reserve labor force, see Chandra Mukerji, *A Fragile Power: Scientists and the State* (Princeton, 1989).

[10] Jean-Jacques Salomon, "Science Policy Studies and the Development of Science Policy," in *Science, Technology, and Society: A Cross-Disciplinary Perspective,* ed. Ina Spiegel-Rösing and Derek de Solla Price (London, 1977), 43–70; Bruce L. R. Smith, *American Science Policy since World War II* (Washington, D.C., 1990).

of industry and the great private philanthropies, such as the Rockefeller Foundation and the Carnegie Corporation.[11] Federal support for R&D was around $75 million in 1940. Five years later it had soared to $1.5 billion. It had fallen back to about $1 billion by 1950 but had shot up to $3 billion by 1953 and doubled again by the late 1950s. Much of this expenditure (90 percent in 1950) was for research sponsored by the Department of Defense and the weapons-related Atomic Energy Commission (AEC).[12] Health was not ignored. The budget of the National Institutes of Health climbed from a mere $3 million in 1946 to $52 million by 1949 and to $71 million by 1953.[13]

This new configuration between science, technology, and the state was expressed not just at the local, national level but also at the international level. Internationalism in science, too, has a long history.[14] Its epistemic premise was that truth was universal, that by virtue of the "scientific method" local practices could produce "objective" knowledge—though exactly what that means or how that transformation is socially achieved has been an abiding concern of historians of science and technology for the past two decades.[15] This assumption was wedded to pragmatic concerns in some fields, the recognition that in sciences such as astronomy, meteorology, and oceanography certain problems transcended national borders and required a coordinated international effort. Prior to World War II, scientists themselves "informally" conducted much of that international effort, as individuals or through their national and international scientific societies. After the war (and sometimes building on foundations that had been prepared for a while),[16] the international bonds between scientists, and the collaborative practices that expressed them, provided a preestablished platform for the integration of science and scientists into foreign affairs. Now they were essential not only for the development and security of the nation but also in its dealings with other states, in its efforts to project and consolidate its power in the international domain and to build a stable world order.[17]

The postwar intersection between science, technology, and foreign policy has been extensively studied.[18] Some authors have focused on the role of scientific experts in

[11] Daniel J. Kevles, "Foundations, Universities, and Trends in Support for the Physical and Biological Sciences, 1900–1992," *Dædalus* 121 (Fall 1992): 195–235.

[12] Paul Forman, "Behind Quantum Electronics: National Security as Basis for Physical Research in the United States, 1940–1960," *Historical Studies in the Physical and Biological Sciences* 18 (1987): 149–229; Daniel J. Kevles, "Principles and Politics in Federal R&D Policy, 1945–1990: An Appreciation of the Bush Report," in *Science, the Endless Frontier: A Report to the President,* by Vannevar Bush, National Science Foundation [NSF] report 90-8 (Washington, D.C., 1990), ix–xxxiii (an introduction to the version of the report reprinted by the NSF on its fortieth anniversary).

[13] Nicolas Rasmussen, "The Mid-Century Biophysics Bubble: Hiroshima and the Biological Revolution in America, Revisited," *History of Science* 35 (1997): 245–93.

[14] Aant Elzinga and Catharina Landström, *Internationalism and Science* (London, 1996).

[15] For a short introduction to the issue as phrased in these terms, see Robert E. Kohler, *Landscapes and Labscapes: Exploring the Lab-Field Border in Biology* (Chicago, 2002), chap. 1.

[16] Clark A. Miller, "'An Effective Instrument of Peace': Scientific Cooperation as an Instrument of U.S. Foreign Policy, 1938–1950" (this volume).

[17] Joseph Manzione, "'Amusing and Amazing and Practical and Military': The Legacy of Scientific Internationalism in American Foreign Policy, 1945–1963," *Diplomatic History* 24 (Winter 2000): 21–55; Allison L. C. de Cerreño and Alex Keynan, *Scientific Cooperation, State Conflict: The Roles of Scientists in Mitigating International Discord,* Annals of the New York Academy of Sciences, vol. 866 (New York, 1998).

[18] For an early overview, see Brigitte Schroeder-Gudehus, "Science, Technology, and Foreign Policy," in Spiegel-Rösing and Price, *Science, Technology, and Society* (cit. n. 10), 473–506; Ronald E. Doel and Zuoyue Wang, "Science and Technology," in *Encyclopedia of American Foreign Policy,* ed. Alexander DeConde, Richard Dean Burns, and Fredrik Logevall, rev. ed. (New York, 2001), 443–59.

presidential decision-making,[19] the ways in which developments in science and technology have extended, or limited, foreign policy options, and the place of science and technology in the projection of American power abroad, and the defense of national interests on foreign soil.[20] Shifting the focus to other nations and regions, there has been extensive work done on the place of science and technology in the postwar reconstruction of Europe and the forging of supranational organizations and identities, as well as their implementation as symbols of modernity and to consolidate regional, if not global, power.[21]

Historians of science and technology have made significant contributions in these areas.[22] However, they have usually been reluctant to situate their studies in a broader historical context or to build bridges with other disciplines (and vice versa). Take the case of reconnaissance satellites. Historians have done excellent research in this domain.[23] Yet it was a political scientist who drew attention to the crucial role of reconnaissance from space in the Eisenhower administration's commitment to launching a scientific satellite in the International Geophysical Year (1957–1958).[24] It was a diplomatic historian who stressed the significance of reconnaissance satellites for monitoring compliance with arms control measures.[25] It is mostly political scientists who have analyzed nuclear test ban negotiations and who are now extremely active in studying the impact of chemical and biological weapons on international relations.[26] There is a similar imbalance if we broaden the scope of international affairs to include issues such as biotechnology and global health; or science, technology, and international economic development; or scientists and engineers as actors in international affairs.

This division of labor between the history of science and technology, on the one hand, and policy-related studies, broadly conceived, on the other, with its compart-

[19] Gregg Herken, *Cardinal Choices: Presidential Science Advising from the Atomic Bomb to SDI*, rev. and exp., Stanford Nuclear Age series (Stanford, 2000); Harold Karan Jacobson and Eric Stein, *Diplomats, Scientists, and Politicians: The United States and the Nuclear Test Ban Negotiations* (Ann Arbor, Mich., 1966); Robert Gilpin, *American Scientists and Nuclear Weapons Policy* (Princeton, 1962); Robert Gilpin and Christopher Wright, eds., *Scientists and National Policy-Making* (New York, 1964); Matthew Evangelista, *Unarmed Forces: The Transnational Movement to End the Cold War* (Ithaca, N.Y., 1999); Sheila Jasanoff, *The Fifth Branch: Science Advisers as Policymakers* (Cambridge, Mass., 1990).

[20] United States Congress, House Committee on International Relations, and Library of Congress, Congressional Research Service, *Science, Technology, and American Diplomacy: An Extended Study of the Interactions of Science and Technology with United States Foreign Policy,* 3 vols. (Washington, D.C., 1977).

[21] From the vast literature dealing with Europe, see John Krige, "The Politics of European Scientific Collaboration," in *Companion to Science in the Twentieth Century,* ed. John Krige and Dominique Pestre (London, 1997); Johan Schot, Thomas Misa, and Ruth Oldenziel, eds., *Tensions of Europe: The Role of Technology in the Making of Europe,* special issue, *History and Technology* 21 (March 2005); Margaret Sharpe and Claire Shearman, *European Technological Collaboration* (London, 1987).

[22] For a survey, see Ronald E. Doel, "Scientists as Policymakers, Advisors, and Intelligence Agents: Linking Contemporary Diplomatic History with the History of Contemporary Science," in *The Historiography of Contemporary Science and Technology,* ed. Thomas Söderqvist (Amsterdam, 1997), 215–44.

[23] David C. Arnold, *Spying from Space: Constructing America's Satellite Command Systems* (College Station, Texas, 2005); Dwayne A. Day, John M. Logsdon, and Brian Latel, eds., *Eye in the Sky: The Story of the Corona Spy Satellites* (Washington, D.C., 1998).

[24] Walter A. McDougall, *The Heavens and the Earth: A Political History of the Space Age* (Baltimore, 1997).

[25] John Lewis Gaddis, *The Long Peace: Inquiries into the History of the Cold War* (Oxford, 1987).

[26] But see Gerard James Fitzgerald, *From Prevention to Infection: Intramural Aerobiology, Biomedical Technology, and the Origins of Biological Warfare Research in the United States, 1910–1955* (Ph.D. diss., Carnegie Mellon Univ., 2003).

mentalization of research questions, is partly due to the disciplinary structure of academia. One of the aims of this volume is to dissolve this boundary and draw attention to the immense importance the history of science and technology has for our understanding of international affairs. Its ultimate aim is to build a bridge between historians of science and technology and diplomatic and economic historians, political scientists and policy analysts who study international affairs, to the advantage of both.[27]

Four major sites of contestation are used to organize the material in this collection: the nuclear, the postcolonial, patronage, and globalization. The changing relationship between the state and science and technology over the past fifty years is the leitmotif that connects them. In the early years of the cold war, and in the first phases of decolonization, independent nation-states were the main actors on the global stage. Scientific and technological capability (dominated by the lure of the nuclear) was coupled with demands for independence and modernization in a context of superpower rivalry. The articulation between science and foreign policy was expressed through bi- or multilateral agreements or in newly established intergovernmental organizations intended to harmonize multiple voices and manage diverse, even conflicting interests. Scientists, initially wary of being enrolled in the foreign entanglements of their governments, soon realized that the international arena provided them with additional resources while enhancing their scientific authority and social capital. Since the 1980s, the global role of the state, spurred by the computer and telecommunications revolutions and the end of the cold war, has been transformed—reduced at the expense of the market and civil society in the view of some, enhanced to the advantage of the single remaining superpower in the view of others.

The themes we have identified are not mutually exclusive and indeed are comingled in some of the articles in the collection. Keeping them separate is analytically useful, however. More significantly, while not pretending to be exhaustive, we believe that these themes will be central to any attempt to theorize the relationship between science, technology, and international affairs. They identify what is at stake in forging international arrangements and agreements concerning scientific and technological matters in the postwar period. They are therefore not simply valuable as a grid through which to read the papers presented here. They can also serve as elements for reflection when dealing with other case studies of the ongoing struggle between multiple actors to define the configuration of an international and global space in which technoscientific practices, scientists, and expertise are embedded.

THE CENTRALITY OF THE NUCLEAR

"One is nuclear or one is negligible." Thus spoke the French minister of defense in 1963, soon after his country had successfully tested its first nuclear device.[28] He expressed the fundamental centrality of the nuclear in the relationships between states during the cold war and, indeed, beyond. Hence our emphasis on nuclear and nuclear-related matters in this volume: more than half of the papers deal directly or indirectly

[27] Diplomatic and military historians have discussed commonalities and differences with political scientists in Colin Elman and Miriam Fendius Elman, eds., *Bridges and Boundaries: Historians, Political Scientists, and the Study of International Relations* (Cambridge, Mass., 2001).

[28] Pierre Mesmer, *Journal Officiel,* Jan. 24, 1963, 1613. For the significance of the nuclear in France, see Gabrielle Hecht, *The Radiance of France: Nuclear Power and National Identity after World War II* (Cambridge, Mass., 1998).

with them. The centrality of the nuclear to the postwar balance of power led the United States to try to maintain its atomic monopoly for as long as it could after 1945; led the Soviet Union, then Britain and France, to quickly acquire nuclear deterrents of their own; and precipitated the global proliferation of the nuclear we have seen over the past fifty years, its forms of expression depending on the economic resources, the scientific and technological skills, and the political wills of the governments concerned.[29] To have a nuclear or nuclear-related capability was to acquire the right to be heard with respect at the high table of international affairs and to have an ability to shape the physiognomy of the international world order, or at least to be considered an interlocutor in the negotiations which defined its changing profile.

The nuclear weapons arsenal was at once a fundamental prop of national security and an instrument of power on the global stage.[30] Parallel, or intertwined, civilian nuclear programs were useful, if expensive, risky, and contaminating sources of energy, as well as of radioisotopes for research, for medicine, and for agriculture. As Krige shows, for the superpowers the civil program also masked the reality of the mighty arsenals they had accumulated; Abraham and Hecht stress that, for countries throwing off the yoke of colonialism, such programs were essential tools for affirming national autonomy and defining national identity. The related infrastructure—the uranium-bearing ores, the trained scientists and engineers, the technologies of the reactors and the production plants, the radioactive wastes produced by the thousands of tons—all these were indifferent to the civil-military distinction that had to be maintained and reinforced by conventions backed with safeguards and inspections.

The successive reconfigurations of the international space framed by the nuclear were determined, first by the polarized relationship between the United States, along with its European allies, and the Soviet Union, and then by incremental additions which, even if deplored or denied, were deemed sufficiently containable not to threaten the fundamental parameters of the existing world order. With the collapse of the Soviet Union, the United States has emerged as the single most powerful nation on the globe, with the power, if it wishes to utilize it, to unilaterally sanction states whose nuclear aspirations it sees as posing a threat to international stability. The nuclear remains a central component in the discussion of science, technology, and international affairs, not simply because it is a major determinant of global power, but also because the threat of a nuclear attack still hangs over us.[31]

The dual-use possibilities of nuclear science and technology demand that negotiators wanting to assess the intentions of governments that have nominally civilian pro-

[29] Starting points include Richard Rhodes, *The Making of the Atomic Bomb* (1986, repr. New York, 1995); David Holloway, *Stalin and the Bomb* (New Haven, 1994); Margaret Gowing and Lorna Arnold, *Independence and Deterrence: Britain and Atomic Energy, 1945–1952,* 2 vols. (New York, 1974); André Bendjebbar, *Histoire secrète de la bombe atomique française* (Paris, 2000); George Perkovich, *India's Nuclear Bomb* (Berkeley, 2001); John Wilson Lewis and Xue Litai, *China Builds the Bomb* (Stanford, 1988); Avner Cohen, *Israel and the Bomb* (New York, 1998). For an overview, see McGeorge Bundy, *Danger and Survival: Choices about the Bomb in the First Fifty Years* (New York, 1988).

[30] Robert Jervis, *The Meaning of the Nuclear Revolution: Statecraft and the Prospect of Armageddon* (Ithaca, N.Y., 1989); Michael Mandelbaum, *The Nuclear Revolution: International Politics before and after Hiroshima* (New York, 1981).

[31] Since the end of the cold war, the nuclear threat has shifted to some extent from states to nonstate actors. Charles D. Ferguson, William C. Potter, with Amy Sands et al., *The Four Faces of Nuclear Terrorism* (Monterey, Calif., 2004); Graham Allison, *Nuclear Terrorism: The Ultimate Preventable Catastrophe* (New York, 2004).

grams, or deny having any intention of developing nuclear weapons, have to be thoroughly familiar with the technical issues involved. One can only seriously monitor the nuclear programs of others if one has access to the facilities concerned and can draw on a pool of scientific and engineering expertise. U.S. ambassador Robert L. Gallucci, a career diplomat who has spent most of his professional life negotiating nuclear nonproliferation agreements on behalf of successive American administrations, described the range of sources he can draw on in an interview conducted in the course of this book-project. As he put it, "The labs are prepared to provide it. You can go to DoE [Department of Energy] Washington, or . . . you make your own connections out at Livermore in the Z Division. You may have connections at Los Alamos or Sandia; depending, it might also be at Oak Ridge. It just depends on the circumstances. The Agency, of course, [the] CIA, has a whole group of scientists in this area who write on these issues." Turning to these experts, and being particularly attentive to their disagreements—for "that's how you can learn the most"—Gallucci "got comfortable enough to know how to use scientists and what to get from them."[32] In this domain, in which a state's "degree of nuclearity" (Hecht) maps directly on to its geopolitical status, the negotiated boundary between the civil and the military is a technopolitical one, and a fusion of scientific knowledge with foreign policy considerations is used to construct and to maintain a regime of control.[33]

The dissemination of nuclear technology abroad has always been controversial in the United States, even when a technology as benign as radioisotopes for medicine and research was concerned, let alone nuclear reactors. Krige's paper describes these two successive phases in the foreign distribution of these technologies, the first embarked upon in 1947, and the second triggered by President Eisenhower's spectacular speech to the United Nations General Assembly in December 1953.[34] Eisenhower's Atoms for Peace program was intended to share nuclear materials and technology with countries that sought them. On the assumption that fissionable material was in scarce supply—an assumption that rapidly proved to be mistaken—it was also supposed to serve as an instrument of nonproliferation by creating a uranium pool under the auspices of the International Atomic Energy Authority (IAEA). Source material drawn for this pool would help Western Europe build up its civil programs and divert scarce national resources in "developed" and "developing" counties into nuclear programs dedicated to "peaceful" and socially beneficial purposes. It would also provide an export market for American business, which was deeply concerned about the economics of nuclear power at home but willing to be encouraged to sell reactors abroad as part of a foreign policy effort. Atoms for Peace was also promoted as a way to seal

[32] Ambassador Robert L. Gallucci, dean of the Edmund A. Walsh School of Foreign Service, Georgetown University, interview with Kai-Henrik Barth, Georgetown Univ., Washington, D.C., Sept. 23, 2005. A copy of the interview is available on request from Barth. See also Joel S. Wit, Daniel Poneman, and Robert L. Gallucci, *Going Critical: The First North Korean Nuclear Crisis* (Washington, D.C., 2004).

[33] Thus Lawrence Scheinman, "Was Atoms for Peace a Mistake?" in *Atoms for Peace: An Analysis after Thirty Years,* ed. Joseph F. Pilat, Robert E. Pendley, and Charles K. Ebinger (Boulder, Colo., 1985), 197–203, remarks that the distinction between peaceful and military uses cannot be made technically except in terms of end-product; they can only be separated by political means (202).

[34] Angela N. H. Creager, "Tracing the Politics of Changing Postwar Research Practices: The Export of 'American' Radioisotopes to European Biologists," *Studies in the History and Philosophy of Biological and Biomedical Science* 33 (2002): 367–88; Martin J. Medhurst, "Atoms for Peace and Nuclear Hegemony: The Rhetorical Structure of a Cold War Campaign," *Armed Forces and Society* 23 (1997): 571–93.

alliances and win "hearts and minds" in the cold war struggle between the United States and the Soviet Union for influence in the many new states of the "third world" that emerged in the 1950s and 1960s.[35]

As it became clear in the late 1950s that uranium in one form or another was widely available, the international weight of traditional suppliers of the source material, like South Africa, was devalued. At the same time, the presence of the substance within their boundaries provided an opportunity for newly decolonized nations in Africa to demand a say in the framing of nuclear nonproliferation safeguards by the IAEA. They, too, wanted to capitalize on their degrees of nuclearity, albeit limited, to reconfigure in their favor the geopolitical world order from which they had been excluded for centuries. Hecht describes the tactics that the South African delegation used to defend its elevated position on the "barometer of nuclearity" and so retain significant influence in the international body. First, it insisted that the IAEA was primarily a technical agency and that, as an international organization, it should ignore the internal politics of its member states and the disagreements between them (including the violent opposition to apartheid). From this point of view, South Africa deserved to be influential, since not only was it a traditional supplier of source material but it also had an "advanced" nuclear program (pursued by its Atomic Energy Board expressly set up to give the country increased technopolitical weight and, eventually, a bomb).[36] While framing its strategy in these terms, the delegation could not ignore that membership on the IAEA Board of Governors was deeply influenced by the geopolitical distribution of power in the cold war and that seats were allocated on a combination of technical and regional considerations. Now embedding the technical in the political, South Africa claimed it was an African country with a shared history of rebellion against colonial rule with its neighbors to the north and so could legitimately speak for the continent in the IAEA's deliberations—seeking thereby to contain the influence of other countries in Africa and the Middle East that objected to South African policies. The contradiction could not be sustained indefinitely, and in 1977 Egypt was selected to represent Africa on the board of the IAEA.

The South African delegation (and some historians of the IAEA, too, Hecht reminds us) disparagingly dismissed the successful attempt to unseat them from the IAEA as signaling the politicization of the IAEA by what it called "the Afro-Asian upsurge." A few years before, that same "upsurge" had led to similar charges being leveled by many western politicians and intellectuals against UNESCO.[37] In the wake

[35] It continues to have this significance today. For example, Gallucci has explained how the United States offered the North Koreans light-water reactors built in South Korea, which "were knock-offs of American, Westinghouse reactors," to stop them from building gas-graphite reactors, which could be used for a weapons program. Gallucci interview (cit. n. 32).

[36] On the South African program, see Peter Liberman, "The Rise and Fall of the South African Bomb," *International Security* 26 (Fall 2001): 45–86; David Albright, "South Africa and the Affordable Bomb," *Bulletin of the Atomic Scientists* 50 (July/Aug. 1994): 37–47; Mitchell Reiss, *Bridled Ambition: Why Countries Constrain Their Nuclear Capabilities,* Woodrow Wilson Center Special Studies (Baltimore, 1995), 7–43; Helen E. Purkitt and Stephen F. Burgess, *South Africa's Weapons of Mass Destruction* (Bloomington, Ind., 2005), chap. 3. For a perspective from those involved in building the bomb, see Hannes Steyn, Richardt van der Walt, and Jan van Loggerenberg, *Armament and Disarmament: South Africa's Nuclear Weapons Experience* (Pretoria, 2003). For an official history of South Africa's early nuclear program (which, of course, doesn't mention nuclear weapons), see A. R. Newby-Fraser, *Chain Reaction: Twenty Years of Nuclear Research and Development in South Africa* (Pretoria, 1979).

[37] Clare Wells, *The UN, UNESCO, and the Politics of Knowledge* (London, 1987).

of the Arab-Israeli War, the 1974 General Conference passed a number of resolutions against Israel and South Africa, and the UN General Assembly invited Yasser Arafat to address it on behalf of the PLO (Palestine Liberation Organization). Some American and Israeli physicists and mathematicians were incensed. As de Greiff explains, not satisfied with simply condemning UNESCO, they launched a boycott of the International Centre for Theoretical Physics (ICTP) in Trieste, Italy. This institute, established in 1964 with the renowned Pakistani theoretical physicist Abdus Salam as its first director (Nobel Prize, 1979), was a mecca for international scientific exchange and a shrine at which physicists from the third world, in particular, could do research along with the best minds in the industrialized countries. In 1974, at the time of the boycott, the center was funded by the IAEA, by UNESCO, and by the Italian authorities. In de Greiff's view, it was boycotted because of its association with the international organization, because it was the only meaningful pressure point at which the boycotters could express their opposition to the third world–led anti-Israeli resolutions in that body (which Salam did not publicly condemn), and because refusing to teach or visit there would not harm their careers. Scientific life at the ICTP was eventually revived thanks to Salam's shrewd use of the boycott to raise more money from UNESCO and the IAEA and thanks to European (and some American) physicists who were lukewarm about the strategy, pointing out that anyway UNESCO had always been a site for political initiatives in the scientific arena.[38] What had changed, of course, was the balance of power in the organization and the ability of the "Afro-Asian upsurge" to impose its will on the minority western nations.

SHAPING THE POSTCOLONIAL WORLD

The determination of the major western democratic powers to maintain their nuclear monopoly, and to limit the proliferation of nuclear weapons as best they could (and then only to their allies), introduced a major imbalance into international relations. The negotiations inside the IAEA over a state's degree of nuclearity were also, as Hecht points out, sites for struggles over national sovereignty and international recognition. Such struggles mattered deeply to relatively marginal states such as South Africa, on whom she focuses her attention. They also mattered to states throwing off the yoke of colonialism. A concept of development promulgated by the West provided elites in some of these countries with a vision of the postcolonial order and of national identity that included science and technology at its core. Having a nuclear capability of some kind was at once a guarantee of international recognition, a symbol of modernity for leaders and their allies among national elites, a bargaining chip with which to affirm national autonomy and to protect national sovereignty and national political agendas, and potentially an invaluable addition to military strength.[39]

[38] Theoretical physicists Leon Van Hove in Europe and Victor Weisskopf in the United States were particularly emphatic about this point. Both were directors general of CERN, the European Organization for Nuclear Research in Geneva, and knew very well that an American-sponsored resolution in UNESCO in June 1950 had been crucial to the establishment of the regional laboratory. For the U.S. foreign policy concerns that inspired this intervention, see John Krige, "Isidor I. Rabi and CERN," *Physics in Perspective* 7 (2005): 150–64. See also idem, *American Hegemony and the Postwar Reconstruction of Science in Europe* (Cambridge, Mass., 2006), chap. 3.

[39] Political scientists have long debated why states seek nuclear weapons. For an overview of the debate, see Tanya Ogilvie-White, "Is There a Theory of Nuclear Proliferation? An Analysis of the Contemporary Debate," *Nonproliferation Review* (Fall 1996): 43–60. For one of the most influential

In this context, any nuclear program had an "ambivalent" character—a term which, as Abraham explains, is thick with meaning. The ambivalence lay both in the technology, which was dual-use, and in the attitudes of those responsible for charting the course of the program. These individuals contested the economic, technological, and political consequences of each quantum jump along the path of nuclearity. Thus Abraham insists that the Indian nuclear reactor program initiated in the 1950s, and the ensuing nuclear test explosion in 1974, should be decoupled from the acquisition of a nuclear weapons capability, which he situates in the late 1980s.[40] Jawaharlal Nehru's initial promotion of a nuclear India has to be understood as one of many gigantic, modernizing state-driven technological projects such as building hydroelectric dams, giant capital-intensive iron and steel plants, or new capital cities. All were symbols of the power, the legitimacy, and the autonomy of the post-colonial state being forged soon after independence. According to Abraham, the first nuclear test was just that, a scientific "experiment," and as such not necessarily the harbinger of an "inevitable" bomb.

Gallucci made a similar point regarding the supply of new light-water reactors to North Korea. Whereas many of his colleagues believed that North Korea was determined, in 1993, to acquire nuclear weapons, he was unconvinced. "I believe they were indeed prepared to give up their weapons program in order to get modern nuclear technology. In other words," Gallucci went on, "in the great debate, 'Would the North Koreans ever give up nuclear weapons?,' I think the answer is yes, if they thought they were getting the political relationship that they needed to assure themselves that we would not engage in regime change."[41] In a country with limited resources and crippled by a dictatorial regime, the need to modernize—and to win an agreement from the United States to be left alone—trumped the decision to acquire nuclear weapons in the early 1990s. In short, each phase of a nuclear program has to be historically contextualized, not read as a logical step on the path to an "inevitable" outcome, the acquisition of nuclear weapons. Indeed, Abraham insists that to believe that it does, as many western diplomats and scholars of international affairs tend to do, is actually to encourage proliferation, precisely the opposite of what is intended. Fetishizing the bomb endows it with a mystical attraction for those who do not have it.

The India of the 1960s is the focus of two other papers in the volume, those by Doel and Harper and by Leslie and Kargon. Both explore the role of experts other than nuclear scientist and engineers in the pursuit of American foreign policy objectives in the region. Doel and Harper also seek to draw scholarly attention to President Lyndon Johnson's deep personal commitment to the physical environmental sciences, especially oceanography and meteorology, which they trace back to his youth in an arid zone of Texas and his success in securing New Deal federal funds to dam the Colorado River for local power production. It is well known, and frequently deplored, that late in 1966 Johnson personally suspended grain shipments to India at a time of famine in an effort to force the ruling Congress party to embark on a program of agricultural reform (and probably also to punish the party for its criticisms of the war in Vietnam).

models of proliferation, see Scott D. Sagan, "Why Do States Build Nuclear Weapons? Three Models in Search of a Bomb," *International Security* 21 (Winter 1996/97): 54–86.

[40] See also Itty Abraham, *The Making of the Indian Atomic Bomb: Science, Secrecy, and the Postcolonial State* (London, 1998); Perkovich, *India's Nuclear Bomb* (cit. n. 29).

[41] Gallucci interview. See also Wit, Poneman, and Gallucci, *Going Critical.* (Both cit. n. 32.)

What Doel and Harper stress, however, is that, in parallel, Johnson also looked for prestigious scientific and technological projects with which to improve agricultural yield. These were intended to relieve human suffering by helping the country to be self-sufficient in food, to deflect the government's attention away from a nuclear weapons program, and to enhance India's political prestige in the region, which had been badly damaged by China's successful nuclear test in 1964. In particular, these authors describe the highly secretive attempt, undertaken behind closed doors in consultation with some leading Indian officials, to modify the weather in Bihar and Uttar Pradesh, using technologies of cloud seeding developed for the military at the Naval Ordnance Test Station at China Lake in California. This was a case of using swords as ploughshares: the original reason for the research was to increase precipitation and so clog with mud the Ho Chi Minh trail through Laos and Cambodia, along which North Vietnamese guerillas moved south. The experiment failed for lack of clouds. Even if it had been a success, it would probably have been abandoned for fear that the military origins of the project would be exposed by astute critics who connected increased rainfall in India with the same phenomenon in Laos. This revelation would have seriously embarrassed both the Indian and the U.S. authorities and incensed many American meteorologists and the world meteorological community.

Leslie and Kargon describe a quite different American initiative to move India (and Iran) along the development path. They tell how Gordon Brown, the dean of engineering at MIT (Massachusetts Institute of Technology), was firmly convinced that, by the late 1950s, he had developed a science-based engineering curriculum without equal that was embedded in, and responded to the needs of, the modern and modernizing industrial state.[42] Brown was aware of the success that his colleagues Max Millikan and Walt Rostow (who was a senior adviser to both Presidents Kennedy and Johnson) had had in turning modernization and "nation building" into foreign policy issues.[43] Believing firmly that MIT had blazed a path that developing countries could only follow to their advantage, MIT engineers first tried to recreate two institutions similar to the Cambridge school in India. One, in Kanpur, turned out to be an "isolated island of academic excellence," producing graduates more attuned to the challenges provided them in the United States than in their home country. The other, sponsored by industrialist G. D. Birla, was only reluctantly supported by MIT, whose staff felt that it was not academic enough. In the event the Birla Institute proved more attuned to the needs of the local context. The export of an MIT to Iran in the early 1970s proved to be a spectacular success, but not in the way Brown or the U.S. administration had foreseen. In 1979, the shah was forced into exile by an Islamic revolution led by the Ayatollah Khomeini from France. The Aryamehr University of Technology, established by imperial decree in 1965, was temporarily closed. It was later split by the revolutionary government into two new technical universities. Both were modeled on MIT, both were organized around research centers, and one of them (in Isfahan) was tightly linked to the needs of the Iranian defense sector. What Joseph S. Nye Jr., the former dean of Harvard's Kennedy School of Government, calls "soft power" is likely to be as ineffective and even as counterproductive as "hard," coercive power if it is not

[42] On the MIT that Brown deemed exemplary, see Stuart W. Leslie, *The Cold War and American Science: The Military-Industrial-Academic Complex at MIT and Stanford* (New York, 1993); idem, "Profit and Loss: The Military and MIT in the Postwar Era," *Hist. Stud. Phys. Biol. Sci.* 21 (1990): 59–85.

[43] Walt W. Rostow, *The Stages of Economic Growth: A Non-Communist Manifesto,* 3rd ed. (Cambridge, 1990).

exercised with due attention to the history, politics, and culture of the nation or region in which it is exercised.[44]

In the first two decades after World War II, the United States was determined to use scientific and technological collaboration as an instrument to contain nuclear proliferation, assisting in building the strength of its allies but also channeling their resources along nonnuclear paths as far as possible. By the 1970s, it was clear to the United States that it would have to accommodate itself to the determination of other countries to go nuclear. With that accepted, the effort to deflect other states away from a nuclear weapons capability could be trumped by other foreign policy concerns deemed more important at the time. For example, with Marxism on the march in all the surrounding territories, a stable white South Africa was deemed crucial to U.S. interests in the region, no matter how odious the apartheid regime had become and no matter what nuclear intentions the regime had. Thus the State Department was willing to blur intelligence suggesting that Pretoria was about to embark on a nuclear test in the Kalahari Desert.[45] The same thing happened with Pakistan. It also denied having a nuclear weapons program, even though, as Gallucci put it, "everybody knows that A. Q. Khan had worked in Almelo [the Netherlands] in the Urenco program, and not only brought out designs, but brought out the list of companies in Europe who produced and could produce the components of a gas centrifuge to the tolerances required by the engineering drawings, and that the Pakistanis, under A. Q. Khan's guidance, went about shopping, using various cut-outs and brass-plate companies, in order to get this shipped to Pakistan."[46] Even though American diplomats were convinced that Pakistani president Zia was lying about his nuclear program, they did not challenge him because, as Gallucci explains, "at that time we wanted very much to support the mujahideen in Afghanistan against the Soviets, and we needed Pakistan for that."[47]

[44] Joseph S. Nye Jr., "Soft Power," *Foreign Policy* 80 (Autumn 1990): 153–71. Nye writes: "Soft cooptive power is just as important as hard command power. If a state can make its power seem legitimate in the eyes of others, it will encounter less resistance to its wishes. If its culture and ideology are attractive, others will more willingly follow. If it can establish international norms consistent with its society, it is less likely to have to change. If it can support institutions that make other states wish to channel their activities in ways the dominant state prefers, it may be spared the costly exercise of coercive or hard power" (167). See also idem, *Soft Power: The Means to Success in World Politics* (New York, 2004).

[45] Gallucci provides an insight into the internal disputes over the interpretation of intelligence. He tells how, on one occasion, he was called over by the CIA to study pictures of a "hole in the Kalahari Desert" detected by Soviet intelligence, which suggested that South Africa was about to test a nuclear weapon. He was told by a senior official in the State Department that this intelligence "was inconvenient politically," and that Galluci had to "find another explanation." This did not mean that the CIA was expected "to twist the intelligence or anything," but the senior official wanted Gallucci "to go back and tell [the CIA] I know they think this is a test site, but I want them to use a little imagination here, because South Africans tell me it's not a test site. So I want you to go back and say, 'If it's not a test site, what else could it be?' " Gallucci and the CIA, very reluctantly, did as they were asked. "So we, in fact, produced a paper on what else could it be. Nobody believed any of it, but we did it." Galluci interview (cit. n. 32).

[46] Gallucci interview (cit. n. 32). On Pakistan's nuclear program see Samina Ahmed, "Pakistan's Nuclear Weapons Program: Turning Points and Nuclear Choices," *International Security* 23 (Spring 1999): 178–204. Urenco is a uranium enrichment consortium established in 1970 by Germany, the Netherlands, and the United Kingdom. The three countries sought advanced enrichment capabilities to guarantee a steady fuel supply for their nuclear power plants. On A. Q. Khan, see William J. Broad, David E. Sanger, and Raymond Bonner, "A Tale of Nuclear Proliferation: How Pakistani Built His Network," *New York Times,* Feb. 12, 2004, A1; David Albright and Corey Hinderstein, "Unraveling the A. Q. Khan and Future Proliferation Networks," *Washington Quarterly* 28 (Spring 2005): 111–28.

[47] Gallucci interview (cit. n. 32).

The rivalry between the United States and the Soviet Union in the cold war opened up a space that other states could exploit to take their own paths to development and to technological autonomy, including acquiring nuclear weapons. With one superpower now dismantled, the control of that space is being increasingly contested.

SCIENCE AND STATE PATRONAGE IN THE INTERNATIONAL DOMAIN

The ways in which patronage refashions scientific identities, promotes particular lines of research, and delimits the research questions addressed, transforms academic curricula, dissolves disciplinary boundaries, and leads to the creation of new forms of intellectual production embodied in "think tanks" and "summer studies" has been extensively studied by historians of science.[48] Particular attention has been given to the transformation in the material practice of strategic natural scientific fields through the pressures of the "military-industrial-academic" complex.[49] The enthusiastic engagement of scholars in the humanities and the social sciences in studies addressing the cold war concerns of the American administration have also been explored.[50] This militarization of the natural and social sciences during the cold war went along with the loss of scientific autonomy and independent, critical thinking. As Herbert Kelman, chair of the Doctoral Program in Social Psychology at the University of Michigan, put it: "Research that is tied to foreign policy or military operations is, of necessity, conceived within the framework of existing policy," and the status quo's "value assumptions are so much second nature to the members of the society that they perceive them as part of objective reality."[51] Scientific internationalism was also sacrificed on the altar of nationalism and patriotism. For Jean-Jacques Salomon, once science was supported, not as an adventure of the mind, but for the promises of its applications, the invocation of a universal, *internationale* of science that transcended narrow national interests was little more than ideology: now "the scientist's double loyalty, to science and to humanity, . . . had to conform to the common law of national loyalties."[52]

This model of scientific patronage is shaped, and limited, by its preoccupation with the place of science and scientists in the construction of the national security state, in

[48] On refashioning identities see Mario Biagioli, *Galileo Courtier: The Practice of Science in the Culture of Absolutism* (Chicago, 1993); David Kaiser, "The Postwar Suburbanization of American Physics," *American Quarterly* 56 (2004): 851–88. On disciplinary transformations, see the references in note 49. On one major think tank, see Martin J. Collins, *Cold War Laboratory: RAND, the Air Force, and the American State, 1945–1950* (Washington, D.C., 2002); David Hounshell, "The Cold War, RAND and the Generation of Knowledge," *Hist. Stud. Phys. Biol. Sci.* 27 (1997): 237–67.

[49] See Michael Aaron Dennis, "'Our First Line of Defense': Two University Laboratories in the Postwar American State," *Isis* 85 (1994): 427–55; Forman, "Behind Quantum Electronics" (cit. n. 12); Daniel J. Kevles, "Cold War and Hot Physics: Science, Security, and the American State, 1945–56," *Hist. Stud. Phys. Biol. Sci.* 20 (1990): 239–64; Leslie, *The Cold War and American Science* (cit. n. 42); Rebecca S. Lowen, "'Exploiting a Wonderful Opportunity': The Patronage of Scientific Research at Stanford University, 1937–1965," *Minerva* 30 (1992): 391–421; idem, *Creating the Cold War University: The Transformation of Stanford* (Berkeley, 1997); Kai-Henrik Barth, "The Politics of Seismology: Nuclear Testing, Arms Control, and the Transformation of a Discipline," *Soc. Stud. Sci.* 33 (Oct. 2003): 743–81.

[50] See, e.g., Allan Needell, "'Truth is Our Weapon': Project TROY, Political Warfare, and Government-Academic Relations in the National Security State," *Diplomatic History* 17 (1993): 399–420. See also Mark Solovey, "Project Camelot and the 1960s Epistemological Revolution: Rethinking the Politics-Patronage-Social Science Nexus," *Soc. Stud. Sci.* 31 (April 2001): 171–206.

[51] Kelman is quoted in Solovey, "Project Camelot and the 1960s Epistemological Revolution" (cit. n. 50), 189. See also Gilpin, *American Scientists* (cit. n. 19).

[52] Jean-Jacques Salomon, "The *Internationale* of Science," *Science Studies,* 1 (1971): 23–42, on 26.

which academia and academic laboratories became "our first line of defense" during the cold war.[53] It is too blunt an instrument to come to grips with the activities of scientists in the international arena. Undoubtedly, as Edwards makes clear in his paper on the globalization of meteorology, some scientists *feared* that state-sponsored international cooperation would rob them of their scientific autonomy and credibility, disrupt the functioning of international scientific collaboration, and force them to toe the political line of their governments. Undoubtedly, too, this did happen to some extent. But it must also be stressed that the enrollment of governments in regional and international scientific and technological ventures had many advantages for scientists. It provided an *additional* source of financial support for science, one that was sometimes supplementary to the distribution of resources on the national level.[54] It was also a *reliable* source of support, less vulnerable to the vagaries of domestic infighting over the science budget and its distribution between different fields. Once governments were enrolled in international ventures, they became locked into institutional structures from which it was extremely difficult to extract themselves for fear of international opprobrium. Intergovernmental organizations also provided the scientists involved with additional prestige and symbolic capital at home, which they could then use to lever further resources for their field at a national level. Finally, since international collaboration entails a *dilution* of national sovereignty, the international arena provided scientists with some scope to detach themselves from the pressures of uncompromising national loyalty and, contrary to what they might have intuitively feared, to chart an "independent" course that, while not orthogonal to their patron's interests, was not necessarily dominated by them either. The boundary between scientific autonomy and state loyalty is contested, particularly in the international arena, and all the more so in democratic societies.

The place of science in U.S. foreign relations is often traced back to the report prepared by a blue-ribbon panel chaired by Lloyd Berkner and delivered to the State Department in May 1950 under the title "Science and Foreign Relations."[55] Miller's contribution to this volume pushes back the date, drawing attention to the role which a little-known committee responsible for coordinating U.S. technical assistance programs in Latin America had in promoting the importance of science and technology for American foreign policy even before the war and, in a new guise, in the first five years after it. The leitmotif running through its programmatic statements saw science working along with U.S. foreign policy for the general betterment of "mankind." American science and technology, by contributing to economic growth and social stability, would promote "a free stable and prosperous democratic order."[56] For these policy makers, science was essentially an instrument of foreign policy used to project and to protect America's power abroad.

[53] Dennis, "'Our First Line of Defense'" (cit. n. 49).

[54] In some European countries, the budget for European scientific and technological collaboration often comes from the foreign office, not the science ministry.

[55] On the report, see Anonymous, "Science and Foreign Relations: Berkner Report to the U.S. Department of State," *Bulletin of the Atomic Scientists* 6 (June 1950): 293–8. See also Allan A. Needell, *Science, Cold War, and the American State: Lloyd V. Berkner and the Balance of Professional Ideals* (Amsterdam, 2000).

[56] These words are from Clark A. Miller, "Scientific Internationalism in American Foreign Policy: The Case of Meteorology, 1947–1958," in Miller and Edwards, *Changing the Atmosphere* (cit. n. 2), 167–217, on 173.

It was for the Berkner panel as well, of course. But faced by the closed world of Stalinist science, and fearful of the dangers that the Soviet Union posed to American security, in a classified annex the Berkner report also suggested that international scientific exchange and meetings could provide an excellent opportunity for informal intelligence gathering by scientists. In a climate of trust and mutual respect, they would be able to glean significant information about the research capabilities of their colleagues in other countries. To establish that trust required not simply that the scientists not be integrated into the formal structure of state surveillance, notably the Central Intelligence Agency (CIA). It also required that they be seen as independent, not simply as loyal servants of their government's policies.[57] In short, there were excellent *pragmatic* reasons for the state to leave scientists a good deal of leeway in international forums so as to protect their credibility in the eyes of their foreign peers and to enable them better to promote their own and their government's interests. This balance between autonomy and loyalty was recently stressed again by President Clinton's secretary of state Madeleine Albright. Speaking to the American Association for the Advancement of Science (AAAS) in February 2001, she insisted that the fundamental values and interests of U.S. diplomacy dovetailed perfectly with those of science. "The purpose of American diplomacy is to protect American interests," Albright said. But America today had "a stake in the stability of every part of the globe" and sought "to bring nations closer together around the basic principles of democracy, liberty and law." This provided a deep bond between U.S. diplomacy and science, "because the best science is driven by a similar impulse to improve not just the American condition but the human condition." More to the point here, Albright went on to stress that scientists could be better ambassadors than professional diplomats, for they had the authority and credibility that came from being "independent" authorities, while official representatives were often misunderstood, "resented by those who confuse leadership with hegemony and distrusted by those who, in our place, would use their power for more selfish ends."[58] Scientists could correct these misunderstandings among the international scientific community.

Krige uses Berkner's advocacy of informal intelligence gathering in his analysis of the famous scientific meeting held in Geneva in August 1955 as part of Eisenhower's Atoms for Peace program. He confirms that, from the point of view of foreign policy, the relative transparency provided by international activities, and the ready informal engagement of qualified scientists in them, was an invaluable point of entry into the scientific and technological programs of friend and foe alike, a "panoptical" instrument of evaluation, of surveillance, and of control.[59] Transparency also boosted

[57] The State Department and the CIA grappled with this issue. A scientist had "to be known internationally in order to have the entrée he needs for collecting information in a foreign country." At the same time, scientists were very wary of being associated with "spying" since they feared it would jeopardize their scientific careers. Hence the importance of *informal* intelligence gathering. See Wilton Lexow, "The Science Attaché Program," *Studies in Intelligence,* CIA Document CSI-2001-00018, published April 1, 1966, released July 30, 2001, available at http://www.cia.gov/csi/kent_csi/author-combine.htm (accessed on February 2, 2006).

[58] See Albright, AAAS speech (cit. n.1).

[59] A task force reporting to the State Department on the dangers of the proliferation of dual-use rocket technology in the mid-1960s made the point quite explicitly. Multilateral programs should be encouraged, it asserted, since "[i]n such a framework rocket programs tend to be more open, serve peaceful uses and are subject to international control and absorb manpower and financial resources

national programs. Once home from Geneva the physicists could play the card of rivalry to extract more resources from their governments to improve their national physics capabilities in the highly competitive cold war atmosphere of the time.

Two papers in the collection, by Hamblin and by Barth, describe how scientists working outside international structures have helped contain irresponsible behavior by nuclear weapons states. Hamblin's paper addresses the disputes in the late 1950s and early 1960s over the risks associated with dumping radioactive waste at sea. In this conflict, oceanographers and marine biologists, with the public support of charismatic figures such as the French explorer of the deep Jacques-Yves Cousteau, took on the formidable scientific establishment in the atomic energy agencies of Britain, France, and the United States. The oceanographers and their allies insisted that, without a better understanding of the nature and velocity of currents on the ocean floor, it was dangerous and irresponsible to use the sea as a graveyard for radioactive waste. This was seen by their opponents in the three AECs as little more than a crude attempt to pander to irrational public fears and to secure patronage for oceanography. To ward off the threat, the agencies developed a joint response to their critics, essentially agreeing to be more prudent about dumping and to enroll oceanographers and marine biologists in discussions and decisions about the disposal of radioactive waste at sea. The oceanographers also got a new international laboratory in Monaco financed by national governments through the IAEA, whose programmatic statement allowed them to do pretty much the kind of research they wanted. In short, in this case the critics of the atomic energy establishment saw the international agency as a new source of patronage to be exploited, and a previously complacent patron was forced to finance research that could expose the dangers of its existing policies of waste management.

Barth focuses on the relentless efforts made by some U.S. and Soviet scientists to install seismic detection devices near underground nuclear test sites in Kazakhstan and in Nevada. In the face of considerable hostility from senior officials in the administrations of both countries, but with the notable support of President Gorbachev who was personally engaged, these "transnational" actors (i.e., *nonstate* actors who *do not* operate on behalf of a government or international organization) succeeded in demonstrating that on-site monitoring by experts from a different, even rival, country was a politically possible and technically reliable arms control measure.[60] Scientific internationalism was here a potent, independent force for the imposition of checks on nuclear testing and was particularly effective because it was deployed outside the usual channels of foreign policy and interstate negotiations.

The paper by Hamblin and that by Gaudillière, who discusses disputes over the risks associated with the use of genetically modified organisms (GMOs), map strategies of patronage on to disciplinary distinctions. The disagreements over risk that they

---

that might otherwise be diverted to purely national programs. National rocket programs tend to concentrate on militarily significant solid and storable liquid fueled systems, are less open, and less responsive to international controls." Cited in John Krige, "Technology, Foreign Policy, and International Cooperation in Space," in *Critical Issues in the History of Spaceflight,* ed. Steven J. Dick and Roger D. Launius (Washington, D.C, 2006).

[60] This term is so defined by Thomas Risse-Kappen, "Bringing Transnational Relations Back in: Introduction," in *Bringing Transnational Relations Back in: Non-State Actors, Domestic Structures, and International Institutions,* ed. Thomas Risse-Kappen (Cambridge, 1995), 3–33, on 3.

describe, and about who can speak with the authority bestowed by "scientific independence" on matters of public concern, arise between researchers from different disciplinary backgrounds: oceanographers and marine biologists against physicists (including health physicists) in Hamblin's case; ecologists and agricultural scientists versus molecular biologists in Gaudillière's example. These confrontations are also a challenge to claims to legitimacy and scientific authority rooted in the appeal by physicists and molecular biologists to be doing "more fundamental" research. And they build on public fears and anxieties, of the nuclear or of the pollution of the food chain in these cases.[61] The more important, epistemological point is that, at the research frontier, policy-making is inevitably underdetermined by the empirical evidence, which is vague, contestable, and contested. Disputes between experts may be amplified by the agendas of specific interest groups, but they cannot be reduced to them: they are also a response to the messiness and complexity of the world.[62] Former U.S. secretary of state Madeleine Albright has said that "good science is vital to good diplomacy."[63] In practice, though, the place of science in diplomacy will be contested, especially since the uncertainties that go with doing "good science" will sometimes require an international posture that is tentative, flexible, and open to change, and so not necessarily congenial to diplomats.[64]

THE STATE AS ONE (DOMINANT?) ACTOR AMONG MANY IN A GLOBAL WORLD

With the emergence of globalization beginning in the 1980s and the formation of strong supranational, regional conglomerates (like the European Union), the notion of the state and its significance in the global world order is increasingly contested. While few will agree on exactly what constitutes "globalization," no one can dispute the importance of science- and technology-based factors, such as communication technologies, transport technologies, and pandemics as embodiments and expressions of globalization. The emergence of a technologically mediated global civil society, comprising transnational actors and institutions, mechanisms of consolidation and control, and an accompanying ideology of "a single world order," has challenged

[61] Europeans have been traumatized by the outbreak of "mad cow disease" and are now far more sensitive than many others about what they eat, although it must be said that the enthusiasm for "organic food" in the United States is rising sharply as well. On public perceptions of the nuclear, see Spencer R. Weart, *Nuclear Fear: A History of Images* (Cambridge, Mass., 1988). More generally, Mary Douglas and Aaron Wildavsky, *Risk and Danger: An Essay on the Selection of Technological and Environmental Dangers* (Berkeley, 1983).

[62] See Stephen Hilgartner, *Science on Stage: Expert Advice as Public Drama* (Stanford, 2000).

[63] Secretary of State Madeleine K. Albright, "Science and Diplomacy: Strengthening State for the 21st Century," *Policy Statement on Science & Technology and Diplomacy,* attachment to memorandum signed May 12, 2000, http://secretary.state.gov/www/statements/2000/000512b.html (accessed Nov. 21, 2005).

[64] Thus, speaking to the American Association for the Advancement of Science in February 2000, Albright claimed that "science does not support the 'Frankenfood' fears of some, particularly outside the United States, that biotech or other products will harm human health." Cited by Jasanoff, "Biotechnology and Empire: The Global Power of Seeds and Science" (this volume). The dismissal of the critics of GMOs (including some British and French scientists) as being irrational and impervious to the supposedly unambiguous dictates of "science" suggests that the secretary of state was here more interested in harnessing science to legitimate U.S. business interests than in framing policies that reflected the scientific uncertainties and public anxieties about the effects of genetic modification. See also Sheila Jasanoff, *Designs on Nature: Science and Democracy in Europe and the United States* (Princeton, 2005).

the foundations of the nation-state, and nationally based conceptions of democracy and identity.[65] How have scholars dealt with this challenge?

Consider, for example, Eugene Skolnikoff's *Science, Technology, and American Foreign Policy,* published in 1967, and his *The Elusive Transformation,* published in 1993.[66] The latter is one of the few books by a political scientist that analyzes the role of science and technology in international affairs on a systemic level. Have advances in science and technology changed the Westphalian system of nation-states? Have such advances reduced a nation's sovereignty in a significant way? Skolnikoff concluded that science and technology have changed international affairs in fundamental ways but have left the nation-state system essentially intact. The transformation of the international system postulated by many analysts, he argued, remained elusive. In other words, while the international system had evolved toward more interdependency, very few elements in international affairs had fundamentally changed. Most important, Skolnikoff pointed out, the idea of statehood was still strong despite growing limitations of a nation's autonomy in the global economy and the global security and global health context.

Other analysts have argued that the changes in science and technology, in particular in information technology, have shifted the power from states to international nongovernmental organizations. An oft-cited article by Jessica T. Mathews, then a senior fellow at the Council on Foreign Relations, argued that in fact the central role of the nation-state was decreasing, in large part as a consequence of science and technology.[67] She emphasized the relocation of power from governments toward nongovernmental organizations (NGOs), which often have the power and resources to set the agenda even for powerful governments. Among the areas in which NGOs had demonstrated their influence, she highlighted issues of border-crossing pollution, women's and human rights, and arms control. In contrast to Skolnikoff, she suggested that the nation-state was on its way out, largely driven by information technology and the associated rise of NGOs.

Diplomatic and economic historians sensitive to the immense military and economic power of the United States in the postwar era, and to the increasing willingness of America to use that power unilaterally in the absence of a superpower rival, see "the state" neither as an irreducible actor in international affairs nor as being bypassed by regional groupings or NGOs. For them one particular state, the United States, has built an "empire," a hegemonic regime sometimes by the use of force, more often by coopting a fraction of a national elite that shared America's conception of democracy and whose fortunes were tied to promoting it at home.[68] These notions merge into a view of globalization that sees the neoliberal political, economic, and cultural

[65] Ann M. Florini, ed., *The Third Force: The Rise of Transnational Civil Society* (Tokyo, 2000); Margaret E. Keck and Kathryn Sikkink, *Activists beyond Borders: Advocacy Networks in International Politics* (Ithaca, 1998).

[66] Eugene B. Skolnikoff, *Science, Technology, and American Foreign Policy* (Cambridge, Mass., 1967); idem, *The Elusive Transformation* (cit. n. 5).

[67] Jessica T. Mathews, "Power Shift," *Foreign Affairs* 76 (Jan./Feb. 1997): 50–66. Jessica Mathews has been the president of the Carnegie Endowment for International Peace since 1997. This is one of the most influential think tanks in Washington, D.C.

[68] John Lewis Gaddis, *We Now Know: Rethinking Cold War History* (Oxford, 1997); Geir Lundestad, *Empire by Integration: The United States and European Integration, 1945–1997* (Oxford University Press, 1998); Charles S. Maier, "An American Empire? The Problems of Frontiers and Peace in Twenty-First Century World Politics," *Harvard Magazine* 105 (Nov./Dec. 2002).

"model" of the United States gradually becoming universalized, backed by the strength of the country's democratic principles, its power to dominate markets, the lure of its consumer society, and behind it all, an immense military force. In this view, the decline of the nation-state, with its distinct national identity and national political agenda, is a symptom of the globalization of American hegemony, of the dominance of one state over the others.

Several papers in this volume address the significance of the changing role of the nation-state as a global actor. Barth's contribution draws attention to the ability of scientific NGOs in a democratic system to impose significant policy changes even in the highly sensitive area of security, in which one would expect the prerogatives of the nation-state to be untouchable. The U.S. scientists who worked along with their colleagues from the Soviet Academy of Sciences were supported by the National Resources Defense Council (NRDC), an NGO based in the United States that claimed to have more than 60,000 members in 1986 and an annual budget of some $7 million. By virtue of its resources, its youth and relative independence from the inertia of a large bureaucracy, and its ability to focus on key disarmament issues unencumbered by other foreign policy concerns and "interdepartmental rivalry," this NGO served as an agile and effective broker of nuclear test monitoring procedures, which would otherwise have been impossible at the time. The NRDC's success nicely illustrates Mathews's point that networked groups of transnational actors in a global world are sometimes "able to push around even the largest governments."[69]

The global ramifications of biotechnology are central to the papers by Gaudillière and Jasanoff. Gaudillière compares regulatory regimes for breast cancer genes and GMOs in France and the United States. He contrasts the *ancien régime* of what its practitioners called "molecular biology" (national, state-funded, legitimized by a rhetoric of basic science and weakly coupled to industry), with the situation that has emerged over the past two decades. Now genes are no longer regarded as "natural," and so discovered, but as laboratory artifacts, and so patentable, and subject to disputes over intellectual property in an increasing global market that intersects with different regimes of regulatory control. By comparing two different biotechnologies in two different countries, Gaudillière concludes that the usual way of characterizing transatlantic differences (elite scientists strongly tied to a centralized state in France, and a more democratic, but market-oriented, regulatory system in the United States), while a useful approximation, is too simple. Indeed, he shows that the structures of regulatory regimes are determined by complex alliances between the state, scientists, industry, and citizens groups and NGOs. The balance between them and the influence of each depend very much on local historical, political, cultural, and ideological constellations. For example, the ability of patients' groups to influence the regulation of tests for breast cancer in the United States and their irrelevance in France is partly a consequence of the very different forms the feminist movement has taken in the two countries. By contrast, the power and determination of French farmers to oppose GMOs, while their American counterparts have willingly accepted them, are deeply rooted in the passionate defense of the small farm in France, a passion with deep historical roots nourished by a hostility to mass production, Americanization, and globalization.

Jasanoff reads the complex tapestries of the debates over genetically modified

[69] Mathews, "Power Shift" (cit. n. 67), 53.

foods through the lens of empire. She structures her argument using a typology of five modes of imperial governance of agricultural biotechnology. Empires, as she stresses, are fragile, heterogeneous constructs, hybrid entities in which cohesion constantly eludes those who build them, and whose stability demands ongoing recourse to various modes of control, with science and technology as prime instruments of ordering and homogenization.[70] While she stresses that some forms of empire (e.g., the constitutional type embodied in the European Union) are more fluid and open to democratic change, she is also deeply disturbed by the collusion of the state and corporate interests in inventing oppressive forms of biopower in our neoliberal age. To destabilize this project, Jasanoff pleads for the development of new institutional arrangements that exploit global advances in technology to enable citizens to participate meaningfully in "political world-making beyond the nation state."

Jasanoff stresses the importance of technologies of standardization for the "legibility" (the term is Scott's) essential to any mode of governance.[71] Edwards's paper addresses the establishment of internationally agreed and enforceable standards for a global system of weather prediction. He describes the successive steps taken by meteorologists since the end of the nineteenth century to coordinate the collecting and sharing of data, from the informal, voluntary cooperation of national weather services in a nongovernmental framework to the formation of an intergovernmental body after World War II that increasingly managed to impose its standards of measurement on all participating states, thereby producing "a shared understanding of the world as a whole." This project was made possible by the building of "permanent, unified world-scale institutional-technological complexes," based on satellites and computers, in which the standards were embedded, and by making those standards quasi-obligatory through peer pressure and by technical assistance programs to developing nations.[72] As a result, weather prediction is no longer a prerogative of the nation-state acting alone but is in the hands of an intergovernmental body whose backbone is a vast technological system, a global network of power.[73]

CONCLUSION

The papers presented here provide an entrée to some of the major concepts and topics that scholars in the field of "science and technology studies" use to explore the intersection between science, technology, and international affairs. The bias in favor of the nuclear, and nuclear nonproliferation, in particular, is also a recognition that, even though the cold war is over, we are still haunted by the dangers of these weapons and deeply disturbed by their implications for personal and national security. That said, this collection also moves beyond these typical cold war concerns to embrace the geo-

[70] See also Shelia Jasanoff, ed., *States of Knowledge: The Co-production of Science and Social Order* (London, 2004).

[71] James C. Scott, *Seeing Like a State: How Certain Schemes to Improve the Human Condition Have Failed* (New Haven, 1998).

[72] For a recent private initiative to build "one (satellite-linked) world" in the telecommunications sector, see Martin J. Collins, "One World . . . One Telephone: Iridium, One Look at the Making of a Global Age," *History Technol.* 21 (Sept. 2005): 301–24.

[73] The allusions to Hughes are deliberate: Thomas P. Hughes, *Networks of Power: Electrification in Western Society, 1880–1930* (Baltimore, 1988); idem, "The Evolution of Large Technological Systems," in *The Social Construction of Technological Systems: New Directions in the Sociology and History of Technology,* ed. Wiebe E. Bijker, Thomas P. Hughes, and Trevor Pinch (Cambridge, Mass., 1987), 51–82.

physical sciences, notably meteorology, oceanography, and seismology, as well as biotechnology and biomedicine. We have given due weight as well to the struggles of states to assert themselves in a postcolonial order marked by major imbalances in economic and military power, itself built on national and regional disparities in the production of scientific and technological knowledge, and we have considered the implications of globalization for state power. Above all, though, we have tried to cross disciplinary and conceptual boundaries and to make the case that people working in the field of "science and technology studies" and scholars of international affairs can interact with one another to the mutual benefit of both. And perhaps to the benefit of building a less dangerous, more just, and more democratic world.

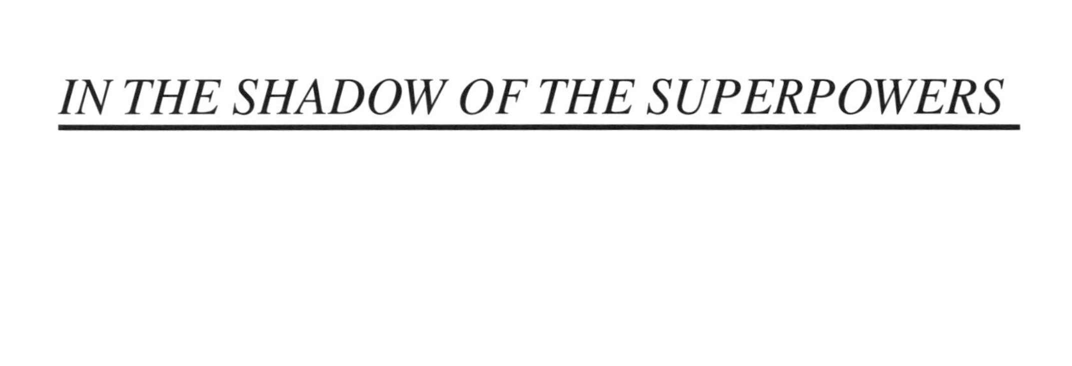
IN THE SHADOW OF THE SUPERPOWERS

Negotiating Global Nuclearities:
Apartheid, Decolonization, and the Cold War in the Making of the IAEA

*By Gabrielle Hecht**

ABSTRACT

Throughout most of its history, the International Atomic Energy Agency (IAEA) has been portrayed as a technical agency in which geopolitics are either extraneous or inappropriate. This chapter argues that this separation of technology and politics was discursive and never enacted in practice. Looking at the role of South Africa in the early history of the agency, this chapter shows that the IAEA's technopolitical regime was the continually contested outcome of negotiations between visions of a hierarchical, bipolar global order structured by cold war tensions and visions of a decentralized global order inspired by decolonization. This chapter also explores how dynamics between the apartheid state, decolonizing nations, and the United States inflected the meanings and implications of the "nuclear" in IAEA technopolitics. "Nuclearity"—that is, the degree to which a nation, a program, a policy, a technology, or even a material counted as "nuclear"—was a spectrum, not an on-off condition. Both nuclearity and its implications emerged in substantive ways from the dynamics between cold war and postcolonial visions of the world.

INTRODUCTION

After years of relative obscurity, the International Atomic Energy Agency (IAEA) has received a lot of public attention over the past decade. International political discourse—and journalistic coverage thereof—portray the IAEA as the most authoritative judge of a nation's ability to produce atomic weapons. In this discourse, the agency's impartiality is guaranteed by both the technical qualifications of its inspectors and the diversity of their national origins. The verdict of IAEA weapons inspectors determines whether the nation under scrutiny has become a "rogue state" or has remained within the bounds of acceptable technopolitical behavior. When the United States ignored IAEA findings about Iraq's nuclear weapons capabilities, international protest grounded itself in results of the IAEA inspections process to insist on the failure (or fabrication) of U.S. "intelligence" and the illegitimacy of its war efforts. At least symbolically, most political leaders treat the IAEA as the ultimate arbiter of global nuclearity, notwithstanding the agency's inability to temper

* Department of History, University of Michigan, 1029 Tisch Hall, Ann Arbor, MI 48109-1003; hechtg@umich.edu.
Thanks to Kai-Hendrik Barth, John Krige, and Paul Edwards for their helpful comments on earlier drafts.

bellicose behavior—be it that of the world's most powerful nation or of the world's worst pariahs.

A complex, historical explanation for the IAEA's failure to control the spread of nuclear weapons would require a volume unto itself, an endeavor attempted by some international relations experts.[1] Although they vary in their approach, these scholars all frame their work primarily in cold war terms. They agree that the IAEA's main mission at its inception was to transcend East-West divisions and (with the exception of the IAEA's official historian) focus the bulk of their attention on the agency's role in negotiating the terms of successive nonproliferation regimes. Some even distinguish between politics "intrinsic" to the IAEA's *raison d'être*—and therefore appropriate subjects of debate within the agency—and those that remained "extraneous," and hence inappropriate.[2] The distinctions in this analytic scheme appear straightforward. "Intrinsic" politics deal directly with "nuclear" matters, such as U.S.-Soviet negotiations over safeguards. "Extraneous" politics introduce "other" (presumably "non-nuclear") agendas—such as those that led to the expulsion of apartheid South Africa from the IAEA's board of governors in 1977 and from its General Conference in 1979.

This approach misses two key elements in the history of the agency. First, the IAEA's technopolitical regime[3]—the political programs, technical and scientific practices, and institutional ideology that constituted and governed its structure and development—was not merely a product of cold war politics. It was also a product of decolonization politics. More accurately, it was the continually contested outcome of negotiations between visions of a hierarchical, bipolar global order structured by cold war tensions and visions of a decentralized global order inspired by decolonization.[4] Second, the meanings and implications of nuclearity in IAEA technopolitics were neither universally agreed upon nor stable in time or space. Nuclearity—that is, the degree to which a nation, a program, a policy, a technology, or even a material counted as "nuclear"—was a spectrum, not an on-off condition. Both nuclearity and its impli-

[1] Lawrence Scheinman, *The International Atomic Energy Agency and World Nuclear Order* (Washington, D.C., 1987); David Fischer, *History of the International Atomic Energy Agency: The First Forty Years* (Vienna, 1997), offers an insider's history of the agency. The most sophisticated historical analysis is Astrid Forland's "Negotiating Supranational Rules: The Genesis of the International Atomic Energy Agency Safeguards System" (Ph.D. diss., Univ. of Bergen, 1997).

[2] Scheinman, *International Atomic Energy Agency and World Nuclear Order* (cit. n. 1), 210.

[3] Gabrielle Hecht, *The Radiance of France: Nuclear Power and National Identity after World War II,* Inside Technology series (Cambridge, Mass., 1998)

[4] Fischer, *History of the International Atomic Energy Agency;* and Forland, "Negotiating Supranational Rules" (both cit. n. 1), both acknowledge the role of "developing nations" in the history of the IAEA and take some elements of postcolonial geopolitics in the agency seriously. Fischer argues that the influence of the "Group of 77" (G-77, the so-called nonindustrialized countries) on agency policies grew stronger beginning in 1973. But the tone of his analysis suggests that this influence was an obstacle to be overcome, rather than its own legitimate force, and that accusations of safeguards as "neocolonial" practices were frivolous, rather than serious, expressions of an alternate vision of global technopolitics. He celebrates the IAEA's role in distributing nuclear technology to developing nations and describes them as "aid programs" without excavating the self-interest of donor nations, the geopolitical conflicts involved in distributing funds and technologies, or the utility of the programs themselves. Forland treats the positions expressed by some "third world" nations—especially India—with greater analytic seriousness, but not in much depth. Her analysis is strongest in showing that nations and institutions had very different ideas about safeguards. Rather than constituting the IAEA's *raison d'être* from the start, she argues, safeguards (as policy and practice) emerged in fits and starts from negotiations among these competing ideas. Still, she and Fischer both ultimately offer teleological narratives depicting the agency as moving inexorably toward more effective safeguards, in which political obstacles posed by India and other nations are steadily overcome on the road to "technical perfection" (Forland, 132).

cations emerged in substantive ways from the dynamics between cold war and postcolonial visions of the world.

The conclusion to this chapter suggests ways in which this double reframing of IAEA history can help to account for the stunning inadequacy of today's international nonproliferation regime. Before we can get there, though, we need to understand more about how cold war and postcolonial politics confronted each other in the agency and how this confrontation shaped negotiations over the substance and meanings of nuclearity. To achieve this, I have chosen a somewhat unorthodox route, via the South African archives.

South Africa's role in the early history of the International Atomic Energy Agency offers a unique lens on early struggles over the meanings of global nuclearity. In the late 1940s, South Africa had signed contracts with the Combined Development Agency (CDA) to supply thousands of tons of uranium for American and British weapons programs. South Africa's importance as a uranium producer led to its inclusion in the original eight-nation group that began drafting the IAEA's statute in 1954. South Africa subsequently secured a position on the agency's board of governors, thereby remaining among the most powerful second-tier nations in the organization until its expulsion in the late 1970s.

The apartheid nation's influence was controversial from the start. As new African nations joined the agency in the early 1960s, the opposition to South Africa's position initiated by Egypt, India, and (more discreetly) the Soviet Union gained momentum. Newly postcolonial countries construed South Africa's centrality in the world's only truly global nuclear institution as a major affront, an indication of the West's insincerity about decolonization and racial equality. South African delegates to the IAEA navigated these storms by triangulating between three points: (1) the relationship between technology and politics that the IAEA should properly define and enact; (2) the degree to which uranium production constituted a nation's level of nuclearity in the global hierarchy defined by the IAEA; and (3) the definition and testing of a dual national identity for South Africa as both "Western" and "African."

FRAMING GLOBAL NUCLEARITIES

The IAEA had its immediate origin in Eisenhower's famous Atoms for Peace initiative. Initial negotiations on a draft statute for the agency began in 1954 and involved eight nuclear weapons states and uranium suppliers in NATO and the Commonwealth: the United States, the United Kingdom, France, Canada, Australia, Belgium, Portugal, and South Africa. In 1956 the drafting group expanded to include the Soviet Union, Brazil, India, and Czechoslovakia. This group of twelve submitted a draft to the UN in September of that year, right around the decolonization of Morocco, Sudan, and Tunisia. In July 1957, the IAEA officially came into existence, and Ghana became independent. By the end of 1960, another eighteen nations had gained independence from colonial empires. Of these, five had—or would soon have—confirmed uranium reserves: the Central African Republic, Gabon, Madagascar, Niger, and Zaire.

The IAEA's board of governors would both reflect global nuclearities and shape their future, so its composition was hugely contentious. In order for the IAEA to have political credibility, both West and East needed adequate representation. In order for the agency to have technical credibility, expertise had to play an important role in selection. But these criteria made the Indian delegation anxious. Wouldn't the IAEA simply end up reproducing the global imbalances perpetrated by the twin processes of colonialism and industrialization?[5] India's solution gained easy acceptance: combine nuclear "advancement" with regional distribution in the selection criteria for board membership. The delineation of regions simultaneously reflected cold war tensions and anticipated decolonizations: North America, Latin America, Western Europe, Eastern Europe, Africa and the Middle East, South Asia, Southeast Asia and the Pacific, and the Far East. In 1956, the twenty-three seats on the board of governors were then defined as follows:

1. Five seats to the member states "most advanced in the technology of atomic energy *including the production of source materials*": the United States, the Soviet Union, the United Kingdom, France, and Canada.
2. Five seats to the "most advanced" states, again "*including the production of source materials,*" in regions other than those covered by the top five: South Africa, Brazil, Japan, India, and Australia.
3. Two rotating seats to uranium producers: Belgium and Portugal would alternate on one seat, and Czechoslovakia and Poland on the other.
4. One seat, designated by the board, to a supplier of technical assistance. This seat seemed likely to rotate among the four Scandinavian countries.
5. Ten seats elected by the IAEA's General Conference (which represented all members), "with due regard to equitable representation on the Board as a whole, of the members in the [eight] areas."[6]

The focus on "advancement" thus made the obsession with technological rankings that dominated cold war discourse a structural feature of the IAEA. But the regional framework—even as it sought to replace a narrative of inequality with one of expertise[7]—conjugated that advancement into a postcolonial tense.

What made a nation count as "most advanced in the technology of atomic energy including the production of source materials"? Indeed, what were "source materials," and to what degree did producing these constitute a significant manifestation of nuclearity?

These were high-stakes questions for South Africa. By 1957, the IAEA was the only international organization in which the increasingly ostracized nation could hope to have major influence. Concretely, South African delegates anticipated that the IAEA would play a critical role in shaping the emerging uranium market. Uranium production had saved many of South Africa's gold mines from collapse and would surely continue to be central to the nation's economy. A permanent seat on the IAEA Board, therefore, seemed vital to the growth of South Africa's uranium industry. Only by invoking technical expertise and imbrication in the West's nuclear systems could

[5] This concern resurfaced at the general conference on the statute. According to the leader of the South African delegation, Arab, Asian, and some Latin American delegations raised concerns parallel to those they were raising at UN General Assembly meetings, concerning the "'undemocratic' nature of the Statute, and the perpetuation of an elite of 'have' nations which would repeat the inequalities of the first industrial revolution." "Conference on the Statute of the International Atomic Energy Agency, First Progress Report, 20th September to 2nd October 1956," Oct. 4, 1956, BLO 349 ref. PS 17/109/3, vol. 2, National Archives of South Africa, Pretoria (hereafter cited as NASA).

[6] Fischer, *History of the International Atomic Energy Agency* (cit. n. 1), 39–40 (my italics).

[7] Beverly Sauer, "The Multi-Modal Character of Disciplinary Knowledge in South African Coal Mine Safety Training" (talk, Univ. of Michigan, Ann Arbor, Sept. 13, 2004).

South Africa hope to secure a spot, since mounting opposition to apartheid would make it impossible to get elected to a board seat.

Much hinged, therefore, on the nuclearity of "source materials." There were (at least) two forums in which this mattered. The first had to do with global hierarchies of nuclearity. If producing "source materials" was a full-fledged nuclear activity, then producers of those materials would sit further up the hierarchy. If the nuclearity of "source materials" was weak, however, then their producers couldn't count on much clout in the IAEA. The second forum had to do with the technopolitics of safeguards: control over the flow and use of nuclear materials and technologies. Again, if "source materials" were strongly nuclear, then they should be subject to strict safeguards; otherwise, controls could be more cursory. In this section, I will address the first forum by looking at the implications of uranium's shifting nuclearity for global hierarchies. The last section of the essay will take up the implications of the nuclearity of "source materials" for early safeguards discussions.

In the decade following World War II, the nuclearity of uranium seemed self-evident. It was constituted by a set of technopolitical assumptions and practices, which went something like this:

- Uranium was the only naturally occurring radioactive material that could (with considerable effort, granted) fuel nuclear weapons.
- Atomic bombs were a fundamentally new kind of weapon, capable of rupturing not only global order but the globe itself. They were (if you believed Western political rhetoric) single-handedly responsible for obviating the old colonial order in favor of a new nuclear order.[8]
- Uranium was a rare ore. If the West could monopolize its supply, it could keep the Communist ogre at bay and make the world safe for democracy. It was, therefore, imperative for the West to find and secure all sources of uranium around the world. Nothing mattered more than this (and certainly not the institution of apartheid on the southern tip of Africa).[9]
- Uranium's position at the core of Western nuclear systems made it imperative to proceed as secretly as possible. Geological surveys, actual and potential reserves, means of production, terms of sales contracts—state secrets one and all.[10]

And if uranium's nuclearity imposed secrecy, that secrecy in turn reinforced its nuclearity. Uranium was the only ore subject to legislation specifically targeted at ensuring the secrecy of its conditions of production.

But global agreement on the nuclearity of uranium would soon fray. By the mid-1950s, it had become clear that while pitchblende (very high-grade uranium) was rare, lower grades of uranium were not. In fact, an astonishing variety of geological conditions could lead to uranium ore—the stuff was everywhere. Meanwhile, the Soviets had tested an atomic bomb and clearly planned to continue doing so. They had found

[8] Gabrielle Hecht, "Rupture-Talk in the Nuclear Age: Conjugating Colonial Power in Africa," *Social Studies of Science* 32 (2002): 691–727.

[9] Thomas Borstelmann, *Apartheid's Reluctant Uncle: The United States and Southern Africa in the Early Cold War* (New York, 1993); Jonathan E. Helmreich, *Gathering Rare Ores: The Diplomacy of Uranium Acquisition, 1943–1954* (Princeton, 1986).

[10] Ibid.

their own sources. So a western monopoly of the "source material"—if defined simply as uranium in whatever raw state it occurred—would be impossible. The challenge lay not in finding uranium but in processing it to weapons-grade quality.

Just as the final discussions on the IAEA statute got underway, therefore, the nuclearity of uranium ore eroded. One tangible manifestation of this erosion was the fact that nations in Western and Eastern Europe whose only claim to nuclear fame lay in their uranium production would have to rotate seats on the IAEA Board. In a blatant attempt to increase its own standing in the hierarchy, India had already tried once to relegate South Africa and Australia to mere "producers" (rather than the "most advanced" in their regions), which would have forced those two nations to share a seat.[11] The United States and the United Kingdom helped their Southern Hemisphere allies resist these efforts, but all signs indicated that more challenges were to come. Anticipating these, the South African delegate pushed mightily, and successfully, to include the "production of source materials" as an indicator of nuclear "advancement."[12] Yet he also foresaw that uranium production wouldn't suffice to keep South Africa's status as "most advanced" in the region. In the long run, South Africa could not "sit back and rely upon [its] importance as a producer or . . . position in the West."[13] How, then, could South Africa ensure long-term influence in the agency? For its delegates, the answer lay in triangulating among three technopolitical positions:

1. advocating a maximal separation between technology and politics within the agency, and asserting that the IAEA should be pursuing technical, *not* political, agendas;
2. rehearsing the history of South Africa's imbrication in western nuclear networks, while simultaneously situating their nation firmly in Africa, thereby testing the national identity cultivated by their government;
3. emphasizing South Africa's non-uranium, nuclear accomplishments, which in turn involved lobbying Pretoria for the establishment of an atomic energy program.[14]

To get a sense of what these positions involved, let us first take a look at some general performances of the first two positions (the third emerged somewhat later). We'll then turn to a series of IAEA debates on nuclearity and "development," in which South African delegates deployed all three positions simultaneously.

The man who would become South Africa's primary representative and policy maker at the IAEA, Donald Sole, was a career diplomat. A native English speaker, he had taken a degree in history from Rhodes University in Grahamstown. Posted in London during World War II, he became the close colleague of several Afrikaner members of the foreign service. These friendships and his own mastery of Afrikaans (as well as a shortage of qualified Afrikaner diplomats) probably helped him survive the elections that brought the Nationalist Party to power in 1948. Sole continued to

[11] "International Atomic Energy Agency," Annex to South Africa Minute No. 79/2, 28/7/56, 4. BLO 349 ref. PS 17/109/3, vol. 2, NASA.

[12] Donald B. Sole, "'This Above All': Reminiscences of a South African Diplomat" (unpublished memoirs, Cape Town, n.d.).

[13] "International Atomic Energy Agency," Annex to South Africa Minute no. 79/2 (cit. n. 11), 10–1.

[14] "Statement by Mr. W. C. du Plessis, Leader of the South African Delegation, on Oct. 1, 1956," 2, BLO 349 ref. PS 17/109/3, vol. 2, NASA.

rise in the diplomatic ranks, representing South Africa at the UN before getting selected for its delegation to the IAEA drafting group. Sole's unpublished memoirs suggest that he deplored (without actively opposing) the increasing racism of apartheid, not the least because the resulting isolation of his nation made his job as a diplomat increasingly difficult. Although his loyalty to the government never wavered, he frequently struggled to find ways to make South Africa more palatable internationally and sometimes tried to temper Pretoria's more excessive isolationist practices.

Early on, Sole decided that the best way to navigate international opposition to apartheid in the IAEA was to depoliticize the terms of negotiations by focusing on technical issues. Although a common strategy in nuclear politics around the world, this approach had special resonance in South Africa, where the apartheid government justified a range of policies by insisting that they emerged from administrative, business, or technical considerations—not political ones. Sole's opening speech at the first IAEA general conference in 1957 declared that "this Agency should be primarily a technical Agency. Political considerations, of course, cannot and should not be excluded from the functioning of an organization of this character, but they should be allowed the minimum possible influence . . . If they are permitted to get out of hand, this Agency can all too easily develop into a propaganda forum—into yet another platform for the wasteful and time-consuming exposition of conflicting ideologies."[15] For most delegates from other nations, this statement appeared to refer to East-West tensions. So who could possibly disagree? For Sole, meanwhile, insisting on the technical functions of the IAEA would lay the foundation for arguing that objections to apartheid were purely political and hence not appropriate subjects for IAEA consideration and debate.

Similarly, Sole fought to keep the IAEA autonomous from the UN. At the 1956 statute conference, Russian, Indian, and Czech delegates had tried to place the IAEA under direct UN control, probably anticipating that the growing influx of postcolonial nations would favor them. Sole "strongly resisted this development" but only got anemic support from the United States and the United Kingdom, which had much less at stake. To rally others to their cause, South African delegates played on fears about the dilution of national sovereignty[16] and "stressed the dangers inherent in the possibility of the Agency being called upon to give reports on the action of members. This, we pointed out, might open the way to United Nations interference in the domestic affairs of individual members." After intense lobbying, Sole won a compromise—namely that only UN actions "in accordance with the [IAEA] statute" could be considered by the agency. In other words, only UN resolutions pertaining to nuclear matters would be relevant to the IAEA.

For most members, this seemed reasonable enough: after all, the IAEA's *raison d'être* surely revolved around managing global nuclearity—not global affairs more generally. Not so for India, which was poised to become the leader of postcolonial geopolitics and the nonaligned movement and had the most to gain from close relations

[15] "Statement Delivered by Mr. D. B. Sole, Leader of the South African Delegation at the Opening of the General Debate of the First General Conference of the International Atomic Energy Agency," Oct. 7, 1957, 3–4, BLO 349 ref. PS 17/109/3, vol. 2, NASA.

[16] As Forland shows in "Negotiating Supranational Rules" (cit. n. 1), the infringement of IAEA actions on national sovereignty was a persistent theme in agency discussions, one raised most often by India and a few other nonwestern nations.

with the UN. The Indian delegate wasn't fooled for a second, and he objected that Sole's amendment was "too restrictive." According to Sole, the Canadian representative "thereupon enquired whether the Group should assume from the Indian representative's reasoning that he wished the United Nations to go 'fishing' in the affairs of member states. The latter, in some confusion, then withdrew his objection." Sole wasn't fooled either, of course. He understood only too well India's "success in giving the impression of speaking on behalf of the new African countries" and feared that "the Afro/Asian alignment may be perpetuated under Indian leadership in the Agency itself."[17] His report to Pretoria concluded that "we should take careful note of the struggle which led to the inclusion of those five words '*in accordance with this Statute.*' This proves conclusively that they were specifically included (unanimously) to prevent such 'fishing expeditions.'"[18]

Sole thus managed to obtain some sort of statutory separation between technology and politics. Of course, the edges of nuclearity remained nebulous. He knew that quarrels about what counted as appropriate actions for the IAEA would continue to erupt. He had succeeded, however, in defining some boundaries for such discussions and would not hesitate to invoke these in the years to come.

Meanwhile, tensions over South Africa's presence in the IAEA had taught Sole that he would have to carefully manage his country's national and geopolitical identity. A discursive strategy with deep roots in the history of the South African state readily presented itself: for decades, Afrikaners had styled themselves simultaneously western and African. Set against the rise of black African nationalism, this approach might seem grotesque. Yet the duality of identity discourse only strengthened during the cold war/apartheid period. On the one hand, apartheid leaders saw theirs as a western nation, qualifying as such because of political heritage, ancestry, corporate connections, and sophisticated industrial infrastructure—not to mention white rule and fervent anticommunism. In this, they were supported by the still-segregated, even more rabidly anti-communist United States.[19] On the other hand, Afrikaner nationalism insisted on its own African-ness: Afrikaners had cut the tie to Europe, pioneered the land, suffered mightily at the hands of British colonialism, and made South Africa their God-given homeland. The National Party's 1948 victory was the culmination of centuries of struggle and offered a unique opportunity to forge the most "advanced" African nation in the history of the world—so advanced, indeed, that it would barely count as African at all. Western-African dualities had woven the institutional and cultural fabric of apartheid; they were central to its logic.[20]

South African diplomats used the IAEA as a laboratory for testing the robustness, credibility, and function of this discursive duality. From the beginning the duality had a specifically nuclear manifestation: South Africa both produced uranium for western nuclear systems (receiving special technical and financial favors as a result) and qualified as the "most advanced" nuclear nation in Africa. But Sole and his colleagues understood that duality would have to be relentlessly performed to maintain its power.

Accordingly, having secured the opening slot in the IAEA's first general debate in October 1957, Sole seized the opportunity to remind the assembly of South Africa's

[17] "Final Report on the Conference on the Statute of the IAEA, 20th September to 26th October 1956," Nov. 6, 1956, BLO 349 ref. PS 17/109/3, vol. 2, NASA.
[18] "International Atomic Energy Agency," Annex to South Africa Minute no. 79/2 (cit. n. 11).
[19] Borstelmann, *Apartheid's Reluctant Uncle* (cit. n. 9).
[20] Hermann Giliomee, *The Afrikaners: Biography of a People* (Cape Town, 2003).

contribution to western nuclear development while simultaneously asserting his nation's African-ness. In particular, he described a 1957 radioisotopes conference held in Pretoria that hosted experts not just from colonial territories but also from the Central African Federation, Ghana, the United Nations, and the World Health Organization.[21] Listing these participants obliquely addressed the question on everyone's mind: Could nonwhite scientists visit or work in South Africa? Sole finessed the point: "We shall do what we can to assist others who may be less fortunately placed than we are . . . In fact we have always been ready to grant research facilities to guest workers from other countries . . . Just before I left South Africa in July, we were happy to welcome two scientists from Japan who were given every facility both in Government Institutes and in the private Laboratories of our uranium industry."[22] Sole thereby implied that foreign nonwhite researchers would receive courteous treatment in South Africa. He glossed over one little-known fact: namely, that Japan's significance as a trading partner had led the government to grant Japanese visitors "honorary white" status.[23] No records survive of how Pretoria handled the visit of Ghanaian and Central African scientists.

As the activities of the IAEA got under way in earnest, Sole would come back to these two themes (technology/politics; western/African) again and again. He would also concern himself ever more intently with the strength of South Africa's own nuclearity. Triangulating among these three points would steer him through one of the most contentious issues in the agency's early history: the role that the IAEA should play in the emerging practices of international "development."

NUCLEARITY AND "DEVELOPMENT"

Debates about the relationship between nuclearity and "development" revealed and reinforced a deep split between the West and the Rest over the *raison d'être* of the IAEA. At least rhetorically, everyone agreed that promoting the "peaceful uses of atomic energy" was the IAEA's central mission. With tiresome predictability, official speeches and press releases the world over endowed this mission with a high moral purpose. In practice, however, the world's nuclear leaders—be they uranium producers or system builders—had more mundane interests in mind: building reactors, making nuclear power commercially viable, and creating a market for technologies and uranium that would sustain their own nuclear industries.[24] Hopes for a global military and economic panopticon embedded in Atoms for Peace[25] thus found even wider expression in the IAEA. The West would sell to the Rest, while somehow (the details remained fuzzy) ensuring that ensuing programs would serve civilian, rather than military, ends. By providing a bigger market, the Rest would help the West commercialize its own nuclear power systems. To work best for the West, the IAEA should channel its resources to nations that could develop nuclear infrastructures quickly, rather than function as yet another international development agency. Perhaps inspired by Sole's

[21] "Statement Delivered by Mr. D. B. Sole" (cit. n. 15), 3.

[22] Ibid.

[23] Jun Morikawa, *Japan and Africa: Big Business and Diplomacy* (Johannesburg, 1997).

[24] In "Negotiating Supranational Rules" (cit. n. 1), Forland sketches out how commercial considerations shaped the ways that the United Kingdom and Canada, in particular, approached safeguards.

[25] See John Krige, "Atoms for Peace, Scientific Internationalism, and Scientific Intelligence" (this volume).

eloquence, many western delegates justified this perspective by insisting that the IAEA should focus on "technical" tasks, augment the world's overall nuclear capacity and expertise, and leave "politics" to the UN. But the Rest had a different view. From the beginning, India had led newly independent, "developing" nations in arguing that the IAEA should ensure that emerging nuclear hierarchies not perpetuate global inequalities. They concurred that the agency should spread nuclear systems but (for equally self-interested reasons) argued for a broader distribution of resources. On no account should politics be left out of discussions.

So, how did these differences manifest themselves in policy negotiations?

The United Kingdom argued that the IAEA should only fund economically viable reactor projects. Of course, this approach would "necessarily limit the construction of power projects to the most advanced countries and the 'have nots' would argue that instead of assisting the underdeveloped areas of the world, the major Agency activity in its earlier years would be concentrated on strengthening countries which already possess comparatively highly developed industrial complexes."[26] To counter this objection, the U.S. Atomic Energy Commission (AEC) proposed that the agency sponsor three "demonstration reactors" to test the viability of nuclear power in different geographic (read: geopolitical) settings. The AEC offered to provide substantial (if unspecified) funding for these provided that the reactors ran on enriched uranium. Conveniently enough, in the late 1950s only the United States designed commercially oriented enriched uranium reactors. Bypassing Soviet, French, and Canadian natural uranium designs, this seemingly generous American offer therefore sought to make U.S. technology the basis for worldwide nuclear development. The Canadians, meanwhile, argued that both the United States and the United Kingdom were jumping the gun. They instead advocated education programs as the first step: "since the provision of adequate training for atomic scientists and atomic engineers is an essential preliminary to the development of a large-scale market in uranium after 1965, the major producers have a direct interest in stimulating the provision for such training whenever possible."[27]

As a uranium producer, South Africa had similar stakes in the IAEA's development policy. South Africans hoped the agency would provide a means of tracking competing uranium producers and of establishing a market price for uranium, which they wanted to influence.[28] They also hoped the IAEA might serve as a marketplace for customers and suppliers to have direct contact. None of these prospects materialized in the first two years, leading Sole to fear in 1959 that South Africa had "considerably over-rated [the IAEA's] importance so far as the immediate future of our uranium industry is concerned."[29] As nuclear programs gained momentum, however, Sole's discreet marketing efforts began to pay off.[30] By 1961, both France and Israel had ap-

[26] "Atomic Energy Agency: Planning of the Operational Programme for the First Few Years," Feb. 1, 1957, BLO 349 ref. PS 17/109/3, vol. 2, NASA.

[27] Ibid.

[28] "First General Conference of the IAEA: Final Report," Nov. 5, 1957, BLO 349 ref. PS 17/109/3, vol. 2, NASA.

[29] "Report of the South African Delegation on the Third General Conference of the IAEA," Oct. 8, 1959, BLO 349 ref. PS 17/109/3, vol. 2, NASA.

[30] In addition to his activities at the IAEA, Sole joined two South African scientists on a nuclear tour of Europe in the late 1950s. The group visited nuclear research and development facilities throughout the continent, in a thinly veiled effort to peddle South African uranium. This tour is described in his unpublished memoirs: Sole, "'This Above All'" (cit. n. 12).

proached Sole about buying South African uranium, and there was good reason to think that other customers would soon follow.[31]

South Africa thus had economic as well as political reasons to advocate a narrow promotional agenda for the IAEA. Sole found the idea of a demonstration reactor seductive and asked his government "whether a bid should be made for the erection of one of the demonstration reactors in Africa, although not necessarily in the Union in view of the difficulties which would arise in accommodation of nonwhites from other parts of the African region."[32] His superiors in Pretoria ignored his question, however. Instead, they specifically instructed him to "apply the brakes" to broadly based initiatives for educating budding scientists in developing nations. But Sole understood the delicate situations produced by South Africa's declining international stature better than Pretoria. He urged caution. "Rather than return a blank negative to any and every attempt on the part of the under-developed countries to secure help,"[33] South Africa should back a few, select proposals. For example, supporting an IAEA fellowship program would constitute "an important gesture vis-à-vis the under-developed countries"[34] that didn't cost too much. Though South Africa shouldn't expect overt thanks from such countries in return, this kind of gesture mattered for retaining the goodwill of nations such as the United States and Canada. The West was well disposed toward South Africa for many reasons but shouldn't be forced into awkward standoffs over such issues.

Budget constraints made questions about resource allocation particularly contentious. For example, in 1959 the agency was considering sponsorship of two possible conferences: one on radioisotopes, and another on small- to medium-size power reactors. The South African secretary for external affairs—who by then seemed to have a better understanding of nuclear systems—instructed Sole to promote the latter topic. "From the Union's point of view, any development which will encourage the increased consumption of atomic fuel and thus help to narrow the widening gap between production and consumption, would of course be of considerable value."[35] In subsequent years, the South African delegation continued to press for concrete steps toward commercial-scale nuclear development, repeatedly asserting that "the development of nuclear power is the main eventual goal in the Agency's work."[36] In this, Sole continued to privilege the technical over the political. The IAEA, he said, "shall attract good scientists only if the Agency has *scientific prestige* and scientific prestige is unfortunately *not* built up on programs of technical assistance."[37] South Africa's

[31] "Mr. Sole's Attendance at IAEA Board Meetings," Dec. 2, 1961, TES 1004 ref. F5/362/5, NASA.

[32] "Atomic Energy Agency: Planning of the Operational Programme" (cit. n. 26).

[33] "International Atomic Energy Agency: Comments on the Report by the Executive Secretary to the Second Session of the Working Group—Document—IAEA/PC/WG.2(S)," n.d. [ca. April 10, 1957], 1–2, BLO 349 ref. PS 17/109/3, vol. 2, NASA.

[34] Ibid., 3–4.

[35] "Meeting of the Board of Governors of IAEA: June, 1959," June 12, 1959, BVV83 13/1 ref., vol. 6, NASA. See also "Report of the S.A. Delegation on the Third General Conference of the IAEA," Jan. 12, 1960, BVV83 13/1 ref., vol. 6, NASA.

[36] "Seventh Regular Session of the General Conference of the International Atomic Energy Agency: Report of the South African Delegation—Part I," Oct. 7, 1963, BVV84 13/1, vol. 7 and annex, NASA. See also "Address Delivered by Mr. D. B. Sole, General Conference," Sept. 19, 1962, BVV84 13/1, vol. 7 and annex, NASA.

[37] "IAEA General Conference 1959: Statement Delivered in the General Debate by the Leader of the South African Delegation, Mr. D. B. Sole, on 23rd September 1959," BLO 349 ref. PS 17/109/3, vol. 2, NASA.

technopolitical strategy thus combined a push to expand uranium demand with an aversion to providing development funds to postcolonial states.[38]

The next step in this logic suggested that a nation's legitimacy within the IAEA rested on its nuclearity: the more nuclear a nation, the louder its voice should be. Of course, the converse applied as well. Sole accordingly expressed disgust at the 1959 general debate:

> Once again very few of the under-developed countries participated in the debate . . . [T]he Agency is of real value to them only in the fields of technical assistance, including fellowships, and the distribution of scientific and technical information. Even in these fields, so far as many of them are concerned, their interests are limited. Nevertheless, many of [these] countries . . . were responsible for transforming one of the issues before the conference—assistance to less developed countries with the production of nuclear power—into a major debate. They have clearly been persuaded by the results of studies and research during the past year that the introduction of small and medium power reactors into some of the less developed countries will be feasible in the fairly near future and the Agency will be under considerable pressure in the coming months to accelerate its program in this field.[39]

Was Sole truly oblivious to the irony that the very conference he'd supported on small and medium reactors had made developing nations think they were feasible? The logic of apartheid had sent him into a tailspin: only technical assistance programs could help developing nations acquire the infrastructure and expertise that would legitimate (according to his reasoning) their participation in the debate.

As a first attempt to resolve the question of what the IAEA should actually do, the board of governors sought a middle road between strengthening existing programs and addressing global inequalities. It decided to sponsor regional training centers in parts of the world already somewhat—but not overly—"advanced" in nuclear research and development. Of course, this in turn involved evaluating the nuclearity of potential candidates. The first mission targeted Latin America. In part because Brazil and Argentina were competing for designation to the board as the region's "most advanced" nation, the mission encountered some controversy over which laboratories should be visited and which experts should participate.

Those squalls paled next to the storm that erupted when the United Arab Republic (UAR)[40] expressed interest in establishing a regional training center in Cairo in 1958. The UAR's initial proposal welcomed research from "friendly nations," implicitly excluding Israel. Undoubtedly hoping to benefit from a probable impasse, Turkey put in a rival bid for a regional training center, with strong support from both Israel and South Africa.

For most delegates, the main source of contention came from conflict in the Middle

[38] "Seventh Regular Session of the General Conference of the International Atomic Energy Agency: Report of the South African Delegation." See also "Address Delivered by Mr. D. B. Sole." (Both cit. n. 36.)

[39] "Report of the South African Delegation on the Third General Conference of the IAEA" (cit. n. 29).

[40] The United Arab Republic was established as a union between Egypt and Syria in February 1958; the hope was that more Arab nations would join. Syria withdrew in 1961, but Egypt continued to use the name until Nasser's death in 1970. Given how rapidly this proposal for an IAEA regional training center followed the founding of the UAR, it seems likely that UAR leaders viewed such a center as a potential technopolitical expression (and creator) of Arab unity that would itself provide incentive for other nations to join. This interpretation would need to be tested in the Egyptian archives, however.

East. But Sole saw the UAR proposal as the first step in an Egyptian plot to usurp South Africa's seat on the board of governors as the region's "most advanced" nation. The Egyptian statement at the 1958 general conference, describing an extensive R&D program that included significant regional cooperation, confirmed his suspicions.[41] Sole dourly reminded his secretary of external affairs that "the Union of South Africa . . . has at present *no* reactor, *no* national isotope centre and is not in a position to accept trainees from other countries in the regional area."[42] (The latter for racist reasons.) Most humiliating of all, the Egyptians seemed disposed to gloat: "the head of the Egyptian Atomic Energy Establishments extended to me a cordial invitation to visit . . . Obviously he would like me to come and see for myself the extent of Egyptian progress."[43]

The UAR challenge made it all the more imperative that South Africa launch its own atomic energy program. For several months, Pretoria had been sitting on two proposals. One involved establishing a separate Atomic Energy Board (AEB). The other involved developing an atomic research program under the auspices of the nation's Council for Scientific and Industrial Research. Sole strongly supported the first proposal, which he felt would make a stronger statement about South Africa's commitment to nuclearity and provide a powerful argument to keep the South African seat on the board of governors.[44] In the end, and partly (though only partly) for this reason, that's precisely what the Pretoria government decided to do.

Meanwhile, discussions about prospective training centers in the Middle East/Africa region continued. When the time came to send an IAEA delegation to visit the region's nuclear installations, Sole jumped in. "The region" included the whole African continent. Surely, the delegation shouldn't be restricted to the Middle East? Working the corridors of the IAEA, Sole reminded Britain, France, Belgium, and Portugal (all colonial powers in Africa) that such a restriction would be "detrimental to the long-term interests of the rest of the region." The Belgians were building a research reactor in the Congo—could that not provide a plausible site for another regional center, "rather than have the Agency focus all its attention and assistance" on Cairo? "It is important," Sole added self-righteously, "to protect the rights and interests in this matter of the African countries and territories as opposed to the Middle East countries and territories."[45] Sole thus deployed South Africa's privileged position as a western nuclear fuel supplier while simultaneously claiming to speak for all Africa. The tactic worked: the board of governors agreed that the IAEA delegation should visit all the countries in the Middle East/Africa region that could claim any substantial nuclear activities.

To determine which countries qualified, the IAEA sent out questionnaires to the possible destinations. The questions were broadly conceived to include all possible manifestations of nuclearity. They pertained to:

[41] "United Arab Republic: Progress in the Atomic Energy Field," Oct 3, 1958, BLO 349 ref. PS 17/109/3, vol. 2, NASA.
[42] Ibid.
[43] Ibid.
[44] Ibid.; "IAEA: Missions to Egypt," Nov. 19, 1958, BLO 349 ref. PS 17/109/3, vol. 2, NASA; "South African Delegation to IAEA Conference," Aug. 24, 1957, 2, BLO 349 ref. PS 17/109/3, vol. 2, NASA.
[45] "IAEA: Proposal to Establish a Regional Centre in Cairo," March 25, 1959, BLO 349 ref. PS 17/109/3, vol. 2, NASA.

- teaching programs
- special facilities for experimental training in nuclear physics
- radiochemical and radioisotope laboratories
- the presence of source materials in the soil and corresponding prospecting programs
- the current and expected state of the nation's atomic energy program
- the presence of chemical and metallurgical ore processing and purification plants
- the availability of research workers in nuclear physics and related fields
- the availability of health physicists and other radiation protection experts
- the degree to which the government was prepared to help finance prospective regional training programs
- the current extent of atomic energy cooperation with other states in the region.[46]

Mightily pleased with how he'd influenced the calibration of this barometer of nuclearity, Sole recommended that Pretoria send "comprehensive replies" to these questions, even though it did "not aspire to provide the headquarters of any regional training centre in the area." In the long run, these replies would inevitably "have some bearing on the rival candidature of the Union of South Africa and the UAR for designation as the most advanced Member State in the area." Given his government's recent approval of an extensive atomic research and development program, Sole felt confident that South Africa would top the region's nuclearity barometer in technical matters. There was, perhaps, just one sticking point: the increasing isolation of the apartheid state meant that claims about regional cooperation wouldn't be too credible. Nevertheless, Sole suggested that under this rubric South Africa emphasize "the degree of co-operation which subsists with the metropolitan States having territorial responsibilities in the area."[47] Here, then, was another test of South Africa's dual identity as western and African.

Ever the creative diplomat, Sole was inspired by debates on the Cairo training center to further develop a vision of South Africa as a regional leader. The Commission for Technical Co-operation in Africa South of the Sahara (CCTA) offered the ideal paradigm, including as it did colonial powers (Belgium, France, Portugal, and the United Kingdom), newly independent nations (such as Ghana and Liberia) and apartheid-infested southern Africa (Nyasaland, Rhodesia, and South Africa). In short, the organization enacted precisely the sort of western-African duality so central to the national identity being crafted by South Africa's Afrikaner government.[48] Sole began a campaign on two fronts: In Vienna, he pushed the IAEA to formalize relations with the CCTA. In Pretoria, he pushed his government to get more involved in CCTA activities.

Ismael Fahmy, the UAR governor, strongly resisted the Vienna push by questioning the political legitimacy of the CCTA. Apparently pursuing its own dual-identity strat-

[46] "Questionnaire: Information Required for Studies Relating to the Establishment of Regional Atomic Energy Training Centers in the Middle East and Africa," June 25, 1959, BLO 349 ref. PS 17/109/3, vol. 2, NASA.

[47] "Establishment of Radioisotope Training Centers in Africa and the Middle East," June 30, 1959, BLO 349 ref. PS 17/109/3, vol. 2, NASA.

[48] Though, in fact, Afrikaners in the South African government didn't necessarily see this, and Sole had to push hard on his own government to participate actively in the CCTA—so on this issue, he was leading a battle on two fronts.

egy, the Egyptian government "was completely opposed to the concept inherent in the CCTA constitution of a distinction between sub-Saharan Africa and the rest of the continent."[49] In a backroom meeting with the "Western powers," Sole suggested that Fahmy just wanted to "divert Agency attention and resources from Africa south of the Sahara in favor of . . . the Mediterranean littoral in general and . . . Cairo in particular."[50] Meanwhile, rumors were spreading "that the African members of the CCTA have never agreed to the conclusion of a relationship agreement with the IAEA. . . . It is alleged that CCTA is a creation of the 'colonialists' and it is recalled that the colonial powers concerned for a long time did their best to keep the United Nations out of Africa."[51] The United Kingdom came to South Africa's rescue. It defended the CCTA's legitimacy by trotting out the good old technology-politics distinction: the functions of the CCTA, argued the British delegate, "were essentially technical, hardly political at all, and very useful." A formal relationship agreement with the IAEA would be no more than an accord between two technical agencies. Besides, geography should calm fears of illegitimacy: the fact that "the Commission was a genuinely African association was proved by the fact that its headquarters were at Lagos."[52]

WHO SPEAKS FOR AFRICAN NUCLEARITY?

As decolonization overtook the continent, the scramble to define and represent African nuclearity became almost surreal. In 1960, for example, Ghana applied for IAEA membership. Delegates fell over each other in a rush to sponsor it. Sole reported that "in order to forestall an anticipated initiative on the part of the United Arab Republic to introduce a resolution approving Ghana's admission to membership which would be co-sponsored by members of the Afro/Asian bloc," the UK delegation had proposed that the Commonwealth bloc introduce and sponsor the resolution. Caught between the imperatives of apartheid and those of western-ness, Pretoria had instructed Sole by secret telegram that his support for Ghana "should go no further than affirmative vote."[53] But Sole feared that "for South Africa to have been the sole member of the Commonwealth *not* to co-sponsor the resolution approving the admission of Ghana would have been regarded as a purely political gesture on the part of a delegation which had been among the foremost in pleading that international politics be kept out of the Agency." He chose to remain true to his own diplomatic strategy. Claiming that his office hadn't deciphered the secret telegram until it was too late, Sole replied that he'd already signed on to cosponsor Ghana's application.

A month later, the IAEA's general conference entertained proposals to increase the representation of the Africa/Middle East region on the board of governors. To maintain its African credentials, Sole decided, South Africa had to retain "such initiative as is possible. Inevitably, even in the IAEA where our reputation stands high, we are suspect in the eyes of other African states, but this year we are the only *African* representative on the Board and we cannot for one moment afford to let the other [elected]

[49] D. B. Sole, "IAEA: Relations with CCTA," June 29, 1960, BLO 581 ref. PS 31/3/13, pt. 1, NASA.

[50] D. B. Sole, "IAEA: Relationship Agreement with CCTA," March 19, 1960, BLO 581 ref. PS 31/3/13, pt. 1, NASA.

[51] D. B. Sole, "IAEA: Relations with CCTA," Sept. 17, 1960, BVV84 13/1, vol. 7 and annex, NASA.

[52] "Report of the South African Delegation to the Fifth General Conference of the IAEA," Oct. 12, 1961, 30–1, BVV84 13/1, vol. 7 and annex, NASA.

[53] Secretary for External Affairs, Pretoria, to S.A. Legation, Vienna, Sept. 19, 1960, BVV84 13/1, vol. 7 and annex, NASA.

representative of the area, Iraq, act as a spokesman for Africa."[54] Sole worked furiously behind the scenes with delegates from Tunisia, Ethiopia, Ghana, Iraq, Morocco, and even the UAR to draft "an amendment to the Statute designed to ensure adequate African representation." The resulting resolution was swiftly and unanimously adopted. For Sole, "the whole episode provided striking evidence of the extent to which East, West, and neutralists are competing for the favor and support of Africa." He hoped his government would capitalize on the "prestige" he was building in the IAEA in order to "improve our relations with the rest of the Continent. In this respect, we [must] exercise a little more imagination than we have displayed in the past."[55] Sole's efforts fell on deaf ears. The apartheid state continued to isolate itself diplomatically: just a few months later, Prime Minister Hendrik Verwoerd withdrew South Africa from Commonwealth membership.[56]

Meanwhile, ongoing decolonizations continued to threaten South Africa's board seat, both by offering alternative candidates and by increasing the number of postcolonial nations likely to oppose it. In 1961, for example, the Republic of Congo announced it was contemplating the reactivation of its uranium industry and the renegotiation of contracts inherited from Belgium. For Sole, this move boded ill:

> This is in line with the ambitions of India (who took the main initiative for the Congo to apply for membership) to secure the replacement of Belgium, as a producer member on the board of governors, by the Congo (Leopoldville). The argument runs that Belgium owes her seat to production in the Congo . . . [I]t can be expected that the Europeans will bring pressure to bear on Belgium not to . . . acquiesc[e] publicly in any demand which may be forthcoming from the Congo, since to do so undermines also the position of Portugal on the Board (already under heavy fire) and would, in the event of this claim succeeding, lead inevitably to the loss of a European seat on the Board.[57]

In the event, Congolese uranium production was not reactivated, and colonizers managed to hold on to their seats.

By 1963, a temporary consensus had emerged on the various issues under debate. To Sole's delight, the IAEA did sign a formal agreement with the CCTA in the context of a range of other agreements (including one with Euratom). Board representation of the Africa/Middle East region had increased by two seats, making South Africa's position considerably more secure. To his dismay, however, the Regional Training Centre for the Arab Countries opened in Cairo, giving a serious boost to Egyptian nuclearity. In addition, 1963 saw the admission of another three new African nations to the IAEA: Gabon (a uranium producer), Ivory Coast, and Nigeria. IAEA director general Sigvard Eklund heralded these developments as "further steps toward close collaboration with African countries" and important examples of the kind of regional collaboration the agency should encourage and support.[58]

[54] "Report of the South African Delegation to the Fourth General Conference of the IAEA," Oct. 4, 1960, BVV84 13/1, vol. 7 and annex, NASA (my italics).

[55] Ibid., 34–5.

[56] See Giliomee, *The Afrikaners* (cit. n. 20), 529, for more on the apartheid state's relations with the Commonwealth.

[57] "Report of the South African Delegation to the Fifth General Conference of the IAEA" (cit. n. 52), 3.

[58] "Address by the Director General of the International Atomic Energy Agency, Dr. Sigvard Eklund, to the Seventh Session of the General Conference," Sept. 24, 1963, 3, BVV84 13/1, vol. 7 and annex, NASA.

But Sole bemoaned these developments, particularly when twenty nations signed a "Declaration on the Incompatibility of the Policies of Apartheid of the Government of South Africa with the Membership of the IAEA." The statement condemned the racism of the apartheid government and urged the IAEA to conduct "a review of South Africa's policy in the context of the work of the Agency."[59] Sole presented a written response accusing the declaration of being "purely political" and insisting that the IAEA was "not a proper forum" for such matters. He categorically denied the accusation that "all Afro-Asian scientists are unacceptable on South African territory and that it is impossible for members of the Afro-Asian group to take part in any seminar organized in South Africa."[60] At the conference itself, Sole lobbied hard against the motion to exclude South African participation. He received promises of backroom support from several delegates who had come to respect him personally. Most important, though, was U.S. support: "the fact that they were prepared to use their lobbying machine on our behalf was in itself a useful factor in its effect on the general attitude of some delegations."[61] The motion did not pass.

Clearly, though, South Africa would have to work continually to maintain the legitimacy of its place near the top of the global nuclear hierarchy. "Only a simple majority of the members of the Board is required to oust us from our position,"[62] Sole remarked. He knew that "the Afro-Asian upsurge, combined with the continued preoccupation with politics, will bring South Africa very much under the harrow." The only solution was to help the IAEA remain narrowly nuclear and not become "just one more international organization for the provision of technical assistance which provides at the same time a platform for the propagation of varying ideologies, Western, Communist, Neutralist, anti-White, anti-colonialist and the rest."[63] At the same time, and regardless of the government's foreign policy in other domains, South Africa had to continue upholding the interests of sub-Saharan Africa as a region.[64] The government's discursive duality was vital for its survival in the IAEA.

SAFEGUARDS AND THE REGULATION OF NUCLEARITY

Early negotiations over safeguards were just as driven by tensions between cold war and postcolonial geopolitical paradigms as those we've examined so far. The United States was the most vocal promoter of safeguards agreements, but most other nations selling nuclear technologies and materials at least paid lip service to their desirability. As historian Astrid Forland argues, commercial considerations strongly shaped the approach that selling nations took toward safeguards.[65] Unsurprisingly, buyers of nuclear systems were less than thrilled at the prospect of controls over end use. India in particular argued that regulating technological access through safeguards perpetuated

[59] "Joint Declaration by a Group of Members in Africa and Asia Regarding South Africa," General Debate and Report of the Board of Governors for 1962–63, agenda item 10, Oct. 1, 1963, BVV84 13/1, vol. 7 and annex, NASA.

[60] "Statement by South Africa," General Debate and Report of the Board of Governors for 1962–63 (cit. n. 59).

[61] Ibid.

[62] "Report of the South African Delegation to the Fifth General Conference of the IAEA" (cit. n. 52), 23.

[63] Ibid., 32.

[64] Ibid.

[65] Forland, "Negotiating Supranational Rules" (cit. n. 1), especially chaps. 2 and 3.

colonial inequalities, defeating what they saw as the very purpose of the IAEA. Of course, no one *wanted* to develop nuclear weapons, Indian delegates asserted, but nevertheless an important principle was at stake. Even more than other nuclear topics, struggles over safeguarding the global nuclear order were saturated with morality talk. But in this domain as in others, more mundane concerns drove the practices that actually emerged. Once again, South Africa's archives reveal not just the country's own aims but also those of other nations. Let us see how by returning to 1956.

For South Africa, the safeguards matter was really a subset of the uranium market issue. South Africans worried that the imposition of safeguards would affect the marketability of their uranium. The Canadians were similarly concerned: "there is little doubt . . . that [the Canadians] regard us as their main potential competitor in export markets for uranium and wish to ensure that their competitive position is not weakened by commitments [i.e., safeguards] which we do not also accept."[66] Evidently, Canadian delegates had approached their South African counterparts to determine where they stood on including a use-control clause in uranium sales contracts.

So where *did* South Africa stand on this question? Sole knew that Pretoria's Department of External Affairs would strongly object to international controls on uranium exports, which necessarily infringed upon the sovereignty so dear to Afrikaner nationalists. But he urged Pretoria to resist its isolationist impulses in order to maintain American goodwill. "To defend a position in which the Union was the only country which refused to enter this system in one way or another . . . would lead to all kinds of difficulties."[67] Besides, South Africa could derive practical benefit by cooperating, notably by making acceptance of safeguards conditional on securing both similar acceptance from its competitors and board membership. Persuaded by these arguments, Secretary of External Affairs G. P. Jooste sought to make controls palatable to the rest of the government by noting that the IAEA's draft statute only committed agency members to accepting safeguards for transactions conducted *through the agency.* Contracts concluded separately would be exempt. "We consider, therefore, that acceptance by the Union of membership in the Agency does not in itself commit the Union, as a producer of source material, to Agency controls or safeguards. Nor would membership, in our opinion, create a moral commitment to apply Agency control over bilateral agreements."[68]

W. C. du Plessis, Sole's colleague on the South African IAEA delegation in 1956, feared that other delegates would recognize this approach as a cop-out and that real dangers inhered in not publicly embracing controls. Besides, he argued, controls wouldn't really affect South Africa's uranium business. First, safeguards would not apply to the existing CDA contracts. Second, controls "would operate in the recipient country only and would not cover production or any processing of source materials in the country of origin." Third, controls of uranium oxide wouldn't be as stringent as controls of "fissionable materials." Meanwhile, the danger of *not* accepting controls

[66] "Peaceful Uses of Atomic Energy: International Safeguards," Aug. 21, 1956, BLO 349 ref. PS 17/109/3, vol. 2, NASA. Forland offers further details on Canadian concerns—see "Negotiating Supranational Rules" (cit. n. 1), 103–4, 135–7.

[67] "Peaceful Uses of Atomic Energy: International Safeguards" Aug. 21, 1956 (cit. n. 66). For more of same, see David Fischer, Secret Correspondence Z4/5/1, Aug, 21, 1956, BLO 349 ref. PS 17/109/3, vol. 2, NASA.

[68] "Peaceful Uses of Atomic Energy: International Safeguards," Sept. 8, 1956, BLO 349 ref. PS 17/109/3, vol. 2, NASA.

was that South Africa could "be accused of having willfully, and in pursuit of its own economic interests, taken the risk of widening the area of international insecurity." Finally, even if the South African government didn't accept a moral obligation to promote safeguards, Canada did. In their eagerness to outbid competitors, the Canadians might seek to persuade the United States that South Africa didn't respect U.S. foreign atomic energy policy: "on one occasion during the Conference, a member of the Canadian delegation, in the presence of some State Department and AEC officials, referred jokingly to what he termed South Africa's apparent unwillingness to show its hand in regard to safeguards in bilateral agreements." Du Plessis couldn't tell whether this was an innocent joke or a deliberate attempt to undermine South Africa's relationship with the AEC, "but the incident illustrates the possibility that the Canadians may bring pressure to bear on Washington."[69]

So South Africa would have to accept safeguards. But what exactly did this mean? What did safeguards consist of? What would be safeguarded, and how? These were all open questions, and there was ample opportunity to shape answers in advantageous ways.

The most basic question involved definitions. Clearly "fissionable materials" required safeguards. Yet how did one define a "fissionable material"? When and how did a "source material" become a "fissionable material"? The statute drafters appointed a technical committee to settle these questions. The report of the scientist who represented South Africa on this committee is worth quoting at length:

> [We] approached the task of definitions, bearing in mind that the determination of what constituted fissionable materials would be strongly related to the inspection and safeguarding provisions of the Statute, that the definitions would have to be essentially practical, rather than "textbook" in nature, that they must be legally watertight and must take account of certain political implications.
>
> The Indian representative was clearly keen to ensure (a) that natural uranium was not defined as fissionable material and (b) that the concentration of plutonium, U235, U233, etc., required before a "source" material became a "fissionable" material were relatively high . . .
>
> [We] adopted the view that natural uranium fitted best into the source material category, although it is of course a fissionable material (but then so is U238 under appropriate conditions), that the term "special fissionable material" should be used, that the line dividing source and special fissionable materials should be at the 0.7% enrichment level . . . and that it was desirable to ensure that source material irradiated only momentarily in a reactor must become "special fissionable materials" in terms of the definitions.
>
> The definitions proved astonishingly difficult to draw up. In the final session of the subcommittee, the fourth new or amended draft was adopted with the Indian representative abstaining on the ground of an inconsistency, in his view, in that if a small amount of say U235 were added to say U238, the resulting materials would fit into the definitions for both "special fissionable material" and "source material."[70]

The difference between "source materials," "special fissionable materials," and "fissionable materials" mattered because each would be subject to different controls. The Indians wanted the greatest possible latitude built into the official IAEA definitions and so preferred the more ambiguous term "fissionable materials." They lost the

[69] "Peaceful Uses of Atomic Energy: International Controls," Nov. 20, 1956, BLO 349 ref. PS 17/109/3, vol. 2, NASA.
[70] "International Atomic Energy Agency," Annex to South Africa Minute no. 79/2 (cit. n. 11).

final vote on this question. The final IAEA statute did not treat "fissionable materials" as a distinct, definable product. Instead, it specified three other categories: "source materials," "special fissionable materials," and "uranium enriched in the isotope 235 or 233."[71] Significantly, there *was* general agreement that uranium oxide with an average 235 or 233 content should count as "source material" for the purposes of safeguards. Each delegation had its own reason for this, ranging from preserving the marketability of uranium oxide to leaving the greatest latitude possible in its use. For our purposes, the point is that these definitions created a barometer of nuclearity, with "source materials" being the least nuclear and therefore subject to the most relaxed safeguards and "special fissionable materials" being the most nuclear and therefore subject to more stringent safeguards.

Definitions, of course, were only the first step. Actual controls remained to be worked out. For uranium producers, the critical question was: *Which* stage of production would be subject to *which* safeguards? The United States and Britain initially wanted to control "every stage from the sale of source material until the manufacture of the end product—plutonium." They soon realized, however, that truly controlling each phase would be highly impractical. Instead, they suggested "work[ing] backwards through the various processes from plutonium towards the original materials in the hope of fixing upon some intermediary points where controls can be effectively applied."[72] Such was the task of the IAEA's new safeguards division.

The Canadian scientist chosen to head this division had been involved in his own nation's studies on safeguards, and the South African delegation actually found his presence reassuring. "Having been so closely concerned with Canadian uranium he is fully seized of the concern of the producers of uranium that if any safeguards are to be applied to source material, they should be simple to operate and not . . . unduly affect the saleability of the product in a competitive market. The system of accountability for source material which he favors is clearly far less elaborate than the American AEC has had in mind and may well be limited to straight-forward book-keeping returns coupled with spot inspections every six or twelve months."[73] Such bookkeeping wouldn't be too onerous. Indeed, it wouldn't represent any extra work beyond what uranium producers did for their own purposes—as long as they were willing to open their books to IAEA inspectors.

Even if the process for safeguarding "source material" was satisfactory, however,

[71] In the end, Article XX of the statute specified the following definitions:

"1. The term 'special fissionable material' means plutonium-239; uranium-233; uranium enriched in the isotope 235 or 233; any material containing one or more of the foregoing; and such other fissionable material as the Board of Governors shall from time to time determine; but the term 'special fissionable material' does not include source material.

2. The term 'uranium enriched in the isotope 235 or 233' means uranium containing the isotopes 235 or 233 or both in an amount such that the abundance ratio of the sum of these isotopes to the isotope 238 is greater than the ratio of the isotope 235 to the isotope 238 occurring in nature.

3. The term 'source material' means uranium containing the mixture of isotopes occurring in nature; uranium depleted in the isotope 235; thorium; any of the foregoing in the form of metal, alloy, chemical compound, or concentrate; any other material containing one ore more of the foregoing in such concentration as the Board of Governors shall from time to time determine; and such other material as the Board of Governors shall from time to time determine."

See appendix 1 in Fischer, *History of the International Atomic Energy Agency* (cit. n. 1), 490–1.

[72] "Safeguards on Sale of Source Materials," Dec. 27, 1957, BLO 349 ref. PS 17/109/3, vol. 2, NASA.

[73] "IAEA: Development of a Safeguards System," Oct. 31, 1958, BLO 349 ref. PS 17/109/3, vol. 2, NASA.

another problem immediately followed: What should be done about small quantities of "special fissionable material" that, in larger quantities, would be subject to stricter safeguards? Could small quantities of (say) enriched uranium be exempt? If so, what was the threshold for exemption? The United States wanted to apply safeguards to quantities of uranium between two and ten metric tons if the uranium-235 content was between 0.5 and 1.0 percent. (Any quantity above ten metric tons would be subjected to full safeguards.) Sole reported that these figures were based on the "contention that more than 200 grams of fully enriched uranium represents a dangerous quantity" and that two metric tons of natural uranium were required to produce 200 grams of enriched product.[74] But here he thought that politics had to trump technology: "whatever the theoretical technical justification for this, it is quite certain that *politically* . . . it is impossible to equate 2 tons of natural uranium with 200 grams of fully enriched material as a potential danger to international security."[75] On a practical level, he argued, a huge amount of knowledge, effort, and money was required to transform two tons of natural uranium into 200 grams of bomb-grade material. The building of the required technical system couldn't fail to escape international notice and could itself be subject to a set of controls. South Africa should firmly insist on exempting ten or fewer tons of uranium with less than 1 percent U235.

As the chairman of the board of governors in 1960, Sole was in an excellent position to lobby for this outcome. He soon secured the support of the UK delegates, who in turn rallied other Commonwealth uranium producers. The UK high commissioner also lobbied the United States, arguing that the "South African proposal would benefit . . . all the producers of source materials. If this concession could be made in their favor, they would be more likely to give active support to Agency safeguards in general—and the support of the producers was a *sine qua non* if safeguards were to be effective."[76] Sole, meanwhile, made it clear that if his ten-ton threshold were rejected, South Africa would not consider the IAEA safeguards system binding on bilateral agreements concluded outside of agency auspices.[77] With a combination of threats and incentives, he won his case.[78]

Despite this victory, Sole remained skeptical about the proposed safeguards system. The good news for South Africa was that he didn't think that "the edifice to be constructed will be more than a façade."[79] While potentially useful for international relations, he doubted—prophetically, as it turned out—that the system would do much to prevent the spread of atomic weapons. Based on what he'd heard from Homi Bhabha, for example, he felt sure that India would eventually build a bomb.[80]

To ensure that even this "façade" would not unduly encumber South Africa's activities, Sole focused his energy on two dimensions of safeguards practices. First, he sought to minimize controls on "source materials." He wanted to make absolutely certain that uranium mines and ore-processing plants would not count as "conversion

[74] Memorandum 137/10/9, Dec. 31, 1960, BPA 25 ref. 31/32, pt. 1, NASA.
[75] Ibid.
[76] Ibid.
[77] "IAEA: Safeguards against Diversion," Feb. 27, 1960, BPA 25 ref. 31/32, pt. 1, NASA.
[78] The ten-ton threshold still applies. See the IAEA's INFCIRC/66/Rev.2, Sept. 16, 1968, as posted on the agency's Web site: http://www.iaea.org/Publications/Documents/Infcircs/Others/inf66r2.shtml (accessed Oct. 18, 2004).
[79] "Report of the South African Delegation to the Fourth General Conference of the IAEA" (cit. n. 54).
[80] Ibid.

plants," and therefore not themselves be subject to safeguards. Since many uranium-producing countries shared this priority, the specific exclusion of mines and ore-processing plants from the definition of a "conversion plant" passed unproblematically. At the same time, he led a largely successful campaign to liberalize other controls over "source materials," arguing that this would prevent a uranium shortage when commercial nuclear power finally became a reality.[81] Second, he worked extensively on the "Inspectors' Document," which governed the selection of IAEA inspectors and their rules of conduct. The selection of inspectors was a source of particular anxiety for several departments in the apartheid state, which all worried about whether South Africa would be forced to subject itself to inspections conducted by Indians or black Africans.[82] Realizing only too well that this racial framing would not fly in Vienna, Sole translated this concern into one about national sovereignty—which most IAEA members shared, and which indeed was a broader issue during the negotiation of personnel for any type of diplomatic mission.[83] It then became easy to work into the "Inspectors' Document" that the inspected country could "reject any individual Agency inspector *without* giving any reasons for its rejection," as long as it didn't reject *all* proposed inspectors.[84]

CONCLUSION

In 1977, Sole's fears proved prophetic. Now comprising seventy-seven nations, the group he'd described as "the Afro-Asian upsurge" demanded that South Africa be ousted from the board of governors and replaced by Egypt. Twenty years later, the IAEA's official historian, David Fischer—who had been Sole's junior colleague on the South African delegation for several years (a biographical detail mentioned nowhere in the book or its jacket[85]) before leaving the South African foreign service to join the IAEA's full-time staff—describes the event as follows:

> In June 1977, the Board decided by a vote of 19 to 13, with one abstention, to uphold the Chairman's nomination of Egypt as the Member State in Africa "most advanced in nuclear technology including the production of source materials." Egypt's nuclear program was very modest and it produced no source materials (i.e., uranium) but worldwide revulsion against apartheid made it politically inevitable that the South African Government would sooner or later lose its seat on the Board. This revulsion also led to the rejection of the credentials of the South African delegation when the General Conference met

[81] "Statement by Mr. D. B. Sole, Leader of the South African Delegation, in the General Debate: 25th September 1963," BVV84, 13/1, vol. 7 and annex, NASA.

[82] "Die Vooregte en Immuniteite van Internationale Atoomkragagentskapinspeketurs," Dec. 9, 1959, TES 1004 ref. F5/362/5, NASA.

[83] "The Rights, Privileges, and Immunities of IAEA Inspectors," Dec. 3, 1959, TES 1004 ref. F5/362/5, NASA.

[84] For more on the "Inspectors' Document," see Fischer, *History of the International Atomic Energy Agency* (cit. n. 1), 247–8.

[85] While I do not mean to suggest any sort of conspiracy by calling attention to this biographical fact, readers may find its elision particularly striking in light of the details that were supplied by Hans Blix, then director general of the IAEA, in his preface to Fischer's 1997 book (*History of the International Atomic Energy Agency* [cit. n. 1]): "David Fischer took part in the negotiation of the IAEA's Statue in 1954–56 and served on the IAEA's Preparatory Commission. From 1957 until 1981 he was the Agency's Director and subsequently Assistant Director General for External Relations. In 1981 and 1982 he was Special Adviser to Director General Eklund and to myself. Since then he has served as a consultant to the IAEA on many occasions."

> in New Delhi in September 1979. After a democratic government had taken power in Pretoria, South Africa, with Egypt's concurrence, regained its seat on the Board in 1995.[86]

Though mild in tone, this description is unequivocal in its interpretation: Egypt's nuclearity could not match that of South Africa, which therefore was ousted for purely political reasons—a fact proven by its reinstatement after the fall of the apartheid government. Fischer was too much the diplomat to comment explicitly on the appropriateness of South Africa's expulsion. In 1987—the year after the South African government declared a state of emergency to quell black insurgency, the year after Mikhail Gorbachev initiated *perestroika*—U.S. political scientist Lawrence Scheinman was far less delicate in his assessment. "The treatment of South Africa by its political opponents in the IAEA starkly illustrates the problem of politicization," he wrote. He concluded his brief account of South Africa's ousting by noting that the United States "while condemning apartheid, has opposed all the actions described above on the basis of the principle of universality and the practical importance of South Africa to nuclear matters."[87] In his view, the treatment of South Africa was an example of "political opportunism," an attempt on the part of "those with a relatively limited interest in the basic purpose of the agency to hold hostage the substantial interest of others in order to secure their own political objectives."[88]

My foray into the early history of the IAEA shows that such interpretations serve merely to reinforce a teleological view of the agency as one whose only true, legitimate purpose was to ease cold war tensions and guard against nuclear proliferation. They unquestioningly deploy the very objects of negotiation in the early years of the agency: the meanings of nuclearity, and the boundaries between the technical and the political. In so doing, they reproduce a vision of global order in which cold war concerns trump postcolonial ones every time—a vision in which postcolonial nations have less right to weigh in on nuclear issues by the simple virtue of being less nuclear.

The distortions effected by this vision have become increasingly clear in the last decade as more nations have acquired nuclear weapons. Designating them "rogue nations" elides the fact that these nations did not acquiesce to the imposition of a global nuclear order that never acknowledged the legitimacy of their voices. As George Perkovich has argued so eloquently in his analysis of India's atomic bomb,[89] it also elides the domestic and regional motivations nations have for developing nuclear weaponry, motivations which—for them—trump the nonproliferation goals of western nations. Condemnations of safeguards and nonproliferation treaties as neocolonial may be intensely self-interested, but they are also genuine expressions of an alternative vision of the technopolitics of global order. In making this point, I am in no way condoning the resulting proliferation of nuclear weapons. Paralleling Perkovich's analysis, I am simply arguing that unless we understand this dimension of the making of global nuclearity, we cannot hope to find workable solutions to proliferation.

This chapter has examined how South Africa—with substantial support from the United States, the United Kingdom, and others—attempted to assert a global consensus on the meanings of nuclearity and the boundaries between the technical and

[86] Fischer, *History of the International Atomic Energy Agency* (cit. n. 1), 93–4.
[87] Scheinman, *International Atomic Energy Agency and World Nuclear Order* (cit. n. 1), 211.
[88] Ibid., 210.
[89] George Perkovich, *India's Nuclear Bomb: The Impact on Global Proliferation* (Berkeley, 1999).

the political. These attempts took heterogeneous forms and sometimes met with considerable success. In setting up a process by which to canvass the African continent prior to establishing a training center in Cairo, Sole helped to define the markings on a barometer of nuclearity that ultimately served to legitimate South Africa's presence on the board of governors for twenty years. His ability to institutionalize a separation between the IAEA and the UN did put a leash on postcolonial politics in the IAEA for twenty years, even if in the end the "G-77" succeeded in ousting the apartheid nation. Sole's relentless insistence on the separation of technology and politics served the United States, the United Kingdom, and other western allies well, and they repaid South Africa handsomely with continued support to the bitter end (when thirteen nations voted to keep South Africa *in* the agency). His success in whittling away at safeguards practices helped make his own prophecy—that these would be little more than façades—come true.

In 1993, the world learned that South Africa itself had been a "rogue nation." It had defied the very global nuclear order it had helped to build, the order from which it was expelled by those it had sought to exclude. The moment of revelation was also cast as a moment of redemption: in the same breath that F. W. de Klerk confessed to South Africa's nuclear weapons capacity, he announced its dismantling. Bombs, enrichment plants, testing grounds—all gone, destroyed, never again to see the light of day. In 1994, South Africa's first democratic elections ensured that global political redemption followed nuclear redemption. In 1995, South Africa was readmitted to the IAEA. The emerging consensus *within* South Africa concerning the reasons for the bombs' destruction (namely, to prevent their falling into the hands of a black government) shimmered only faintly on the edges of global discourse, which basked in the glow surrounding the world's first undoing of nuclear sin. IAEA inspectors rushed to verify the destruction of the weapons and turned South Africa into the poster child for nuclear nonproliferation. In a final, jaw-dropping irony, tenacious officials at the South African Nuclear Energy Corporation (NECSA, formerly the AEB) insist that the IAEA itself has forbidden them from releasing whatever archives remain concerning the South African weapons program, for fear that these might fall into the hands of other "rogue nations."

It is difficult to imagine how this latest erasure of history will help to dismantle nuclear weapons elsewhere.

The Ambivalence of Nuclear Histories

*By Itty Abraham**

ABSTRACT

This chapter argues that a discourse of "control," authored by the overlapping narratives of academic proliferation studies and U.S. anti-proliferation policy, has come to dominate our understanding of nuclear histories. This discourse, with its primary purpose of seeking to predict which countries are likely to build nuclear weapons and thereby to threaten the prevailing military-strategic status quo, has narrowed the gaze of nuclear historians. Among its effects has been to *minimize* the importance of the discovery of atomic fission as a "world historical" event and to *impoverish* our recognition of the fluidity of international affairs in the decade following the end of the Second World War. This chapter concerns the tendency to see nuclear histories as, above all, *national* histories and to privilege concerns about the development of nuclear *weapons* over a fuller and more nuanced understanding of what nuclear *programs* mean and why they matter. Paying attention to the scientific-technological underpinnings of nuclear programs offers an alternative path, opening up new archives and insights into the making of "national" nuclear programs that might have important other, even nonbelligerent, ends. This chapter points to the varieties and importance of international collaboration in the making of "national" programs, and shows how weapons building is by no means a universal end of all nuclear programs.

"GOING" NUCLEAR

There can be little disagreement that the development and use of nuclear weapons by the United States in 1945 changed the nature of the international system, and the academic study of international relations, fundamentally.[1] Even though the firebombing of Tokyo using conventional munitions a few months before the nuclear destruction of Hiroshima and Nagasaki may have been approximately on the same scale in terms of destruction to place, property, and persons, from the moment of their first use there was near universal appreciation of the massive destructive power of these new weapons.[2] The first thermonuclear tests, eight years later, only strengthened this feeling, especially as they were conducted aboveground with relatively large numbers of onlookers. The complete disintegration of an entire Pacific island from the test of a hydrogen bomb and the widely reported death of a Japanese crew member from radioactive

* East-West Center, 1819 L Street NW, Washington, D.C. 20036; abrahami@eastwestcenter.org.
Earlier versions of this chapter were presented at the following workshops: "Science, Technology and International Affairs: Historical Perspectives," held at Georgetown University, Washington, D.C., March 26–28, 2004, and "Atomic Sciences," held at Princeton University, November 5–7, 2004. My thanks to participants at both meetings and to Kai-Henrik Barth, P. R. Chari, Sanjay Chaturvedi, John Krige, M. Susan Lindee, M. V. Ramana, Scott Sagan, and Achin Vanaik for their detailed comments.
[1] Robert Jervis, *The Meaning of the Nuclear Revolution: Statecraft and the Prospect of Armageddon* (Ithaca, 1989).
[2] Richard Rhodes, *The Making of the Atomic Bomb* (New York, 1986), 599, 734, 740–1.

exposure on the unfortunate tuna boat *Lucky Dragon Five* made it vividly clear that the world had never seen the destructive equal of this class of weapon. Reflecting the new awareness was nuclear strategist Bernard Brodie's pithy assertion that "[t]hus far the chief purpose of our military establishment has been to win wars. From now on its chief purpose must be to prevent them. It can have almost no other useful purpose."[3]

In January 1946, the young United Nations created an Atomic Energy Commission. A few months later, members made the first proposals seeking collectively to manage the spread of fissile materials and nuclear weapons. The U.S. proposals (the Acheson-Lilienthal/Baruch plans) were soon followed by the Soviet response presented by Andrei A. Gromyko. The former sought to create an international body that would control fissionable materials, the latter to ban the "stockpiling, production and use" of nuclear weapons altogether.[4] As early as 1950, there was debate at the highest level of the U.S. government on whether the conflict in Korea justified the use of nuclear weapons.[5] In other words, within the first decade of the start of the nuclear age, questions of international control and the potential use of nuclear weapons were openly considered and debated. Fifty years later, these two themes—control and use—are still very much with us, marking an ongoing and widespread concern with the *ends* of nuclear programs. What is surprising, however, is how little agreement exists on a foundational question we might have thought would precede a discussion of ends: Why do states develop nuclear weapons in the first place?

Political scientist Scott Sagan offers us the most comprehensive statement on why states "go" nuclear.[6] Sagan posits that there are three primary models explaining nuclear acquisition: a "security," or realist, model, which argues that states build weapons for security and because others do; a "domestic politics" model, which sees nuclear weapons development as the outcome of actions by powerful coalitions within states that seek institutional power via this end; and, finally, a "norms" model, which argues that "weapons acquisition, or restraint in weapons development, provides an important normative symbol of the state's modernity and identity."[7] Sagan's article argues in conclusion that both "[n]uclear weapons proliferation and nuclear restraint have occurred in the past, and can occur in the future, for more than one reason: *different historical cases are best explained by different causal models.*"[8] In an important rejoinder to positivist approaches to international relations, Sagan argues that parsimony in explanation is unlikely to grasp the full range of the "proliferation" problem, a conclusion with important policy implications.

This is an important finding, but it may leave other scholars of nuclear affairs a little puzzled about its significance. Scholars of foreign policy, national security strategy,

[3] Bernard Brodie, ed., *The Absolute Weapon: Atomic Power and World Order* (New York, 1946), 76, cited in Jervis, *Meaning of the Nuclear Revolution* (cit. n. 1), 7.

[4] David Holloway, *Stalin and the Bomb: The Soviet Union and Atomic Energy, 1939–1956* (New Haven, 1994), 154–66, on 161.

[5] John Lewis Gaddis, *What We Now Know: Rethinking Cold War History* (New York, 1997), 103–7.

[6] Scott D. Sagan, "Why Do States Build Nuclear Weapons? Three Models in Search of a Bomb," *International Security* 21 (Winter 1996/97): 73–85. See also Stephen M. Meyer, *The Dynamics of Nuclear Proliferation* (Chicago, 1984); Bradley A. Thayer, "The Causes of Nuclear Proliferation and the Utility of the Nuclear Non-Proliferation Regime," *Security Studies* 4 (1995): 463–519; Tanya Ogilvie-White, "Is There a Theory of Nuclear Proliferation?" *Nonproliferation Review* 4 (Fall 1996): 43–60.

[7] Sagan, "Why Do States Build Nuclear Weapons?" (cit. n. 6), 55. It should be noted that Sagan uses the term "norms" in three distinct ways: as an ideational-symbolic form, as a form of international mimicry, and as a reflexive constraint on autonomous actions.

[8] Ibid., 85 (my italics).

and international relations have long identified the moment states "go" nuclear—a techno-political event involving the planned (and hopefully controlled) explosive release of nuclear energy—as a moment of the greatest importance. Whether identified as a "bomb," a "test," a "peaceful nuclear explosion," or a "demonstration," this event is of primary significance in setting analytic calendars in these fields as it is seen to mark the unambiguous moment when a country has crossed over a particular set of political and technological boundaries. The nuclear explosion is taken to mark a shift in the international distribution of power, leading to new scales of international threat and casting into question existing regimes of nuclear control. It is easy to see why, for realist analysts of foreign policy and for governmental policy makers, this event matters.

However, is trying to understand why countries conduct their first nuclear tests the same as explaining why countries begin nuclear programs? By identifying the first nuclear test as the moment when a threshold has been crossed—the historic moment—analysts have effectively reduced the variety of histories of any nuclear program to the path that led to this particular outcome. The multiple meanings of nuclear power are shrunk into one register—the desire to produce weapons—an analytic shortcoming with both real world and conceptual implications. From a practical standpoint, this approach highlights the prevailing bias that countries seeking to develop nuclear weapons are of primary interest to scholars, thereby conflating scholarly interests with those of policy makers who necessarily have to be worried about new weapons. Taking this assumption as a starting point reinforces the particular aura of nuclear weapons as objects to be coveted and desired, the very opposite effect sought by policy makers concerned with nuclear proliferation.

The intent of this chapter is threefold. First is to explore how the language of nuclear "control"[9] has helped to narrow our analytic vision. Due to the substantial overlap of two streams of analysis[10]—academic studies of nuclear proliferation and U.S. anti-proliferation policy measures seeking to reduce the spread of nuclear weapons worldwide—a discourse of "control" has come to dominate our understanding of nuclear issues. By examining two key concepts in nuclear proliferation studies, we come to realize how a singular focus on a single techno-political event, the nuclear explosion, distorts our understanding of the course of nuclear programs; an alternative approach is suggested. Second, to set the study of nuclear histories on a more productive path, this approach draws on a concept derived from colonial discourse studies—ambivalence—to show that there are remarkable and largely unacknowledged similarities between all the "early" nuclear states. Finally, this paper argues that nuclear programs are best understood as one of a larger family of public technology projects, not all of which are weapons related or have destructive ends. The larger point here is to propose that without a careful appreciation of the political and historical context within which decisions are made to develop nuclear *programs,* it is not possible to get closer to understanding the desire for, likelihood of potential use of, and possibility of international control of nuclear *weapons.*

[9] Using "control" rather than the more common "proliferation" reflects our understanding of proliferation as "political language." "Proliferation" indexes an international-legal discourse in which five countries are given a special status as "nuclear weapon states" and the intent of the law is to prevent other states from acquiring the same *de jure* status. For other examples, see Murray Edelman, *Political Language: Words That Succeed and Policies That Fail* (New York, 1977).

[10] Steve Smith, "The United States and the Discipline of International Relations: 'Hegemonic Country, Hegemonic Discipline,'" *International Studies Review* 4 (Summer 2002): 67–85.

OPACITY, AMBIGUITY, AND INDIA'S "PEACEFUL NUCLEAR EXPLOSION"

The centrality and, by extension, the limits of the first nuclear test in analytically determining the "true" course of a country's nuclear program is best appreciated by considering two concepts central to proliferation studies: ambiguity and opacity. India tested a "peaceful nuclear explosion" (PNE) in 1974. The PNE was officially termed a "demonstration," a word that recurs in Indian technological history.[11] Once India had tested, based on the experience of every other country that had conducted a nuclear test since 1945, it could be considered a nuclear power. But was it? India "did nothing" for the next twenty-four years, that is, it didn't test again or overtly weaponize until 1998. This "expected absence" came to be called a state of "nuclear ambiguity."

Nuclear ambiguity is usually defined as uncertainty in the presence of suspicion about the existence of a nuclear weapons program. However, the term "nuclear ambiguity," as Frankel and Cohen point, out "is [itself] ambiguous": it could either mean a lack of clarity on the part of others' knowledge of the extent and abilities of a country's nuclear program—do they have a weapons program or not?—or could mean a multiplicity of views on the part of a country's leadership about the utility, efficacy, and morality of nuclear weapons possession.[12] The conceptual weakness of this term is clear when we realize that when taken to the limit, all nuclear-capable countries could be said to be in a state of ambiguity until they explode a nuclear device. Ambiguity, however, is to be distinguished from "opacity."

Avner Cohen defines opacity as a "situation in which the existence of a state's nuclear weapons has not been acknowledged by the state's leaders, but in which the evidence for the weapons' existence is strong enough to influence other nations' perceptions and actions."[13] The best example of this case is Israel, which has not officially declared its possession of nuclear weapons but has institutionalized opacity at the highest level of national strategy. Ambiguity, in other words, is about uncertainty and lack of knowledge for the outsider; by this definition, so is opacity, but here the uncertainty is "actionable" from a policy point of view.

Opacity can be understood variously as the outcome of (a) indecision at the highest levels of political decision-making (e.g., India), (b) a deliberate strategy of information denial (e.g., Israel), or (c) an effort to finesse executive authority via calculated deception by a government agency or coalition of agencies (e.g., Fourth Republic France). Nuclear opacity on both sides of a dyadic rivalry might even lead to an equilibrium state of mutual tacit (nuclear) deterrence (e.g., India and Pakistan from the late 1980s). We realize that the only possible resolutions to this uncertainty are a nuclear explosion or the public dismantling of the program, à la South Africa. Given the small likelihood of the latter in most cases, ambiguity and opacity become threshold terms describing a liminal stage between intention and a yet-to-happen event, the long moment between the Fall and the Second Coming. In the case of nuclear ambiguity, a nuclear test is taken to mean that the technical means to do so has been converted

[11] That a techno-political event is only a "demonstration" is an official hedge. It should be taken to mean a technical capability to do something that (a) stops short of defining national policy and (b) provides cover to the technologists in case of failure. Central to the meaning of the word are the various audiences—domestic and foreign—who are presumed to be seeking unambiguous meaning from this event.

[12] Avner Cohen and Benjamin Frankel, "Opaque Nuclear Proliferation," in *Opaque Nuclear Proliferation: Methodological and Policy Implications,* ed. Benjamin Frankel (London, 1991), 14–44.

[13] Avner Cohen, *Israel and the Bomb* (New York, 1998), ix.

into formal ability—whether expected by analysts or not; in the case of opacity, a test is taken to show that the decision has been made to "come out of the nuclear closet" and openly declare a nuclear power. Given these shades of meaning, there is little surprise that analysts turn to the material proof of a nuclear test to confirm their concerns about the direction of a country's nuclear program; by the same token, once a test has taken place ambiguity and opacity are no longer meaningful categories.

The narrowing of vision embodied in these terms, built around the expectation that an explosion is inevitable and forthcoming, reinforces the idea of how limited the purposes and meaning of a nuclear program are assumed to be and how devalued is the importance of the political processes that ultimately make these decisions. Yet if it is important to establish when a country has decided to develop nuclear weapons, the moment of a nuclear explosion is convenient but may not necessarily be meaningful. If the counterexample of Israel—a country recognized as having a nuclear weapons program but which has never openly tested a nuclear device—is not sufficient, and we seek to establish whether a country has a "real" nuclear weapons program, an alternative approach might be based on a closer examination of the technical means to nuclear explosive potential. Under this approach, however, the evidence of a single test is neither necessary nor sufficient.

It is not unreasonable to think of a country's first nuclear test explosion as very much an experiment, with all the uncertainties that term implies. Although the feasibility of the fission process has been known for more than half a century, setting off a first explosive device anywhere is still an act of scientific ability, combined with considerable engineering skills, involving trial and error, chance and luck—and not inconsiderable means. To successfully produce a single nuclear explosive device requires, at the minimum, the following expertise: mathematical and statistical modeling skills, the means to obtain sufficient amounts of fissile material, sophisticated materials handling abilities, expertise in conventional explosives, electronics and instrumentation abilities, and the organizational skills to bring all these different elements together effectively. Needed also are adequate finances and a place to explode the device.

However, for this first test to translate into a weapons program and a nuclear arsenal that can be used at will, two things must happen. First, a political decision to proceed has to be made, and second, ad hoc scientific procedures have to be replaced with an organized, ends-oriented technological process. The technologization of the nuclear explosive building process is a discrete step necessary to convert a latent scientific ability to make nuclear explosives into a tangible and reliable process. Every step of the process—fissile material extraction, weapon design and testing, and delivery—has to be converted into an industrial process, built around repetition, with uncertainty minimized, in which laboratory practices are converted into industrial routines, and safety codes and internal security practices are regularized and institutionalized. It may not always be possible for the same organization that produced the first explosive device as a one-off scientific event to industrialize the process. Certainly, new forms of industrial and organizational management have to be employed and the process routinized sufficiently to reduce levels of error to those at which the explosive device meets the standards of military reliability. In other words, if you want to build a reliable nuclear weapons program, a number of tangible, material, organizational objectives have to be put into place, and these can be observed.

How do these clarifications help us better understand India's nuclear history?

Volumes have been written about the 1998 tests, seeking above all to explain why India did what it did, when it did.[14] To many, the still unresolved question is why, following the 1974 PNE, India did "nothing" until 1998, when it set off five more explosions and proclaimed itself a nuclear power. Of course, India didn't actually "do nothing" for twenty-four years. Under five different prime ministers, a very high-level and public debate went on about the larger purpose of the country's nuclear program, the costs of nuclear power versus other sources of energy, the threat to the world from nuclear weapons, the likelihood of global and limited disarmament, the significance and implications of the Non-Proliferation and Comprehensive Test Ban treaties, and finally, whether to build a nuclear weapons arsenal.[15] In 1998, a newly elected government, operating in great secrecy, and, as in 1974, ahead of a political consensus that this was necessary for India's security, decided India should "go" nuclear. It should be noted, however, that this decision was by no means predetermined, nor are 1974 and 1998 necessarily the dates that best reflect the changed status of India's nuclear capabilities.

Based on the technological criteria referred to above, India probably became a nuclear "power" around 1986, when Rajiv Gandhi was prime minister. From this point onward, India was certainly capable of using nuclear weapons in war and could be considered to have an effective, if crude, nuclear deterrent capability vis-à-vis Pakistan.[16] Considerable evidence now exists that there were at least two attempts to test before 1998, though these were stymied by internal political disagreements and U.S. pressure.[17] Certainly, India's nuclear scientific establishment had been keen to push ahead with more tests for some time, but the political leaders had not made up their minds about the value of doing so. It was not until the ascent to power of the right-wing Bharatiya Janata Party (BJP), a radically new political dispensation in government, that the political decision to "out" India's capabilities was reached, well after conditions on the ground existed. While the decision to test again was the outcome of particular political changes, the siren song of a nationalist government finally in power, for all practical purposes India was already a nuclear power. Previous governments, quite unlike the BJP nationalists ideologically, had ensured that India had converted a latent ability into a viable weapons option a decade before. Crossing the test threshold, however, was symbolically significant *as it sought to signal identity with dominant international norms of nuclear meaning.*

In other words, I argue that framing the decision behind the May 1998 tests was

[14] Most important, Praful Bidwai and Achin Vanaik, *South Asia on a Short Fuse* (Delhi, 2000); M. V. Ramana and C. Rammanohar Reddy, eds., *Prisoners of the Nuclear Dream* (Delhi, 2002); George Perkovich, *India's Nuclear Bomb* (Berkeley, 1999); Raj Chengappa, *Weapons of Peace: The Secret Story of India's Quest to be a Nuclear Power* (Delhi, 2000).

[15] Perkovich, *India's Nuclear Bomb* (cit. n. 14), 226–433.

[16] Chengappa, *Weapons of Peace,* 291–305; Perkovich, *India's Nuclear Bomb,* 293–9. (Both cit. n. 14.) Both countries recognized each other's nuclear capabilities no later than December 1988. This date marks the signing of the first bilateral agreement between India and Pakistan not to attack each other's nuclear facilities.

[17] Ashok Kapur in *Pokhran and Beyond* (Delhi, 2002) asserts that Indira Gandhi wanted to test in 1982, but as Perkovich notes, other than the formal request by the nuclear scientists, the rest of the story (Gandhi's agreement to go ahead that was rescinded after twenty-four hours) has never been fully corroborated. *India's Nuclear Bomb* (cit. n. 14), 242–4. More reliably, Raj Chengappa in *Weapons of Peace* (cit. n. 14), 390–5, reports that in 1995 Prime Minister Narasimha Rao ordered a series of tests that were canceled following internal disagreements and U.S. pressure. Of his two successors, Atal Behari Vajpayee also ordered tests, but his government fell in thirteen days, while H. D. Deve Gowda felt that other matters were more pressing than nuclear tests, even though the test site was ready and explosives were in place. Perkovich, *India's Nuclear Bomb,* 375–6.

the desire to reduce the multiple meanings of a "peaceful" nuclear program, to force nuclear ambivalence into a more familiar register. The desire to discipline these excesses of meaning—via nuclear explosions—comes from the intersection of the discourse of control with that of the domestic nuclear scientist seeking "sweet" solutions, more resources, and intellectual bravura in the name of national pride. Each nuclear explosion sought to reduce further the range of meanings of the Indian nuclear program, bringing it closer into line with received interpretations of what a "typical" nuclear program does. In its rejection of postcolonial difference, this event mimicked the simultaneous transformation of India's unique state-led economic development model into a more familiar path, the now orthodox global model of neoliberal, private sector–led economic growth.

NUCLEAR AMBIVALENCE

However, even once a state has "gone" nuclear, seemingly setting to rest doubts whether it is a proliferator, the meaning of what has emerged continues to be unstable. Are these weapons for deterrence, for waging war, for arms control? Do they work? Earlier meanings of the nuclear revolution—atoms for peace and for electricity—do not disappear; they can even gather new force. In what follows, we see the expression of ambivalence in more than one setting, seemingly in contradictory fashion, but only if we consider the expression of polysemic forms a violation of our preferred epistemology.

Scholarly interest in weapons production is usually located within a conceptual framework that isolates the nuclear industry from the larger political economy of the state—occluding the family resemblances of a class of modern technologies both destructive and nondestructive—and that prevents us from appreciating the flow of ideas, rules, procedures, and techniques between the nuclear industry and the rest of the state apparatus.[18] This tendency to isolate individual states and to examine their unique motives for going nuclear prevents us from giving due importance to the varieties of international collaboration that were common and indispensable to all early developers of nuclear programs (and which, by extension, gives us another history of nuclearism). Focusing on the reasons behind the acquisition of nuclear weapons reduces the number of cases that might be part of our analytic universe by focusing primarily on the bomb makers. It also reduces the search for the multiple factors that influence why countries develop nuclear programs by narrowing analytic gaze to the causes underlying weapons acquisition.[19] Putting these together, one can appreciate why there is still little agreement on the far more vexing question of why countries that could "go" nuclear *don't* or, as suggested below, appear not to.

Rather than forcing the analysis down one path exclusively, I prefer to use the term "ambivalence" to discuss the nuclear condition, in order to highlight the simultaneous presence of more than one meaning of nuclear practices, whether during the stage of ambiguity, before, or after. Ambivalence is a permanent feature of the nuclear

[18] One example of these flows is the adoption of highly restrictive procedures originally developed for the Manhattan Project to guard institutional secrets in a variety of settings quite removed from U.S. national security. See Daniel Patrick Moynihan, *Secrecy: The American Experience* (New Haven, 1998), 154–77.

[19] This issue is explicitly recognized but not fully addressed in Benjamin Frankel and Zachary S. Davis, "Nuclear Weapons Proliferation: Theory and Policy," in *The Proliferation Puzzle: Why Nuclear Weapons Spread (And What Results)*, ed. Zachary S. Davis and Benjamin Frankel (London, 1993), 2.

condition, not simply a question of narrow political choice. This semantic excess is not a sign of conceptual weakness, but a recognition of the inability to wholly control the meaning of nuclear events. As the postcolonial cultural critic Homi K. Bhabha puts it (in the context of colonial discourse), ambivalence does not emerge from "the contestation of contradictories [or] the antagonism of dialectical opposition."[20] Ambivalence is rather a "splitting" of discourse, a denial of the possibility of either one or the other side of familiar binaries (e.g., security/insecurity, war/peace), resulting in "multiple and contradictory belief"; splitting is a "strategy for articulating contradictory and co-eval statements of belief."[21] The "strategy" of ambivalence, as Bhabha uses it, is not an instrument of policy under the control of the proliferating state, to be used to deceive or confuse, but rather an effect of the inability of discourse to fix itself unambiguously on one or another nuclear meaning. "Splitting" the discourse of nuclear control is crucial if we are to open up calcified nuclear histories to see what else they can tell us.

One way of doing this is by closer examination of two related and familiar thematics[22] in the telling of nuclear histories: nuclear programs as *national* programs, and the choice of *either destructive or peaceful ends* as natural objectives of all nuclear programs. By demonstrating that no national program can claim to be truly so, and by showing that both war and peace are always present in the meanings attributed to nuclear programs, the discussion opens up nuclear history to explore its intimate relation with the state project of legitimacy in the modern era.

ORIGINS

One of the most enduring tropes of nuclear histories is the idea that atomic energy programs are always *national* programs. The close relation between nuclear power and national power has led to the assumption that, for reasons of security especially, nuclear programs must be uniquely identified with particular countries. Official histories and scientists encourage this belief, for obvious parochial reasons, but it is rarely true. No atomic program anywhere in the world has ever been purely indigenous, nor is it sensible to attribute singular national origins to the scientific efforts to create nuclear fission in laboratories. Given the continental scale of nuclear physics research in prewar years, when scientists from a dozen countries worked together in four or five different countries, it is difficult, and indeed intellectually pointless, to attribute either origins or original successes to one country over another. The scientific importance of nuclear-related discoveries all through the 1930s, in England, in Italy, in Soviet Russia, in Denmark, and in Germany, which culminated in the discovery of nuclear fission by Otto Hahn and Fritz Strassman in December 1938 (published in 1939), guaranteed a wide interest in the latest news from nuclear physics among physicists around the world.[23]

The first effort to create a "national" atomic energy program, the U.S. atomic en-

[20] Homi K. Bhabha, *The Location of Culture* (London, 1994), 131.

[21] Ibid., 132

[22] Another amazing parallel across practically all national narratives of nuclear energy (not developed here) is the figure of the "Father of the [put country name here] Nuclear Program"—a male scientist-bureaucrat-politician who is able to achieve great success in all three domains.

[23] See Atomic Scientists of Chicago, *The Atomic Bomb: Facts and Implications* (Chicago, 1946), 18, for a list of the key publications announcing the discovery of fission and the diverse nationalities of their authors.

ergy bomb project, was inherently a multinational project, with important contributions from British, Canadian, French, and Italian scientists, not to mention the extensive efforts of expatriate German refugees.[24] The Canadian and British atomic energy projects, the latter of which began with the loan of French uranium oxide, derived some of their legitimacy and expertise from experiences gained in the multinational U.S. program. John Lewis and Xue Litai remind us that Chinese scientists worked with Max Born in Edinburgh, in the Joliot-Curies' lab in Paris, and at Pasadena's Jet Propulsion Lab during the war years and, after returning to China, helped build the Chinese nuclear program.[25] The Chinese program began with Soviet help, and scores of Chinese engineers were trained in the schools and labs of the Soviet Union before relations between the two countries broke down. In the early years of their program, the French approached both the Norwegians and the Canadians for help.[26] The Soviet program was built largely through the indigenous efforts of Russian scientists, supplemented by the clandestine work of British and American spies working in the U.S. program.[27] The Norwegians supplied the Israeli program with heavy water and worked closely with the French and the Swedes in the early postwar years and later with the Dutch, a relationship that would lead to the formation of the European nuclear consortium, Urenco, in 1970.[28] The Israeli program was closely tied to the French and Norwegian efforts,[29] and all this before the formation of the International Atomic Energy Agency and a legal regime governing multilateral traffic in nuclear knowledge and materials. It cannot be denied that to some extent international collaboration, especially for the French, was a self-help strategy driven by the legal exclusions of the postwar American nuclear program, notwithstanding the many contributions of non-Americans in its creation.[30]

In 1951, India and France signed an agreement to collaborate, but the agreement did not lead to much by way of practical accomplishments. A few years later, India's nuclear scientists, facing increasing political pressure at home for their lack of manifest achievements, turned, at Sir John Cockcroft's suggestion, to a British swimming pool reactor design that had been published in the trade magazine *Nucleonics.* In addition to design and engineering details, enriched uranium fuel rods were also supplied by the United Kingdom.[31] India's second reactor, the CIRUS (Canada-India-U.S.), was based on a Canadian design, moderated by heavy water supplied by the U.S. Atomic Energy Commission. In India, however, this multinational history would remain largely invisible. At the inauguration of the swimming pool reactor Apsara in January 1957, Nehru would say:

> We are told, and I am prepared to believe it on Dr. [Homi Jehangir] Bhabha's word, that this is the first atomic reactor in Asia, except possibly [in] the Soviet areas. In this sense,

[24] Rhodes, *Making of the Atomic Bomb* (cit. n. 2).
[25] John Wilson Lewis and Xue Litai, *China Builds the Bomb* (Stanford, 1988), 44–5
[26] Lawrence S. Scheinman, *Atomic Energy Policy in France Under the Fourth Republic* (Princeton, 1965).
[27] Holloway, *Stalin and the Bomb* (cit. n. 4).
[28] Astrid Forland, "Norway's Nuclear Odyssey: From Optimistic Proponent to Nonproliferator," *The Nonproliferation Review* 4 (Winter 1997): 1–16.
[29] Cohen, *Israel and the Bomb* (cit. n. 13).
[30] Margaret Gowing, *Independence and Deterrence,* vol. 1, *Policy Making* (New York, 1974).
[31] Itty Abraham, *Making of the Indian Atomic Bomb: Science, Secrecy, and the Postcolonial State* (London, 1998), 84–5.

> this represents a certain historic moment in India and in Asia. . . . We are not reluctant in the slightest degree to take advice and help from other countries. We are grateful to them for the help which they have given—and which we hope to get in future—because of their longer experience. *But it is to be remembered that this Swimming Pool reactor in front of you is the work, almost entirely, of our young Indian scientists and builders.*[32]

A local product, in other words, "almost entirely." The Indian Atomic Energy Commission press release following reactor criticality had Cockcroft grumbling to his colleagues, "Did you see the press release from Delhi? . . . [This characterization of India's achievement] seems rather ungracious in view of the advice and help we have given and are asked to give. Presumably, detailed plant designs and drawings do not constitute outside help!"[33] Why was it so important to insist on the purely national origins of atomic energy?

Modern technology, especially in the postcolony, was always marked with the trace of the foreign. Yet true independence required self-reliance and indigeneity, especially in relation to technology. Seeking approval to set up an atomic energy commission in 1948, Nehru would remind the Constituent Assembly of India that in spite of "its many virtues," India had become a "backward" country and "a slave country" because it had missed earlier technological revolutions, namely those of steam and electricity.[34] This approach defined technological achievement as one of the primary meanings of national independence and elevated the idea of self-reliance to the highest levels of national strategy. In spite of this considerable ideological need for the local and the indigenous, India's large technological projects were almost always the outcome of international collaboration, exchange, aid, and technology transfer. Recognizing the limits of Indian resources and means, Nehru turned to the world for help in building modern India. No matter what the public thought, or was told, the technology being harnessed to transform India was almost always produced in collaboration with foreign countries. No cold war blocs here—the Soviet Union, the United Kingdom, France, West Germany, Poland, and the United States all contributed directly to the building of independent India's dams, steel mills, fertilizer factories, engineering colleges, and cities. Atomic energy was no exception. Even considering that this was the one technology that, given the overwhelming concern with security, one might have expected to be the most privileged and restricted—in a word, nationalized—the Indian atomic energy project was from the outset built in collaboration with multiple foreign partners.

ENDS

Those responsible for the Indian nuclear program had long been aware of the possibility of atomic energy being used to build weapons. In yet another example of the intertwined histories of nuclearization across many sites, we find that both Indian and Soviet scientists became aware of the Manhattan Project before Hiroshima.

David Holloway, in his authoritative study of the Soviet nuclear program, writes:

[32] Nehru, "Apsara," from *Jawaharlal Nehru's Speeches,* vol. 3, *March 1953–August 1957* (Delhi, 1958), 504–5.

[33] Internal memo, n.d., File AB6/1250, Public Records Office, U.K.

[34] *Constituent Assembly of India (Legislative) Debates,* 2nd sess., vol. 5 (Delhi, 1948), 3319–20.

> Early in 1942 Lieutenant [nuclear physicist Georgii] Flerov's unit was stationed in Voronezh, close to the front line. The university in Voronezh had been evacuated, but the library was still there. "The American physics journals, in spite of the war, were in the library and they above all interested me," Flerov wrote later. "In them I hoped to look through the latest papers on the fission of uranium, to find references to our work on spontaneous fission." When Flerov looked through the journals he found that not only had there been no response to the discovery that he and [Konstantin] Petrzhak had made, but that there were no articles on nuclear fission [at all]. Nor did it seem that the leading nuclear physicists [in the West] had switched to other lines of research, for they too were missing from the journals.[35]

Flerov, Holloway reports, concluded that "the Americans were working to build a nuclear weapon."[36] The story told by Flerov, of the "dog that didn't bark," finds an uncanny parallel in India. Govind Swarup, the radio astronomer, reported in an interview some years ago that Homi Jehangir Bhabha, a Cambridge-trained physicist who would become the founder of the Indian nuclear program, had told him that, by 1944, Bhabha, too, had become convinced that the Americans had started a nuclear weapons program.[37] Bhabha's reasoning was similar to Flerov's. He had been in close contact with a number of physicists around the world, largely by letter, through the war years, when he was stuck in India, unable to travel. Letters from colleagues in the United States, always slow because of distance, and made worse by the war, had practically dried up by 1943. Bhabha thought little of it at the time, assuming that the obvious reasons, distance and war, had slowed his mail down. In 1944, still not having heard from his colleagues in spite of a number of letters written by him, Bhabha sat down and made a list of the people who would be likely candidates for a nuclear program. He then made a list of his silent correspondents—the two lists were almost exactly the same.

The near-simultaneous realization by Flerov and Bhabha (and undoubtedly others) that the United States was engaged in a highly secret process to build an atomic weapon should come as no surprise. The potential military implications of these discoveries were also no secret to anyone who had a basic understanding of the fission process, though there was less than unanimity on the exact outcome of a process of nuclear fission.[38] The nuclear physics community in the interwar years was small, close-knit, and multinational. New discoveries were emerging from a relatively small numbers of labs in Europe and the United States and communicated immediately via letter and travel to a transnational epistemic community that eagerly discussed the implications of each new finding.[39]

In the unsettled first decade after Hiroshima, with Europe divided and a hot war breaking out in Korea, many feared that nuclear weapons would be used again. The horror of nuclear weapons led the UN General Assembly to express its "earnest desire," in a resolution introduced by India in 1953, to urge the "Powers principally involved" to sit down and thrash out a means to "eliminate and prohibit" weapons of

[35] Holloway, *Stalin and the Bomb* (cit. n. 4), 78.
[36] Ibid., 48.
[37] Govind Swarup, interview by author, Washington, D.C., Oct. 15, 1995.
[38] Spencer R. Weart, *Nuclear Fear: A History of Images* (Cambridge, Mass., 1988), 77–102.
[39] Daniel J. Kevles, *The Physicists: The History of a Scientific Community in Modern America* (1971; repr., Cambridge, Mass., 1995), 200–86.

"war and mass destruction." Lester Pearson, the influential foreign minister of Canada, spoke for many when he noted: "A third world war accompanied by the possible devastation by new atomic and chemical weapons would destroy civilization."[40] The need to restrain the superpowers, seeing them as the primary source of world insecurity, became for many, aligned and nonaligned alike, the driving consideration of international affairs in the 1950s.

The use of nuclear weapons in Japan had a considerable impact on Indian elites. Mahatma Gandhi, of course, denounced it in no uncertain terms. Responding in typical fashion to the suggestion that atomic weapons were so horrific that they would end war, he wrote:

> This is like a man glutting himself with dainties to the point of nausea and turning away from them only to return after the effect of nausea is well over. Precisely in the same manner will the world return to violence with renewed zeal after the effect of disgust is worn out. . . . The atom bomb . . . destroy[ed] the soul of Japan. What has happened to the soul of the destroying nation is yet too early to see . . . A slaveholder cannot hold a slave without putting himself or his deputy in the cage holding the slave.[41]

The widespread public revulsion against nuclear weapons, especially once the effects of the hydrogen bomb became more widely known, and a desperate need to consider new roads to international peace and development helped shift the discourse around nuclear power. Only a short decade after average Americans polled in a 1946 survey glumly confirmed that "atomic energy means the atomic bomb,"[42] the combination of Atoms for Peace (1953), the first UN-sponsored conference on the peaceful uses of nuclear energy (1955), the Plowshares project, an effort to develop peaceful uses of nuclear explosions, and international competition in the sale of nuclear reactors broke the link between nuclear power and nuclear weapons, at least temporarily. What we should be surprised about is in spite of the intense and repeated association of the nuclear revolution with the use of ever-greater forces of destruction, a divergent but parallel discourse of nuclear power for development and economic growth did emerge. While few questioned the nostrum that the nature of war was now substantially altered as a result of the destructive potential of these weapons, weapons acquisition did not become the only or even primary consideration for countries now facing the real possibility of a global holocaust.

Even countries that began nuclear programs with an explicit intent to develop nuclear weapons, the United States and the Soviet Union in particular, sought to expand the scope of these programs beyond narrowly defined military ends after the war. In both cases, "civilian" technologies were borrowed directly from the military effort. The transfer of technology from the U.S. nuclear submarine project led to the building of civilian, private sector, light water reactors, while Soviet electric power reactors were based on designs taken from a military reactor designed to maximize the availability of plutonium. Not surprisingly, these new civilian programs struggled with their redefinition due to the weight of existing popular sentiments about the destructiveness of atomic power. In David Nye's discussion of the American "techno-

[40] Quoted in Nehru, "The Hydrogen Bomb," in *Speeches* (cit. n. 32), 248.

[41] Mohandas Karamchand Gandhi, "The Atom Bomb, America, and Japan," originally published in *Harijan,* July 7, 1946. Reprinted in *The Gandhi Reader: A Source Book of His Life and Writings,* ed. Homer A. Jack (New York, 1956), 349–50.

[42] Weart, *Nuclear Fear* (cit. n. 37),162.

logical sublime," he develops a genealogy of American technological development that links the U.S. space program with the nuclear program. Nye notes dryly that "[c]onvincing the public that atomic energy was friendly proved difficult, but the space program was popular."[43]

To make the difficult case that nuclear power could be used for peaceful ends, it was necessary to utilize the discursive mediation of other modern technological marvels. Spencer Weart reminds us of some of the remarkable possibilities offered by the peaceful use of atomic power, including "new lands flowing with milk and honey," transforming Africa into "another Europe," and deserts into irrigated land, which led "some Americans [to look] forward to a government operated civilian atomic energy program, an 'atomic TVA' . . . After all, projects already underway, such as the monumental dams of the Tennessee Valley Authority, were scarcely less astonishing."[44] These linkages were not merely rhetorical flourishes: *wunderkind* head of the TVA, David Lilienthal, would be appointed the first chairman of the U.S. Atomic Energy Commission in 1946. Who better to combine, as Weart puts it, the "White City of technology with the green hills of Arcadia"?[45] The Soviets, too, had begun to believe in the possibilities of nuclear power for nonmilitary ends. Given a history of promoting the virtues of Communism through modern technology projects, including "the most ambitious programs in hydro-electric power and canal building in the 20th century, as well as the largest nuclear power plants ever built,"[46] the Soviets discussed using nuclear explosions to change the course of major rivers for irrigation and electricity-generating purposes. "Along with Marxism, a fierce national pride urged Russians to stand second to none in modern technological projects; huge reactors would join huge dams, rockets and steel mills as proofs of [international] pre-eminence."[47] Even in a country where security imperatives would seem to override all others, Israel, atomic energy was more than just that. Avner Cohen quotes Shimon Peres as saying: "Ben Gurion believed that Science could compensate us for what Nature has denied us," and "Ben Gurion's romantic, even mystical faith in science and technology sustained his utopian vision of a blossoming Negev desert and the use of nuclear power to desalinate sea water."[48]

Contradictions abounded. Even as international demands to control nuclear weapons grew, led by countries with large civilian nuclear programs such as Canada and India, the same countries sought to affirm their own national sovereignty and atomic autonomy. Nehru would say:

> [T]he use of atomic energy for peaceful purposes is far more important for a country like India whose power resources are limited, than for a country like France, an industrially advanced country. Take the United States of America, which already has vast power resources of other kinds. To have an additional source of power like atomic energy does not mean very much for them. No doubt they can use it; but it is not so indispensable for them as for a power starved or power hungry country like India or like most of the other countries in Asia and Africa. I say that because it may be to the advantage of countries which have adequate power resources to restrain and restrict the use of atomic power because

[43] David E. Nye, *American Technological Sublime* (Cambridge, Mass., 1994), 225.
[44] Weart, *Nuclear Fear* (cit. n. 37), 158–9.
[45] Ibid., 160.
[46] Loren R. Graham, *Science in Russia and the Soviet Union* (Cambridge, Mass., 1993), 166.
[47] Weart, *Nuclear Fear* (cit. n. 37), 165.
[48] Cohen, *Israel and the Bomb* (cit. n. 13), 11, 353 n. 9.

> they do not need that power. It would be to the disadvantage of a country like India if that is restricted or stopped.[49]

Nehru's ambivalence, expressed through the simultaneous demands for international control over nuclear weapons and domestic sovereignty over India's nuclear development, would be resolved by a discursive shift in the meaning of nuclear energy, aligning it not with destruction but the history of technology and India's colonial past.

> Often our people fail to realise what the modern world is all about. How did Europe and the United States of America advance? Why were they able to conquer us? It is because they had science through which their wealth and economic and military strength grew. Now they have even produced the atom bomb. All these things stem from science and if India is to progress and become a strong nation, second to none, we must build up our science.[50]

NUCLEAR POWER

The ambivalence of the meaning of atomic energy in postcolonial India is demonstrated by the inability to represent this object in terms of either war or peace. Even though India appeared to argue for a peaceful orientation to this new technology, in contrast to the belligerent views present elsewhere, atomic energy was neither one nor the other, "but something else besides." Seeing atomic energy as a necessary means for preventing recolonization and setting newly gained independence on a solid foundation was much more than parliamentary rhetoric to get the 1948 Atomic Energy Act passed. For Nehru and the Indian elite, the central political problem, postindependence, was to create a new basis for Indian nationalism, to project India's strength, and to be taken seriously on the international stage: *to create political legitimacy for the postcolonial state.*

The closest parallel to the Indian program with regard to the larger national-technological meaning of atomic power is probably the French program. Gabrielle Hecht reminds us that "[t]he fundamental premise of discussions about a future technological France was that, in the postwar world, technological achievements defined geopolitical power." She goes on to quote de Gaulle as saying, "A State does not count if it does not bring something to the world that contributes to the technological progress of the world."[51] Both to recapture the "radiance" of France and to offset American dominance in postwar Europe and the world, France needed technology, especially nuclear technology. In the discussion of the first French Five Year Plan, atomic energy was justified by noting the country's lack of traditional energy sources (coal, oil, hydroelectric power). The planners noted: "there is no doubt that in a few years the energy sources put at the disposition of people would so profoundly and radically transform their economic activity that the nations that do not have it will appear as helpless as the most backward nations of the world today appear in the face of modern nations."[52] Given the French image of themselves, what choice did France have?

[49] Nehru, "Control of Nuclear Energy," Speech in the Lok Sabha, May 10, 1954, *Speeches* (cit. n. 32), 255.

[50] Nehru, *Selected Works of Jawaharlal Nehru,* 2nd ser., vol. 28, *1 Feb.–31 May 1955* (Delhi, 2001), 31.

[51] Gabrielle Hecht, *The Radiance of France* (Cambridge, Mass., 1998), 39.

[52] Scheinman, *Atomic Energy Policy in France* (cit. n. 26), 75.

Establishing the base for an Indian atomic energy program was much more than a scheme for building weapons. The urgent political need for national development and state legitimation was intimately wrapped up in the technological success of atomic energy, defined in terms of national strength, uniqueness, and security. "So what should our role be in this dangerous and fast changing world? It is obvious that the first thing is to make ourselves strong and better off to face any danger."[53] The constant iteration of themes of self-reliance, autonomy, independence make it impossible to separate atomic energy from a host of other techno-political projects also begun by the Indian state soon after independence. These included the building of large-scale electricity-generating, flood control, and irrigation management systems, congealed into the sign of the high dam; the urgent creation of a modern industrial base, including a capital-intensive heavy goods industry, steel mills, and iron ore extraction and milling plants, all positively reinforced by repeated images of industrial furnaces and billowing smoke stacks generated by the Ministry of Information and Broadcasting.[54]

The centrality of monumentality and novelty in the representation of these techno-political projects points to their legitimation function for the postcolonial state. Note the similarity in the rhetorical tropes used in Nehru's speeches inaugurating two seemingly very different postindependence technology projects: the massive Bhakra Nangal hydroelectric dam project and the new planned city of Chandigarh, designed by the French architect, Le Corbusier:

> *I do not think that there is any project on such a grand scale being undertaken anywhere else in the world.* The leading countries of the world have many huge schemes, but a project as gigantic and difficult as Bhakra-Nangal is not being undertaken anywhere else. . . . It is a symbol of a nation which is alive and on the move. . . . [T]he biggest advantage is that in the process of accomplishing them, the nation gains vastly in strength.[55]

Speaking in Chandigarh in 1955, Nehru said:

> When you see a new city coming up, you wonder what shape it will take, for no city can be a mere collection of buildings made of brick and mortar. There has to be something more. *It gives a hint of the shape a society will likely take in the future.* So I was especially interested in Chandigarh. I am happy that the people of Punjab did not make the mistake of putting some old city as their capital. It would have been a great mistake and foolishness. . . . If you had chosen an old city as the capital, Punjab would have become a mentally stagnant, backward state.[56]

Taken together, Bhakra Nangal and Chandigarh, the dam and the city, constitute a techno-political genus, more related to each other than their immediate physical surroundings. This family of artifacts was the technological expression of a new form of secular reason (modern technology), dedicated to massive change (national development), authored by the sovereign independent state, epitomizing the desired future in the mundane present. State power and legitimacy was expressed through these techno-political artifacts, representing the means by which social transformation would take place as well as the ends of that change.

[53] Nehru, *Selected Works* (cit. n. 50), 30.
[54] Abraham, *Making of the Indian Atomic Bomb* (cit. n. 31).
[55] Nehru, *Selected Works* (cit. n. 50), 29 (my italics).
[56] Ibid., 26 (my italics).

CONCLUSION

This chapter has argued that a discourse of "control," authored by the overlapping narratives of academic proliferation studies and U.S. antiproliferation policy, has come to dominate contemporary understandings of nuclear histories. This discourse, with its primary purpose of seeking to predict which countries are likely to build nuclear weapons and thereby to threaten the prevailing military-strategic status quo, has narrowed the analytic optic of nuclear historians considerably. Among the effects of this discourse has been to bury important historical details, minimize the importance of the discovery of atomic fission as a "world historical" event, and impoverish recognition of the fluidity of international affairs in the decade following the end of the Second World War. The discursive means that have led to these outcomes are the tendencies to see nuclear histories as, above all, *national* histories and to privilege concerns about the development of nuclear *weapons* over a fuller and more nuanced understanding of what nuclear programs mean and why they matter. Paying attention to the scientific-technological underpinnings of nuclear programs is another analytical path to follow, offering new archives and insights into the making of "national" nuclear programs that might have other, even nonbelligerent, ends. Such an approach recognizes the varieties and importance of international collaboration in the making of "national" programs and shows how weapons building is by no means the only or even the most common end of all nuclear programs.

Returning now to a question that has not received a complete answer: Why don't all countries that could build nuclear weapons do so? My answer: they only appear not to.

The discussion above elaborated the multiplicity of meanings encompassed by the nuclear condition, meanings that might be in contradiction with each other but that continue to be available to different audiences *at the same time*. In particular, it pointed to the familial identity of nuclear programs with other kinds of state-led public technology projects. New cities, enormous dams, soaring skyscrapers, ballistic missiles, space programs, and nuclear power are universal techno-political means by which modern states seek to visualize their power and express their authority. Put simply, modern states have always sought popular legitimacy through massive technology projects: nuclear programs are one of the prime sites for the expression of that desired political relationship.

By exploring the history of the Indian nuclear program, atypical from the vantage point of the first countries to build nuclear weapons, the United States and the Soviet Union, we see the simultaneous presence of "military" and "civilian" programs, arguably from the inception of the nuclear program in the late 1940s. While the political decision to come "out of the nuclear closet" and create a nuclear arsenal in 1998 had a number of proximate causes, it was also importantly influenced by the power of the discourse of control. It is extremely important to postcolonial decision makers not to (be seen to) lose autonomy over this program for all the other meanings signified by nuclear prowess—at the very least, national sovereignty and a claim to universal modernity. If not losing autonomy meant making a decision that would reduce the level of uncertainty of what this program meant, to outside observers and in relation to the prevailing discourse about nuclear programs, it was worth the immediate and corresponding decline in the country's net security. Nuclear explosions may not tell us whether a country is developing a nuclear weapons program, but they do signal a desired dialogue with dominant discourses.

If nuclear programs carry this ideological weight, at the same time as they may (or may not) be a means to produce weapons, then a policy that seeks to reduce the spread of nuclear weapons—counterproliferation—must take that "fact" seriously. If reasons of national sovereignty and the desire to make a unique claim to modernity help us understand why Malaysia builds the Petronas Towers, why Taiwan and China follow suit with even taller buildings, why Japan has a space program, why Brazil, Ghana, and Indonesia, each, at various points in their respective national histories, claimed to be building the largest dam in the world, why China wants to host the Olympics, why countries as different as Brazil, Pakistan, and Nigeria all built expensive, new, technologically sophisticated (if unaesthetic and antisocial) capital cities, why France identifies her atomic reactors in genealogical relation to the Eiffel Tower and why Sydney's Opera House is much more than a building in which to see *Tosca,* then we can see why getting rid of a nuclear program is extremely difficult.

In the 1950s and 1960s, when the discourse of control was far less determinate (and determined) than it is today, countries such as Australia, Norway, and Sweden could decide to close down their fledgling nuclear weapons programs with little ideological pain. Other technological marvels could take their place. Today, it takes a radical reformation of the state—the end of the Soviet Union, the end of white racist rule, and the end of two decades of military rule—for Ukraine, Kazakhstan, South Africa, and Brazil and Argentina, respectively, to give up their nuclear weapons and weapons-building programs. The degree of reformation gives us clues as well to the likelihood of reversibility, to wit, recent stories about the possible return of Brazil to the nuclear ranks.

What countries are giving up, especially democratic ones, when they dismantle their nuclear programs is a claim to a form of national modernity that they once took pride in and took for granted. Little wonder that nuclear "control" is so difficult, especially when the unevenness of the demand to dismantle is as visible as it is today. If there needed to be another reason given as to why the process of global disarmament needs to begin from the top—from those who have the most weapons—it is because the country with the greatest access to the highest forms of modernity is also the best starting point to disabuse the world of the common sense of the relation between nuclear weapons and international prestige.

Prometheus Unleashed:
Science as a Diplomatic Weapon in the Lyndon B. Johnson Administration

*By Ronald E. Doel and Kristine C. Harper**

ABSTRACT

Scholars who have examined science policy within the Johnson administration have generally argued that science played a limited role in U.S. foreign policy in the mid- and late 1960s. Most point to the President's Science Advisory Committee (PSAC), which reached its zenith of influence late in the Eisenhower administration then declined through the Kennedy and Johnson years before being abolished by Richard Nixon in 1973. These accounts, however, have overlooked Lyndon Johnson's determination to employ science and technology as tools in foreign policy and the rapid growth of the State Department's international science office early in his administration. They also overlook the singular importance that Johnson-era officials placed on the physical environmental sciences—especially oceanography and meteorology—as tools of foreign policy. This article, based on archival sources, examines how Johnson administration officials embraced science in diplomatic policy from 1964 through 1968, when rising tensions over Vietnam limited these efforts. Our study includes a detailed examination of one such instance: a secret administration effort to employ weather modification in India and Pakistan as a technological fix to mitigate the Bihar drought and famine of 1966–1967 and to achieve U.S. policy goals in this strategically important region.

INTRODUCTION

Science and technology have long been used as tools of the state. Their use intensified in the second half of the twentieth century—particularly after the launch of Sputnik in 1957—when scientific and technological achievements came to symbolize the strength and vitality of nations.[1] In 1960, Paul Nitze, the noted nuclear arms negotiator, de-

* Ronald E. Doel: Department of History, Oregon State University, 306 Milam Hall, Corvallis, OR 97331; doelr@geo.oregonstate.edu. Kristine C. Harper: Humanities Department, New Mexico Institute of Mining and Technology, Socorro, NM 87801; kharper@nmt.edu.

We gratefully acknowledge support from the Johnson Presidential Library travel grants program. This work was also supported by the National Science Foundation (Grants No. SBR-9511867 and DIR-9112304) (Doel) and the American Meteorological Society Graduate Fellowship in the History of Science and the Dibner Institute for the History of Science and Technology, Cambridge, Mass. (Harper). Finally, we are grateful to participants at the Georgetown workshop, particularly its organizers, Kai-Henrik Barth and John Krige, and to Paul Forman, Zuoyue Wang, and Aaron Wolf.

[1] Ronald E. Doel and Zuoyue Wang, "Science and Technology in American Foreign Policy," in *Encyclopedia of American Foreign Policy,* rev. ed., ed. Alexander DeConde, Richard Dean Burns, and Fredrik Logevall (New York, 2001), 443–59. For a Soviet perspective, see Nikolai Krementsov, *The Cure: A Story of Cancer and Politics from the Annals of the Cold War* (Chicago, 2002).

clared: "the most important tool of foreign policy is prestige."[2] A high-level report on U.S. science declared the following year: "Our scientific 'prestige' is an increasingly important component in our international bargaining power, perhaps even more important with those who know little of science than with those who do."[3] White House aides determinedly sought new ways to promote their administration's foreign policy goals through scientific achievements and technological applications.

Until recently, most studies of the role of science as an element of U.S. foreign policy have largely focused on physics. In August 1945, the atomic bomb created a dramatic new role for science in foreign policy, and physicists became visible figures in efforts to negotiate treaties, shape world opinion, and articulate models of international governance.[4] In the United States, the Eisenhower administration—seeking a peaceful application of nuclear weapons—advocated "atoms for peace," while the Atomic Energy Commission (AEC) emphasized through Project Plowshare the applications of atomic energy toward civil engineering problems.[5] Post-Sputnik, the U.S. space program also served to demonstrate the importance of American science and technology as symbols of the West's vitality and to persuade newly independent nonaligned nations to follow the West's lead.[6]

Largely missing from these accounts, however, are the roles played by members of the geophysical sciences—the realm of the earth sciences that might best be termed the "physical environmental sciences" to distinguish them from the biological environmental sciences.[7] Their fields were among the most militarily strategic, since advanced weapon systems critical to U.S. defense depended on geophysical knowledge. As field scientists, dependent upon globally collected data, geophysicists were also internationally minded, active players in national security and foreign policy networks. Finally, some geophysicists held out the promise of controlling nature—including

[2] Quoted in Eugene B. Skolnikoff, *Science, Technology, and American Foreign Policy* (Cambridge, Mass., 1967), 209.

[3] Joseph Pratt, *Science in UNESCO: United States Interest in Science Abroad,* Aug. 16, 1961, Box 9, Frank Press Papers, MIT Archives, Cambridge.

[4] Relevant literature is voluminous. See, e.g., Lawrence Badash, *Scientists and the Development of Nuclear Weapons: From Fission to the Limited Test Ban Treaty, 1939–1963* (Atlantic Highlands, N.J., 1995); Gregg Herken, *Cardinal Choices: Presidential Science Advising from the Atomic Bomb to SDI* (New York, 1992); Daniel J. Kevles, "Cold War and Hot Physics: Science, Security, and the American State, 1945–56," *Historical Studies in the Physical and Biological Sciences* 20 (1990): 239–64; Melvyn P. Leffler, *A Preponderance of Power: National Security, The Truman Administration, and the Cold War* (Stanford, Calif., 1992); Spencer R. Weart, *Scientists in Power* (Cambridge, Mass., 1979).

[5] On "atoms for peace," see Bruce W. Hevly and John M. Findlay, *The Atomic West* (Seattle, Wash., 1998); Richard G. Hewlett and Jack M. Holl, *Atoms for Peace and War, 1953–1961: Eisenhower and the Atomic Energy Commission* (Berkeley, 1989); and Martin J. Medhurst, "Eisenhower's 'Atoms for Peace' Speech: A Case Study in the Strategic Use of Language," *Communication Monographs* 54 (1987): 204–20. On Project Plowshare, see, e.g., Barton C. Hacker, *Elements of Controversy: The Atomic Energy Commission and Radiation Safety in Nuclear Weapons Testing, 1947–1974* (Berkeley, 1994); and Peter Coates, "Project Chariot: Alaskan Roots of Environmentalism," *Alaska History Magazine* 4 (Fall 1989): 1–31.

[6] Rip Bulkeley, *The Sputnik Crisis and Early United States Space Policy: A Critique* (Bloomington, Ind., 1991); Robert A. Divine, *The Sputnik Challenge* (New York, 1993); W. Henry Lambright, *Powering Apollo: James E. Webb of NASA* (Baltimore, 1995); Roger D. Launius, *NASA: A History of the U.S. Civil Space Program* (Malabar, Fla., 1994); Roger D. Launius, John M. Logsdon, and Robert W. Smith, *Reconsidering Sputnik: Forty Years Since the Soviet Satellite* (Amsterdam, 2000); John M. Logsdon, *The Decision to Go to the Moon: Project Apollo and the National Interest* (Cambridge, Mass., 1970); Walter A. McDougall, *The Heavens and the Earth: A Political History of the Space Age* (New York, 1985).

[7] Ronald E. Doel, "Constituting the Postwar Earth Sciences: The Military's Influence on the Environmental Sciences in the USA after 1945," *Social Studies of Science* 33 (2003): 635–66.

tools for exploiting the oceans and deliberately modifying the weather—that tempted state leaders. Assessing the contributions of the physical environmental sciences is thus an important step in reassessing the interplay between science and foreign policy in mid-twentieth-century America.

Lyndon Johnson's fascination with the physical environmental sciences—particularly his faith in weather modification and the control of nature—is thus especially revealing. For Johnson, the physical environmental sciences were more than national security tools: he sought to apply these sciences to alleviate the suffering and raise the standard of living of the domestic population and U.S. allies, finding them a political tool of the utmost importance. Yet until now his secret use of science and technology to "fix" emerging environmental problems while bolstering foreign regimes has remained hidden.[8] Johnson's attempt to control the weather in India to aid food production challenges existing historical accounts that emphasize extraordinarily tense relations between the United States and India created by Johnson's food policy, which left India begging for grain as monsoon rains failed.[9] But while he very publicly withheld grain shipments, behind the scenes Johnson was deploying a secret diplomatic weapon: a highly classified method for augmenting rainfall. By taming nature, the president hoped to tame the world.

NATIONAL SECURITY, U.S. FOREIGN POLICY, AND THE ENVIRONMENTAL SCIENCES

From the start of the cold war, the physical environmental sciences began gaining influence within U.S. foreign policy as the result of two distinct, yet overlapping, developments. The first was the growing importance of issues at the intersection of natural resource policy, international law, and U.S. foreign relations. The spread of nuclear fallout made environmental pollution an international concern, stimulating efforts to forge a nuclear test ban treaty.[10] Finding a political solution to competing claims for the Antarctic continent—a key issue for U.S. diplomacy since the end of World War II—came to hinge on making Antarctica a "continent for science." The International Geophysical Year, 1957–1958, involving extensive Antarctic studies, cemented this thrust and helped form the framework of the Antarctic Treaty, signed in 1959.[11] Oceanography played a role as intensified fishing practices—including the development of factory trawlers capable of operating at unprecedented distances from their home ports—and growing interest in seafloor mining helped spawn the first U.N. Conference on the Law of the Sea in 1958. Debates over the extent of international

[8] Modern presidential historians generally have not found science and technology compelling themes. For instance, Robert Dallek's massive LBJ biography contains just one reference to Johnson's science adviser; see Robert Dallek, *Flawed Giant: Lyndon Johnson and His Times, 1961–1973* (Oxford, 1998), 424.

[9] See, e.g., Francine R. Frankel, *India's Political Economy 1947–1977: The Gradual Revolution* (Princeton, 1978); H. W. Brands, *India and the United States: The Cold Peace* (Boston, 1990); Lloyd I. Rudolph and Susanne Hoeber Rudolph, *The Regional Imperative: The Administration of U.S. Foreign Policy towards South Asian States under Presidents Johnson and Nixon* (Atlantic Highlands, N.J., 1980). For broader context, see Robert J. McMahon, *The Cold War on the Periphery: The United States, India, and Pakistan* (New York, 1994).

[10] Robert A. Divine, *Blowing in the Wind: The Nuclear Test Ban Debate, 1954–1960* (Oxford, 1978); Kai-Henrik Barth, "The Politics of Seismology: Nuclear Testing, Arms Control, and the Transformation of a Discipline," *Soc. Stud. Sci.* 33 (2003): 743–81.

[11] Aant Elzinga, "Antarctica: The Construction of a Continent by and for Science," in *Denationalizing Science: The Contexts of International Scientific Practice,* ed. Elisabeth Crawford, Terry Shinn, and Sverker Sörlin (London, 1992), 73–106.

waters, nuclear waste disposal, and deep-sea mining claims all highlighted the increasing relevance of the environmental sciences for U.S. foreign policy.[12] As a time of intense geographic exploration, it is hardly surprising that the physical environmental sciences were resurgent.[13] All emerged as significant issues for the White House.

The second development—less visible at the time because of secrecy and national security concerns—was the importance of the physical environmental sciences for national defense and military operations. By the mid-1950s, almost all fields of the earth sciences, including meteorology, upper-atmospheric research, ionospheric studies, solar-terrestrial relations, geodesy, terrestrial magnetism, and oceanography, were specifically identified as critical to the national military establishment—particularly the development of ballistic missile systems and antisubmarine warfare.[14] Understanding the operating environment became one of the most pressing areas of military research. This need drove an expansion of geophysical research and the creation of new geophysical and oceanographic research institutions.[15] International atomic issues predominated in public and classified discussions, but the physical environmental sciences were a close second in funding and influence on foreign policy. When Washington senators Henry M. "Scoop" Jackson and Warren G. Magnuson protested the U.S. position at the Second U.N. Conference on the Law of the Sea, in 1960, arguing that it harmed fishing interests, they learned that submarine nuclear defenses trumped fishing concerns.[16]

For many geophysicists, international policy concerns and military applications were integral parts of a unified, seamless continuity. Unlike the work of their laboratory-based colleagues, the work of geophysicists in many cases depended upon the immediate, global sharing of data. Seeking global programs, meteorologists developing increasingly sophisticated numerical weather prediction models (to meet both military and domestic civilian needs) pointed out that they were totally dependent upon surface and upper-air weather observations shared by all nations—no matter what their ideologies.[17] Highlighting their contributions to national security, many geophysicists became active in efforts to expand the role that earth scientists could play in international affairs. They began intensive political outreach efforts to persuade the White House and Congress of the strategic value of developing cooperative international earth sciences programs.[18]

The launch of the IGY-connected Sputnik in October 1957 earned the physical

[12] Ann L. Hollick, *U.S. Foreign Policy and Law of the Sea* (Princeton, 1981), 144–59; Jacob Darwin Hamblin, *Oceanographers and the Cold War: Disciples of Marine Science* (Seattle, 2005).

[13] On geography's relationship to the state, see Steven J. Harris, "Long-Distance Corporations, Big Sciences, and the Geography of Knowledge," *Configurations* 6 (1998): 269–304.

[14] For overviews, see John Cloud, "Introduction," and Michael A. Dennis, "Earthly Matters: On the Cold War and the Earth Sciences," in *Soc. Stud. Sci.* 33 (2003): 629–34 and 809–19.

[15] Doel, "Constituting the Postwar Environmental Sciences" (cit. n. 7), 635–66; and Naomi Oreskes, "A Context of Motivation," *Soc. Stud. Sci.* 33 (2003), 697–742.

[16] Secretary of State Christian Herter to Sen. Henry Jackson, March 10, 1960, Jackson 3560-3, Box 58, Folder 18a, Henry Jackson Papers, Special Collections, University of Washington, Seattle.

[17] Data requirements for successful numerical weather prediction are discussed in Kristine C. Harper, "Boundaries of Research: Civilian Leadership, Military Funding, and the International Network Surrounding the Development of Numerical Weather Prediction in the United States" (Ph.D. diss., Oregon State Univ., 2003); see also idem, "Research from the Boundary Layer: Civilian Leadership, Military Funding, and the Development of Numerical Weather Prediction (1946–55)," *Soc. Stud. Sci.* 33 (2003): 667–96.

[18] See, e.g., Joseph Kaplan to Lloyd Berkner, Sept. 25, 1953, Box 1, USNC/IGY, International Geophysical Year collection, National Academy of Sciences Archive, Washington, D.C.

environmental sciences additional stature and authority within the White House and the State Department. The election of John F. Kennedy as president in 1960 further served to spotlight these fields. More than Eisenhower, Kennedy made the nuclear test ban treaty negotiations a national priority.[19] He supported large increases in the public-enthralling space program. In oceanography, the so-called wet space program, Kennedy recognized the growing dilemma of looking to the oceans for food while using them as a dumping site for radioactive wastes. He also supported plans for the large-scale Indian Ocean expedition to gather detailed information about the biology, meteorology, and physical oceanography of this politically sensitive region, allowing researchers to "show the flag" while promoting international cooperation.[20]

The Kennedy administration became interested in the environmental sciences for yet another reason: the potential of large-scale physics-based experiments to tarnish the U.S. reputation abroad. Kennedy seemed quite gratified by the Limited Test Ban Treaty's eventual passage in 1963, since it indicated the nation's commitment to limiting radioactive fallout. Yet the United States had been embarrassed in October 1962 when the *New York Times* leaked that the government had tested a nuclear bomb in space more than a hundred miles above Hawaii. Aware, too, that the AEC plan to detonate several nuclear weapons in Alaska to create an artificial harbor (code-named Project Chariot) was sparking intense controversy, Kennedy sought to limit governmental programs with environmental impacts.[21] On April 17, 1963, the president issued a secret directive from the National Security Council, National Security Action Memorandum 235, to all members of his cabinet, insisting that any large-scale scientific or technological experiments with possible adverse environmental effects (either physical or biological) be reviewed in advance for their potential harm.[22] Seeking to reassure the American scientific community, Kennedy addressed the National Academy of Sciences one month before his death, acknowledging concern about deliberate environmental modification. While in the past such modification had mainly been inadvertent, Kennedy noted, for the first time science "could undertake experiments with premeditation which can irrevocably alter our physical and biological environment on a global scale."[23]

Kennedy's speech touched on the principal environmental sciences concerns that had come into focus since the late 1950s, including conservation policy as well as understanding and exploiting the sea and atmosphere. One final issue that Kennedy raised—much in the news at that time—was deliberate weather modification. In his

[19] Zuoyue Wang, "American Science and the Cold War: The Rise of the U.S. President's Science Advisory Committee" (Ph.D. diss., Univ. of Santa Barbara, 1994), 250–4.

[20] Oceanography concerns were reported in H. K. Bourne, "Biological Implications of Radioactive Isotopes in the Sea," FV 7/52, The National Archives, Kew, UK. The Indian Ocean expedition is discussed briefly in Skolnikoff, *Science* (cit. n. 2), 62; on ocean pollution, see Jacob Darwin Hamblin, "Environmental Diplomacy in the Cold War: The Disposal of Radioactive Waste at Sea during the 1960s," *International History Review* 24 (2002): 348–75.

[21] Divine, *Blowing in the Wind* (cit. n. 10); on Alaska's Project Chariot, see Dan O'Neill, *The Firecracker Boys* (New York, 1995).

[22] National Security Action Memorandum No. 235/1, April 17, 1963, Foreign Relations of the United States (FRUS), *Organization of Foreign Policy; Information Policy; United Nations; Scientific Matters,* Vol. 25 (2001), item #352. This memorandum directed the CIA and the Department of State (among other agencies) to undertake advance reviews of potentially controversial programs in the environmental sciences; it emerged from discussions within PSAC, reported in "Notes on International Cooperation in Science," Box 16, Detlev Bronk Papers, Rockefeller Archive Center, Sleepy Hollow, N.Y.

[23] See "Text of Kennedy's Address to Academy of Sciences," *New York Times,* Oct. 23, 1963, 24.

October academy speech, the president noted that the state needed to work to ensure that the potential benefits of weather control were not outweighed by their risks—"against the hazards of protracted droughts or storms."[24] Just one month before his speech, the *New York Times* had editorialized on "controlling the weather," declaring that ambitious schemes to "improve the weather in one area" might well come "at the expense of that in another area. When control of the weather actually becomes possible, argument about who should gain and who should lose could become significant sources of international tension."[25]

Thus when President Johnson later began thinking seriously about weather modification as a tool of the state, he was engaged not in marginal scientific undertakings but in issues that, like the broader environmental sciences themselves, had become central to U.S. foreign policy and to the practice of international law.

LYNDON JOHNSON'S AIMS FOR U.S. SCIENCE AND TECHNOLOGY POLICY

As president, Lyndon Johnson had a clear idea of how he wished to utilize science policy. On one of his first days as science adviser to the president, Donald Hornig, a Harvard-trained chemist, listened as Johnson spelled out his aims for science and technology policy. Several impressions stayed with him. He later recalled Johnson's emphasis on the *applications* of science: "for $18 billion a year," Johnson told him, referring to the total federal R&D budget, "there ought to be something to say at least once a week." Johnson, Hornig also quickly grasped, "saw everything in political terms."[26]

Johnson's attitude toward science and technology had its origins in—indeed was deeply shaped by—experiences in his childhood.[27] Applications of technology became key issues in his earliest political campaigns. Born in the arid Hill Country of central Texas in 1908, Johnson came of age as radio, movies, and electricity rapidly spread across the nation. In this poor and sparsely populated region, however, neither electricity, radios, nor paved roads existed through the 1930s. After his election to the House of Representatives in 1937, Johnson successfully brought the New Deal to central Texas by securing federal funds to dam the Colorado River for local power production.[28] Celebrating his victory, Johnson had declared that Texans could now "turn the vicious Colorado, which for centuries had gone whooping and snorting down the valleys on its sprees of destruction, into the quiet ways of work and peace."[29] A dedicated New Dealer, Johnson embraced the early twentieth-century ethos that technology could improve the material lives of ordinary citizens.

Another closely related and equally characteristic attitude—Johnson's drive to use technology to modify and improve the natural environment—also stemmed from his childhood experiences. While growing up, Johnson never forgot the anxiety of worrying whether vital rain would come. His faith in technological progress was linked

[24] Ibid.

[25] Editorial, "Controlling the Weather," *New York Times,* Sept. 25, 1963, 42.

[26] Donald Hornig, "The President's Need for Science Advice: Past and Future," in *Science Advice to the President,* ed. William T. Golden (New York, 1980), 42–52, on 47, 50; Wang, "American Science and the Cold War" (cit. n. 19), 258.

[27] "For all the patterns of his life have their roots in this land," stated by Robert A. Caro in *The Years of Lyndon Johnson: The Path to Power* (New York, 1982), xxiii.

[28] Ibid.

[29] Jordan A. Schwarz, *The New Dealers: Power Politics in the Age of Roosevelt* (New York, 1994), quoted on 274.

to a perception that water policy was fundamental to domestic politics, an issue he later came to see as global. Journalist Hugh Sidey, visiting the president at his family homestead in the mid-1960s, wrote that Johnson was "quite convinced that adequate water in the areas of shortage and control of the water in the areas of surplus could do more for peace than just about any technological breakthrough."[30]

Science and technology policy retained a fascination for Johnson as he rose politically. As Senate majority leader, Johnson championed legislation that established the National Aeronautics and Space Administration. Initially seeing space as a matter of national defense, Johnson quickly grasped its potential for showcasing U.S. science and technology around the world. He similarly encouraged an applied role for science and technology when he backed the proposed East-West Center in Hawaii, arguing that the United States needed to step up efforts to train young Asian intellectuals and researchers.[31] As vice president, he became Kennedy's point man for the space program.[32] These commitments foreshadowed two themes of Johnson's future presidency: a deep-rooted New Deal enthusiasm that state-backed technological systems could be used to improve the living standards of all Americans (indeed, of individuals around the globe), and a faith that technological centers, including the military-industrial complex, were critical for promoting the economic growth of disadvantaged citizens as well as advancing U.S. military strength.[33]

Until now, most scholars have argued that Johnson primarily sought to use science and technology to further domestic aims.[34] For instance, in his history of the President's Science Advisory Committee (PSAC), founded after the 1957 launch of Sputnik, Zuoyue Wang argued that compared with Eisenhower and Kennedy, Johnson "liked to focus on domestic and not international affairs."[35] Environmental historians such as Samuel P. Hays have emphasized Johnson's commitment to addressing environmental pollution and improving the environment—perhaps because Johnson wished to extend Kennedy's environmental interests (illustrated by his public embrace of Rachel Carson's *Silent Spring* and making natural resources conservation a high administration priority).[36] There is no doubt that Johnson did care deeply about natural resource and environmental issues. PSAC's 1965 report on environmental pollution—largely confirming Carson's concerns about the overuse of pesticides—was one of the Johnson administration's most influential science reports.[37]

[30] Hugh Sidey, *A Very Personal Presidency: Lyndon Johnson in the White House* (New York, 1968), on 16; see also 137.

[31] Office of the President, University of Hawaii, memorandum for the files, Laurence H. Snyder, Aug. 13, 1959, Chancellor's Records, Univ. of Hawaii Coll., 1991:003, University of Hawaii Archives, Manoa.

[32] Dallek, *Flawed Giant* (cit. n. 8), 22–3; Robert A. Caro, *The Years of Lyndon Johnson: Master of the Senate* (New York, 2002), 1028–30. On the recent historiography of Johnson's foreign policy, see Thomas Alan Schwartz, *Lyndon Johnson and Europe: In the Shadow of Vietnam* (Cambridge, Mass., 2003), especially 1–8.

[33] Schwarz, *New Dealers* (cit. n. 29), 266, 270, 283.

[34] For instance, A. Hunter Dupree, "A Historian's View of Advice to the President on Science: Retrospect and Prescription," *Technology and Society* 2 (1980): 175–90; and William G. Wells Jr., "Science Advice and the Presidency: An Overview from Roosevelt to Ford," in Golden, *Science Advice* (cit. n. 26), 191–220, 206.

[35] Wang, "American Science and the Cold War" (cit. n. 19), 258.

[36] Samuel P. Hays, *Beauty, Health, and Permanence: Environmental Politics in the United States, 1955–1985* (New York, 1985); Dallek, *Flawed Giant* (cit. n. 8), 83.

[37] Wang, "American Science and the Cold War" (cit. n. 19), 260; Dallek, *Flawed Giant* (cit. n. 8), 229–30.

Nevertheless, other scholars have noticed that Johnson was no less interested in employing technology and science to aid U.S. foreign policy.[38] Certainly like his immediate predecessors, he was concerned with ballistic missile development, basic research and graduate education, assessments of Soviet science, scientific exchanges with the Soviet Union, and the problems of international science.[39] But Johnson's interests ranged considerably beyond these points. For instance, Johnson demanded that Hornig find ways to provide technical assistance to developing countries he visited on formal state visits. Working with State's science experts, Hornig crafted cooperative programs in space technology, typhoon damage control, and oceanography to offer Philippine president Ferdinand Marcos when Johnson visited Southeast Asia in 1966; on that same trip, in South Korea, Johnson announced U.S. support for a new Korean Institute for Industrial Technology and Applied Science.[40] Johnson also demanded that his science advisers work to ameliorate the perceived "technological gap" involving Western Europe. Growing fears of Western European nations that they were losing the battle to regain their prewar technological footing made the technological gap a potent, but short-lived, flash point in U.S. foreign policy.[41] Johnson ordered his White House scientists to address the "gap" issue, determine its accuracy, and take appropriate steps as necessary. Later, State's science officials remembered it as one of the most time-consuming challenges of the Johnson presidency.[42]

Johnson's fascination with the physical environmental sciences—and their implications for foreign policy—is less well known. In part this had to do with secrecy: Johnson's most incisive use of environmental sciences applications in India and Pakistan, and later in Vietnam, as the Vietnam War escalated, was carried out entirely in secret. Another critical factor was the tendency of contemporary outside observers to not recognize the significance of the environmental sciences as a cohesive field, even if key Johnson advisers did.[43] (Hornig referred to "environmental sciences" as shorthand for "aeronomy, geology, geodesy, seismology, hydrology, meteorology, oceanography, and cartography," placing biological environmental sciences in a separate mental category.)[44] One of the first initiatives that Johnson sought to promote—not surprising, given his upbringing—was international water policy. In 1964, Johnson sought to focus efforts on nuclear-powered desalination. He launched a major international program—Water for Peace—and lobbied Congress hard to adequately fund

[38] W. Henry Lambright, *Presidential Management of Science and Technology: The Johnson Presidency* (Austin, 1985).

[39] On PSAC's assessment of key national security concerns, see Hornig to president, memorandum, Feb. 17, 1964, Box 1, Hornig Papers, Lyndon Baines Johnson Presidential Library, Austin (hereafter cited as Hornig-LBJ).

[40] The Korean Institute idea had first been broached during a 1965 state visit to the United States by the South Korean president; see Hornig, memorandum for the files, Oct. 7, 1965, Box 3, Hornig-LBJ.

[41] Jean-Jacques Servan-Schreiber, *Le Défi américain* (Paris, 1967); Johnson to department secretaries, Nov. 25, 1966, Confidential File Container 85: SC Sciences, Lyndon Baines Johnson Presidential Library, Austin (hereafter cited as LBJ); Eugene V. Kovach, interview with Ronald E. Doel, June 28, 2001, American Institute of Physics, College Park, Md.

[42] Kovach interview (cit. n. 41).

[43] In 1965, the Environmental Science Services Administration (ESSA) was created under the Department of Commerce, a step urged by geophysicist Lloyd V. Berkner to make the environmental sciences as visible as physics. It is not clear, however, what role Johnson played in this consolidation; see Hornig to J. B. Lucke, Sept. 17, 1965, Box 2, Hornig-LBJ; see also Luther J. Carter, "Earthquake Prediction: ESSA and USGS Vie for Leadership," *Science* 151 (1966): 181–3.

[44] Hornig to Joseph Califano, memorandum, Jan. 25, 1966; see also Hornig to Charles Schultze, Director, Bureau of the Budget, memorandum, Jan. 10, 1966, both Box 3, Hornig-LBJ.

the project.[45] Two years later, the president also accepted a proposal to speak at the dedication of a new research vessel (R/V *Oceanographer*) from his White House advisers, who saw this as a chance for Johnson to "talk not only about the resources of the sea but in broader terms about worldwide peace, higher standards of living throughout the world, feeding a rapidly growing population, etc."[46] In 1966 as well, Johnson gave strong backing to the World Weather Watch, a World Meteorological Organization cooperative program first proposed and vetted in the early 1960s, that promised the United States critical data while allowing the state to proclaim its commitment to international scientific cooperation and nation building.[47] None of these individual efforts was individually surprising, but as a whole they formed a pattern: Johnson devoted more effort to practical environmental sciences initiatives than to initiatives in any other scientific fields, showing little inclination to promote pure science efforts with less certain practical benefits.[48] If anything, Hornig wanted Johnson to broaden his efforts in environmental sciences research, urging him to pay more attention to marine policy and to deep-ocean operations, particularly following the U.S. inability to recover its sunken nuclear submarine *Thresher* in 1963 and its embarrassing loss of an atomic bomb off Spain.[49]

Each of these presidential activities—Johnson's public speeches and presidential initiatives—represented typical ways that White House occupants sought to push their policies and views. What is extraordinary was the extent to which Johnson sought to use science and technology, particularly the environmental sciences, to micromanage U.S. foreign policy and the internal affairs of other nations. Seeking results—aware that he had inherited from Kennedy a world that was increasingly unresponsive to U.S. leadership[50]—Johnson also undertook actions that bypassed and ignored the views of his science advisers in his pursuit of desired political outcomes. He particularly did so in two countries where the United States was already promot-

[45] On desalination, see Hornig, memorandum for the record of conversation with president, July 9, 1964, Box 1; on Water for Peace, see Hornig to Hon. Joseph J. Sisco, Dept. of State, memorandum, Oct. 25, 1965, Box 3, both Hornig-LBJ; and Bernard J. Rotklein to Mr. Shaver, memorandum, Oct. 12, 1966, RG 59, Bureau of International Scientific and Technological Affairs, Box 17, National Archives and Records Administration II (hereafter cited as NARA II), College Park, Md.

[46] Robert E. Kintner to Will Sparks and Bob Hardesty, memorandum, July 6, 1966, Confidential Files RA, Container 85, Oceanography, LBJ. In this same year, the Johnson administration decided to give India the oceanographic research ship *Anton Bruun;* see Hornig, memorandum for president, March 11, 1966, Box 3, Hornig-LBJ. At the conclusion of a dry dock period prior to transfer to India, however, *Anton Bruun* sank, sustaining damage beyond economical repair.

[47] See Charles E. Johnson to Bromley Smith, memorandum accompanying Draft Press Release on World Weather Watch, Sept. 22, 1966, NSF Subject File, Container 52, Folder: "World Weather Watch," LBJ. In 1962, under the Kennedy administration, meteorological projects involving data exchanges had been vetted by the Pentagon, the CIA, and key members of Congress and found "quite safe," worth promoting because they aided U.S. foreign policy goals while strengthening international cooperation; see FRUS (cit. n. 22) #387, cover note dated July 13, 1962.

[48] While Johnson certainly spent considerable energy promoting space exploration and space flight, here, too, he especially emphasized such developments as meteorological satellites, which overlapped directly with the environmental sciences.

[49] See, especially, Hornig, memorandum for president, April 1, 1966, Box 3, Hornig-LBJ; on these incidents see Oreskes, "Context of Motivation" (cit. n. 15). Hornig reiterated this point several months later, writing Johnson that "National Security poses the most urgent requirement for effective use of the sea." Hornig, memorandum to president, June 7, 1966, Box 3, Hornig-LBJ.

[50] On Johnson's concern about the limits of U.S. effectiveness see Warren I. Cohen, introduction to *Lyndon Johnson Confronts the World: American Foreign Policy, 1963–1968,* ed. Warren I. Cohen and Nancy Bernkopf Tucker (New York, 1994), 1–8, on 1.

ing more traditional forms of science and technology support, including reactor design, science education programs, and advanced cardiac clinics: India and Pakistan.

JOHNSON, SCIENCE, AND FOREIGN POLICY IN THE INDIAN SUBCONTINENT

The primary objective of cold war U.S.-India foreign policy was to pull nonaligned India—the world's most populous democracy and a vital linchpin in U.S.-Asia policy—into the Western sphere as a bulwark against the Soviet Union and Communist China. It was in South Asia that the environmental sciences and U.S. foreign policy came to intersect when Johnson decided to use environmental sciences applications as a foreign policy tool in India and Pakistan in early 1967. The science was meteorology. The tool was weather control.

Weather control fulfilled two disparate foreign policy goals toward this objective. First, State wanted to dissuade India from becoming a nuclear nation. Sporadic border disputes with neighboring China (Aksai Chin plateau—1962) and Pakistan (Rann of Kutch and Kashmir—1965), from which India emerged politically diminished, had threatened to escalate into major destabilizing conflicts. However, it was the 1964 explosion of China's first nuclear device that had most seriously damaged India's self-esteem. By joining the nuclear nations, China had usurped India's claim to a prestigious position in science and technology—a position that needed to be reclaimed if India were to maintain influence with Asian and African nations. Pressure began to build within India to start a nuclear weapons program.[51]

During the cold war, in India, as in other developing countries, industrialization and weapons programs primarily served a political purpose: to reach equity with more powerful nations. As political scientist Hans J. Morgenthau noted, these potent symbols of modernity and power were, and remain, crucial to the self-image of emerging nations. Similarly, as the historian George Perkovich has argued, too often U.S. policy makers in the 1960s tended to see these issues through a security framework, when in fact they were efforts to achieve increased national prestige and status and the economic benefits that accrue from them. Nevertheless, most Western policy makers did assume that all nations wanted to improve their lots through the application of advanced science and technology.[52]

Certainly, that was foremost in the minds of State Department officials in mid-1966 as they grappled with India's perceived motivations for "going nuclear." State ultimately recommended to Johnson that the India desk make a special examination of "more specific steps that might be taken to enhance India's political prestige, including scientific and technical projects." Johnson approved.[53] Of particular value would be "dramatic uses of modern technology to attack India's basic problems of food, population, health and education."[54] Hornig, with State's science office, sought to identify ways that U.S. scientific programs could both aid India and bolster U.S. foreign

[51] Itty Abraham, *The Making of the Indian Atomic Bomb: Science, Secrecy, and the Postcolonial State* (London, 1998), 124–5; George Perkovich, *India's Nuclear Bomb* (Berkeley, 1999), 6–7.

[52] Hans J. Morgenthau, *A New Foreign Policy for the United States* (New York, 1969), 93. See also Perkovich, *India's Nuclear Bomb* (cit. n. 51), 8–13.

[53] W. E. Gathright to Garthoff, Schneider, Coon, and Weiler, July 1, 1966, Box 15, Folder: Def 18-2, State Department Records, Entry 5255-NARA II.

[54] Raymond A. Hare to Herman Pollack, Sept. 7, 1966, Box 15, Folder: Def 18-1, State 5255-NARA II.

policy.[55] India's strong history of outstanding achievement in meteorology, for example, might make a cooperative effort to place a geosynchronous satellite in orbit over the Indian Ocean a viable possibility.[56]

It was against this backdrop of military aggression that India suffered a serious crop failure in 1965 when the summer monsoon—India's rainy season—failed. Hardest hit was the state of Bihar, with a population of more than 51 million and a primary grain crop heavily dependent upon water: rice. Without water, the paddies dried up and the rice shriveled and died. Although crops can be saved with irrigation, Bihar was dependent almost exclusively upon rainfall. During a good monsoon season, the rice crop was good. When the monsoon failed, the rice crop failed with it.[57] India had depended upon U.S. grain shipments since the mid-fifties, using imports to provide cheap food to the masses while its capital was invested in heavy industry instead of a strong agricultural base. India's attempts to reform its inefficient agriculture program through the use of increased amounts of chemical fertilizer, high-yield seeds, and privatization, had largely failed because low government-imposed prices eliminated the financial incentive to incur the costs associated in producing higher-yield harvests.[58] However, even fertilizer and high-yield seeds are of no use when there is no water. In 1965, India was without water and without grain.

Already by 1964, Johnson had taken the extraordinary step of wresting control of food aid from the U.S. Agency for International Development (USAID)—which he did not trust—and became the de facto "desk officer" controlling Public Law 480 food aid to India and ten other countries under the auspices of the more easily managed U.S. Department of Agriculture.[59] In 1965, the Bihar drought compounded preexisting food problems. So while State continued to address nuclear weapons issues, Johnson personally took the reins of the second foreign policy goal: to make India self-sufficient in food. Indeed, as historian Paul Hammond has argued, Johnson's role in the Bihar famine was an extreme example of exercising presidential influence on another government.[60] Although the president publicly argued that the "first obligation of the community of man is to provide food for all its members,"[61] his political instincts told him that drought and impending famine in India was a situation to be exploited to force the government to reform its agricultural program by making it a higher economic priority. And that is exactly what he did.

In June 1965, Johnson put his "short-tether" policy into place, releasing just enough

[55] Hornig, memorandum to president, March 11, 1966, Box 3, Hornig-LBJ.

[56] J. Wallace Joyce to Hare, memorandum, Oct. 10, 1966, Box 17, State 3008D, NARA II.

[57] Paul R. Brass, "The Political Uses of Crisis: The Bihar Famine of 1966–1967," *Journal of Asian Studies* 45 (1986): 250.

[58] James Warner Bjorkman, "Public Law 480 and the Policies of Self-Help and Short-Tether: Indo-American Relations, 1965–68," in Rudolph and Rudolph, *The Regional Imperative*, 229; Frankel, *India's Political Economy*, 280. (Both cit. n. 9.)

[59] Dennis Kux, *India and the United States: Estranged Democracies* (Washington, D.C., 1993), 243. PL-480, the Agricultural Trade Development and Assistance Act of 1954 (later called the Food for Peace Program), allowed surplus U.S. agricultural products to be shipped abroad to aid humanitarian and geopolitical aims.

[60] Paul Y. Hammond, *LBJ and the Presidential Management of Foreign Relations* (Austin, 1992), 226.

[61] Lyndon Baines Johnson recommending to Congress steps in an international effort in the War against Hunger, Feb. 2, 1967, quoted in James MacGregor Burns, ed., *To Heal and to Build: The Programs of Lyndon B. Johnson* (New York, 1968), 32.

grain to India to arrive "just in time." Monthly, he personally evaluated the situation, and then, and only then, permitted grain shipments. Otherwise, the supply line was closed. "Short tether" made Indian officials very nervous—and extremely resentful of what they considered a heavy-handed and demeaning tactic. It also spurred heavy investment in agriculture, coming at the same time that a strong supporter of radical agricultural reforms—Chidambaram Subramaniam—was finally making headway as the agriculture minister.[62]

As India's grain harvest plummeted due to lack of rain in early 1966, U.S. agriculture secretary Orville Freeman told Johnson that India's food situation would be desperate by fall. To alleviate catastrophe, Freeman wanted more fertilizer shipped to India. "The weather for next year's crop cannot be controlled," he wrote, "but the amount of fertilizer to be used can be."[63] Freeman was right about the fertilizer. He was wrong about weather control—or at least the lengths to which the Johnson administration was about to go in an attempt to control nature—and the weather.

Throughout 1966, Johnson was keeping his finger on India's agricultural, and weather, pulse. He pored over the detailed weekly rainfall maps,[64] later recalling that he knew "exactly where the rain fell and where it failed to fall in India."[65] Meanwhile, Secretary of State Dean Rusk was reading a point paper outlining the foreign policy implications of weather modification. Although no nation could as yet threaten the economy or security of another by controlling the weather, it seemed only a matter of time before it would be possible. The State Department needed to develop a policy on weather control. As the scale of weather modification research increased, the effects would stray outside national borders. The paper's author, Bureau of Intelligence and Research staffer Howard Wiedemann, continued, "Further research may lead to opportunities for using weather modification techniques for common benefit, including technical assistance to less developed countries," or it could be used to inflict "massive" damage on enemies. Some small-scale programs, including those that resulted in modest rain enhancement, could be a "meaningful way" to render assistance to less developed countries. In fact, Weidemann argued that "in attempting to assist less-developed countries, it may be essential to stress the limits of weather modification in order to keep their hopes within reasonable bounds; in collaborating with other countries on international projects, it may be difficult to strike a neat balance between healthy skepticism and an imaginative approach."[66]

In late 1966, stubbornly determined to make India self-sufficient in food, Johnson turned off the U.S. grain spigot to India. Public outrage both within and outside of the United States at the specter of the potential starvation of millions of people made this

[62] Carleton S. Coon Jr. to Carol Laise, March 2, 1966, Box 15, Folder: Unlabeled, State 5255-NARA II; Brands, *India and the United States* (cit. n. 9), 118.

[63] Orville Freeman to Lyndon B. Johnson, March 22, 1966, WHCF CO 113, Box 38, CO 121 India 3/19/66–3/29/66, LBJ.

[64] Kux, *India and the United States* (cit. n. 59), 255.

[65] Lyndon Baines Johnson, *The Vantage Point: Perspectives of the Presidency, 1963–1969* (New York, 1971), 226.

[66] Noted in Thomas L. Hughes to the secretary of state, April 14, 1966, State Department Records, Entry 3008D, Box 21, NARA II (hereafter cited as State 3008D-NARA II). The Bureau of Intelligence and Research is part of the U.S. intelligence community, providing analysis to State's policy makers on issues of importance to foreign policy.

an extremely unpopular decision. Despite the criticism, the president continued his stranglehold on grain shipments.[67]

However, the continued drought was standing in Johnson's way of forcing India to complete the job of overhauling its agricultural system. The State Department was still looking for that illusive scientific project that would fill its foreign policy requirements for meeting India's development needs while enhancing the country's national scientific and technological prestige. Weather control—with its promise of rainfall to break the drought and provide a cushion against future weather vagaries, and its promise of increasing scientific prestige—was seen as the solution.

PROJECT GROMET

In the cloudless, still-dark early morning of Monday, January 23, 1967, a large unmarked U.S. military transport plane, several small disassembled aircraft resting in its belly, landed just outside New Delhi at 5 a.m. On board were atmospheric scientist Dr. Pierre St. Amand and his associates from the Naval Ordnance Test Station (NOTS), China Lake, California. They were in India to undertake a secret mission: the breaking of the crop-damaging Bihar drought, one of the greatest humanitarian crises of that time, by classified, military-developed weather modification techniques.[68]

NOTS researchers had spent several years perfecting weather control techniques. An unclassified version involved seeding Caribbean hurricane systems with aircraft-dispensed silver iodide (Project STORMFURY). Its classified counterpart involved testing an advanced pyrotechnic dispensing technique (codenamed POPEYE).[69] POPEYE's purpose was to test the feasibility of artificially lengthening the naturally occurring monsoon season in Laos and Vietnam, thus disrupting North Vietnam's extensive supply routes. This new technique targeted large, high-altitude, cold clouds (tops below 25 degrees Fahrenheit) with specially formulated silver iodide. Seeded clouds would "blow up" distinctively and drop large amounts of rain.[70] The Pentagon proposed using this advanced, classified method in India. Its unclassified code name: GROMET.

In late 1966, Defense Secretary Robert McNamara broached the possibility of a "Joint U.S.-India Precipitation Experiment" to U.S. ambassador to India Chester Bowles. Cautioning Bowles that the new techniques had only been tested in limited geographic areas and under special conditions, McNamara stressed the need to avoid raising the level of expectation in the Indian government until the method proved efficacious in the target states: Bihar and Uttar Pradesh. Despite the limited winter monsoon cloud cover, the Department of Defense (DoD) was willing to begin the project in January 1967 in hope of improving India's spring harvest. Seeding would be most advantageous between May and October—the cloudy summer monsoon season. McNamara asked Bowles to immediately explore the possibility with Prime Minister

[67] Hammond, *Presidential Management* (cit. n. 60), 227.

[68] AMEMBASSY New Delhi to U.S. Naval Ordnance Test Station, China Lake, California (Code 50), 270425Z JAN 67 [i.e., 27 Jan. 4:25 AM Zulu, or Greenwich time], NSF Country File India, Box 131, India Memos and Misc. 1 of 2, Vol. 8, 9/66–2/67, Lyndon Baines Johnson Presidential Library, Austin (hereafter cited as India/Memos-LBJ).

[69] John K. Rouleau to Herman Pollack via J. Wallace Joyce, Dec. 16, 1966, State 3008D-NARA II.

[70] Hornig to LBJ, Feb. 20, 1967, NSF Country File Vietnam, Box 41, Vietnam Memos (B), Vol. 66, 2/17 – 28/67, Lyndon Baines Johnson Presidential Library, Austin (hereafter cited as Vietnam/Memos-LBJ).

Indira Gandhi. If she concurred, a technical team would meet with Indian scientists to develop a plan.[71]

Pentagon officials were enthusiastic. State Department personnel were extremely wary. They had just been discussing the necessity of conducting a preliminary study of weather modification's legal issues. However, they did not have time for a study. The decision needed to be made now.[72] Concerned that offering rainmaking to India without making the same offer to West Pakistan could lead to diplomatic problems, State Department officials queried DoD about that possibility. Defense's representative was pessimistic—not because DoD was unwilling, but because cloud cover would be meager. However, he would check into it.[73] In mid-December, State's Science Office advised Bowles of its concerns: the project was classified and associated with the U.S. military, they were facing the possibility of raising false hopes in an adverse climatological environment, and there could be legal problems if the effects of seeding crossed an international border. Despite these issues, DoD remained positive. State's Science Office—aware that Hornig had requested an environmental impact review of STORMFURY in accordance with Kennedy's National Security Action Memorandum 235 directive of 1963—hoped they were right.[74]

Less than a week later, Hornig informed Johnson's national security adviser, Walt Rostow, that POPEYE testing had been successfully concluded. DoD was ready to go operational and needed the president's approval.[75] As 1966 drew to a close, Rostow advised Johnson that the rainmaking experiment in India was going forward on a "highly classified basis." If it worked, the additional rain would "materially improve the chances that [the] spring's crop will produce something in the worst affected areas." Participants would fly in commercially marked planes, and all occupants would wear civilian clothes. In case there were questions from the media, a contingency press release explaining that this was an "agro-meteorological survey" had already been prepared. Rostow finished with a flourish: "May the rain makers succeed!"[76]

A joint State-Defense communiqué advised Bowles that GROMET was proceeding. The memo of understanding stressed the project's classification. There would be no publicity. The Indian government was fully responsible for any resulting claims for personal injury or property damage. There would be no public release of information without the consent of both countries.[77]

Bowles quietly made arrangements in India. The project had to remain secret. The Indians, long suspicious of U.S. military and diplomatic intentions based on previous U.S.-Pakistan military aid, did not want it known that the U.S. military was involved, nor did project participants want their activities known.[78] They were concerned that local residents might conclude that the Americans were trying out this technique in

[71] McNamara to Chester Bowles, 091624Z Dec. 1966, India/Memos-LBJ. The estimated resources: three contract seeding aircraft, one weather reconnaissance aircraft provided by the navy, and seventeen people for a total cost of $300,000.

[72] Rouleau to Joyce, memorandum, Dec. 14, 1966, State 3008D-NARA II.

[73] Rouleau to Pollack via Joyce, Dec. 16, 1966, State 3008D-NARA II.

[74] Pollack to Bowles, memorandum, Dec. 17, 1966, India/Memos-LBJ; Hornig to U. Alexis Johnson, June 13, 1966, Hornig chronological files, Box 4, LBJ.

[75] Hornig to Rostow, memorandum, Dec. 22, 1966, Box 41, Vietnam/Memos-LBJ.

[76] Rostow to Johnson, Dec. 29, 1966, India/Memos-LBJ.

[77] State and Defense to Bowles, Dec. 29, 1966, India/Memos-LBJ.

[78] Bowles to State, 240512Z Jan. 1966, India/Memos-LBJ; Perkovich, *India's Nuclear Bomb* (cit. n. 51), 25.

India because it was illegal in the United States. Indeed, the Defense Department *was* the only U.S. government agency not required to notify Congress before undertaking weather modification experiments.[79] However, the mutually agreed-upon reason for keeping GROMET secret was to avoid raising false hopes for rain.[80] Some members of the Indian government urged abandoning secrecy. The potential for political damage, in their opinion, would be greater if information leaked out requiring "defensive action." However others in the Indian government as well as Bowles wanted firm results first.[81] Their agreed-upon statement:

> Scientists from the United States and India are cooperating in a joint agro-meteorological research project, localized in the Eastern Uttar Pradesh and Bihar to study the cloud physics and rain producing mechanism over these areas of India which have incurred several droughts during the last few years.[82]

The Indian government insisted that any comments on this project connect it to agriculture, not military objectives.[83] Thus, India and the United States were playing a high-stakes game of diplomacy with GROMET. If the rains came, and the crops were saved, India would be able to claim a scientific and agricultural breakthrough. If the project failed, and it later came out that the United States had been using classified military techniques under the cover of an "agro-meteorological survey," both governments could be severely embarrassed.

The weather did not cooperate; skies were clear. Clouds were, however, starting to appear in the northern Punjab. State was hesitant to extend the operational area due to the recent shootdown of a Pakistani aircraft near the border.[84] Bowles was fully aware of the risks. When the primary target areas remained cloudless, the Indian government identified additional areas in Uttar Pradesh.[85] State Department personnel remained uneasy. They wanted to retain control over the seeding areas due to the "sensitivity of the GROMET team activity." Any alternate sites had to be cleared with them first.[86] Bowles argued, and Lakshmi Kant Jha, secretary to Prime Minister Gandhi, agreed, that they needed the flexibility to take advantage of every cloud formation that did not run the risk of provoking an international incident. It was of the utmost importance to prove the efficacy of the rainmaking technique. "Both we and the Indians want to demonstrate that if we can [make rain] India's food and agriculture need not be entirely at the mercy of weather vagaries," Bowles wrote. The Bihar-UP area had been chosen because it needed rain, and there was an outside chance that the cloudless conditions would break. However, the skies had remained cloudless. If the cloud-seeding project were to have any chance of succeeding, it had to move to an area where there were clouds.[87] State finally relented but insisted seeding had to have "some legitimate agricultural use beyond demonstration of the GROMET technique."[88] Furthermore,

[79] Rouleau to Pollack via Joyce, Dec. 16, 1966, Box 21, State 30008D-NARA II. Reporting requirements were addressed by Senate Bill 2916 of Oct. 13, 1966.
[80] Bowles to State, 240521Z Jan. 1967, India/Memos-LBJ.
[81] Bowles to State, 071255Z Feb. 1967, India/Memos-LBJ.
[82] Ibid.
[83] AMEMBASSY New Delhi to State, 301340Z Jan. 1967, India/Memos-LBJ.
[84] State to AMEMBASSY New Delhi, Feb. 8, 1967, India/Memos-LBJ.
[85] AMEMBASSY New Delhi to State, 091256Z Feb. 1967, India/Memos-LBJ.
[86] State to AMEMBASSY New Delhi, Feb. 10, 1967, India/Memos-LBJ.
[87] Bowles to Rusk, 131300Z Feb. 1967, India/Memos-LBJ.
[88] State to AMEMBASSY New Delhi, Feb. 14, 1967, India/Memos-LBJ.

there would be no publicity until the military was out of the picture and civilian agencies were firmly in control.[89]

Seedable clouds finally appeared in mid-February. The outcome was mixed. Some clouds produced heavy rain, others light-to-moderate rain. Large clouds responded better than small clouds did. Team members believed that "economically valuable amounts of rain" could be produced over much of India during and after the monsoon season when nonraining cloud cover was more abundant. The embassy reported that agencies throughout India were now aware of the project and were extremely enthusiastic.[90] Interestingly, the GROMET team did not report how much rain hit the ground—an important measure of the project's success. The dry air evaporated the falling rain. Clearly, rain that failed to land on the parched earth would not aid plants. It would be difficult to call the project "successful."[91] Reporting these events to Johnson, Rostow concluded, "State and the scientists are sorting out what kind of statement to issue—if any."[92] It is unclear if any public statement was made. But absence of GROMET's mention in later books on weather control by Indian authors casts doubt on its having been discussed outside of government circles.[93]

Despite this lack of success, the State Department still wanted to include Pakistan in GROMET. Desiring to ensure regional stability, the United States needed to take an evenhanded approach to aid for the two adversaries, while assuring both India and Pakistan that neither was being given an advantage. As the summer monsoons approached, the wind would blow from east to west. The effect of seeding could carry over into Pakistan. The biggest fear: that rain would fall in India, robbing Pakistan of water. With Pakistan's "almost psychotic fear of India," it would not be a good idea for Pakistani leaders to become convinced that India was trying to steal its water.[94] It was now March, and May, the arrival of the summer monsoon, would bring good "cloud hunting." Time was growing short, and the arrangements needed to be made.[95]

While overtures were being made to both governments, this plan hit a snag.[96] By mid-May, a frantic Bowles had still received no "green light" from State, or the White House, to continue GROMET. He thought his agreement with McNamara was to continue seeding as the monsoon clouds streamed in. Indeed, he had thus sold Gandhi on the project. Bowles had invested considerable time serving as the go-between for State, the Pentagon, and the Indian government while arranging GROMET and creating a suitable cover story. For Bowles, always committed to finding new ways to reinforce a stable India, "the hour for Indian democracy" was late. If the crops failed

[89] Ibid.

[90] AMEMBASSY New Delhi to State, 201254Z Feb. 1967; AMEMBASSY New Delhi to State, 270916Z Feb. 1967, India/Memos-LBJ.

[91] AMEMBASSY New Delhi to State, 281256Z Feb. 1967, India/Memos-LBJ.

[92] Rostow to Johnson, Feb. 28, 1967, India/Memos-LBJ.

[93] Neither P. Koteswaram, *Water from Weather* (Waltair, India, 1976) nor N. Seshagiri, *The Weather Weapon* (New Delhi, 1977) mentions governmental rainmaking efforts in India in 1967. Kux, *India and the United States* (cit. n. 59) discusses grain shipments to India during the 1966–1967 drought years, but not rainmaking efforts.

[94] Walter P. McConaughy, U.S. ambassador to Pakistan, quoted in Robert J. McMahon, "Toward Disillusionment and Disengagement in South Asia," in Cohen and Tucker, *Lyndon Johnson Confronts* (cit. n. 50), 140.

[95] State to AMEMBASSIES New Delhi, Rawalpindi, March 8, 1967, NSF Country File India Box 31 India Cables Vol. 9, 3/67–7/67, Lyndon Baines Johnson Library, Austin (hereafter cited as India/Cables-LBJ).

[96] State to AMEMBASSIES New Delhi, Rawalpindi, March 15, 1967, India/Cables-LBJ.

for a third year in a row, the "fragile Indian democracy" could be in jeopardy, as its large restive population scrambled for food. The clouds were starting to move in. Bowles asked Hornig for help.[97]

Bowles's difficulty in extending seeding to India's rainy season had less to do with U.S.-India foreign policy than it did with the deployment of the "weather weapon" in Laos. Apparently having new doubts about the rainmaking project in South Asia, Rostow spelled out for Johnson the potential problems of launching the weather weapon there. Security was a serious issue. Although a "leak" was unlikely, if GROMET continued in a public venue people would soon make the unwanted connection between enhanced monsoon rainfall in India and the increased rainfall in Laos.[98]

There were also ethical and moral issues at stake. Rostow wrote Johnson: "The fact that we are going ahead with the Indian program on the basis of apparently flimsy back-up evidence has led to speculation that we 'know something' which has not yet appeared." Furthermore, he warned, the administration should not underestimate the "degree of revulsion to be expected in the domestic and international meteorological circles."[99]

Indeed, domestic "meteorological circles" had made it abundantly clear that using cloud seeding for military purposes was unacceptable. University of Washington meteorologist Robert G. Fleagle recalled his service during the 1960s on the National Academy of Sciences' Committee on Atmospheric Sciences. Committee members had rejected a March 1963 recommendation by atomic bomb physicist Edward Teller that they should propose a NATO study of weather control because such a study would jeopardize international cooperation in the atmospheric sciences and there was no scientific basis for making it. Despite this rebuff, a few years later Teller recommended that cloud seeding be used for military purposes during the Vietnam War—relating that NOTS personnel claimed they could muddy up the Ho Chi Minh trail. Several of the committee members opposed Teller, and numerical weather prediction pioneer Jule Charney of MIT spoke out strongly against it. According to Fleagle, Teller "knew when he could not win and [withdrew] his proposal."[100] However, cloud seeding *was* used as a weapon despite opposition—moral, ethical, and scientific—from the National Academy.

Two weeks later, in early June 1967, Hornig informed Johnson that a team sponsored by USAID was heading to India to set up a permanent weather modification program despite potential legal and international complications.[101] However, the rest of the files related to India, Pakistan, and the office of the president's science adviser contain no further references to GROMET or weather modification in South Asia. The arrival of abundant monsoon rains ended the drought in the summer of 1967. Combined with more fertilizer and improved seeds, the result was a bumper grain crop. The specter of famine faded away. State did not need GROMET to keep India's hopes

[97] Bowles to Hornig, May 11, 1967, India/Cables-LBJ.

[98] Rostow to Johnson, May 22, 1967, Box 88, Vietnam-LBJ.

[99] Ibid. For reactions within the scientific community to the possibility that weather had been used as a weapon in Vietnam, see Deborah Shapley, "Rainmaking: Rumored Use over Laos Alarms Arms Experts, Scientists," *Science* 176 (1972): 1216–20.

[100] Robert G. Fleagle, *Eyewitness: Evolution of the Atmospheric Sciences* (Boston, 2001), 76.

[101] Hornig to Johnson, June 5, 1967, CF Box 85, Folder: SC Sciences, LBJ. In this memorandum, Hornig warned Johnson that the international implications were severe—adding that the state of Maryland had declared any form of weather modification to be a crime.

for a better harvest alive, and Defense did not need their cover blown in Laos. GROMET quietly died.

In the end, the lack of a positive outcome combined with the risk of exposing the use of weather modification as a weapon in the Vietnam conflict doomed the secret "agro-meteorological survey." However, despite its failure to produce rain during a normally dry season, the attempt to solve India's water problems, and hence its food problem, with advanced weather control methods was in keeping with Johnson's desire to use environmental sciences as one of many tools intended to bring a kind of "global New Deal" to Asia. As Walt Rostow remembered, "The India food question went right to where he lived. It was part of Johnson's fundamental concern for human beings and his hatred of poverty."[102]

CONCLUSION

"Science" is not usually the first word one associates with Lyndon Johnson or with his presidential administration despite his early championing of the space program and his efforts on behalf of the outer space treaty. Indeed, the Johnson administration is more likely to bring to mind the failed Great Society and the disastrous Vietnam War.[103] But what drove his campaign for the Great Society and his Asian foreign policy was a desire to increase global stability and to improve the quality of life for people around the world, however flawed his vision was in practice, and however much conceived with U.S. interests in mind. Johnson saw science—*applied* science—as the means to see that desire realized.

Unlike his postwar predecessors, for whom physics was the scientific tool of choice as the cold war heated up, Johnson looked to the physical environmental sciences to achieve his aims. They gave him the perfect opportunity to bring geophysical scientists into the fold by supporting their need for global data, the requirements for which were growing rapidly with increases in numerical modeling supported by advancing computer capability. Scientists got their data and strengthened international scientific exchanges, and Johnson got the opportunity to show the flag, increase the prestige of the United States throughout the world, and make diplomatic overtures to other nations. Furthermore, he was able to accomplish all those tasks while meeting national security needs for improved weather forecasts, oceanographic knowledge in support of antisubmarine warfare, and geodetic data in support of ballistic missile programs. Johnson saw the opportunity to apply these sciences to controlling nature: providing water where there was little, preventing flooding where there was too much, and exploiting the oceans and their "unlimited" supply of food for the world's hungry. The physical environmental sciences were not just about understanding the Earth and its atmosphere—or about strengthening national security—but also about improving the quality of life. From the taming of the Colorado River in his congressional district in the thirties to modifying weather in the sixties, controlling nature through the application of science and technology was central to his programs.

In recent years, presidential scholars and diplomatic historians have called for a reevaluation of Johnson's foreign policy. Thomas Alan Schwartz has argued that the

[102] Lloyd Gardner, *Pay Any Price: Lyndon Johnson and the War for Vietnam* (Chicago, 1995), 95; Walt Rostow quoted in Kux, *India and the United States* (cit. n. 59), 243.

[103] Caro, *Years of Lyndon Johnson: Master of the Senate* (cit. n. 32).

Vietnam War "should not be allowed to block a more dispassionate assessment of Johnson's foreign and domestic policies," while Robert Dallek has declared that, where Johnson had genuine choices, "we need to consider whether they were wise or shortsighted, beneficial or destructive to our national and the world's international well-being."[104] Scholars critical of Johnson's South Asia foreign policy have argued that his "real" reason for interrupting the routine flow of grain to India was his anger with Indira Gandhi over her public criticism of U.S. involvement in Vietnam. Since India was already reforming its agriculture, Johnson's attempts to justify his actions as a means of forcing such a reform were disingenuous and a cover for punishing Gandhi's opposition to the war.[105] However, the archival evidence revealing GROMET supports Hammond's conclusion that these critics were wrong. While India had been a fairly low priority for U.S. diplomacy, Johnson nevertheless raised it to a critical level. Instead of staying above such mundane matters as scheduling food shipments, Johnson made a conscious decision to actively intervene. Unlike his subordinates, he recognized that the ruling Congress Party held power in impoverished rural India because of inefficient agriculture that made wheat patronage an effective political strategy: no substantive agricultural reform in India would occur without intense outside pressure. The 40 percent of India's population living in abject poverty written off by the government otherwise would never see an improvement in their way of life. Johnson intervened to improve their way of life in advance of the Green Revolution and was largely successful.[106]

Yet Johnson's decision to keep his weather modification program secret ultimately seems shortsighted and destructive, for it kept a vital policy issue out of public view.[107] In a poignant lament on the trajectory of U.S. science policy, A. Hunter Dupree expressed his frustration that in 1983 President Ronald Reagan informed only six people (including physicist Edward Teller) about his decision to proceed with "Star Wars," the Strategic Defense Initiative, one of the most significant and costly science-technology initiatives of his administration.[108] For Dupree, Reagan had failed to honor the traditional pattern of science advising in the United States and bypassed talented experts available to him. Yet it is now clear that this was not the first time that a strong president in the imperial presidency had conducted an operation with just a few highest-level confidants, against the wishes of a research community that did not believe the tool worked. A single scientist can have tremendous impact: in this instance Pierre St. Amand, a relatively unknown scientist with a passion for employing

[104] Schwartz, *Lyndon Johnson and Europe* (cit. n. 32), 225; Robert Dallek, "Lyndon Johnson as a World Leader," in *The Foreign Policies of Lyndon Johnson: Beyond Vietnam,* ed. H. W. Brands (College Station, Texas, 1999), 6–18, on 17.

[105] Hammond, *Presidential Management* (cit. n. 60), 66. See also Bjorkman, "Public Law 480" (cit. n. 58), 229, 234.

[106] Hammond, *Presidential Management* (cit. n. 60), 67, 98, 106, 225–6. Scholars have begun to reexamine the perceived benefits of the Green Revolution, revising earlier glowing assessments; see Alexis De Greiff and Mauricio Nieto, "What We Still Don't Know about South-North Technoscientific Exchange: North-centrism, Scientific Diffusion, and the Social Studies of Science," in *The Historiography of Recent Science, Medicine, and Technology: Writing Recent Science,* ed. Ronald E. Doel and Thomas Söderqvist (London, 2006), 239–59.

[107] Johnson's penchant for secrecy "has made it difficult for historians to recognize the personal control he exercised over many of his non-Vietnamese polices." Schwartz, *Lyndon Johnson and Europe* (cit. n. 32), 7; see also Dallek, "Lyndon Johnson as a World Leader" (cit. n. 104), 9.

[108] A. Hunter Dupree, "Science Policy in the United States: The Legacy of John Quincy Adams," *Minerva* 28 (1990): 259–71, on 267–68.

weather modification techniques, gained backing from Johnson political appointees because of the benefits his research promised. Lyndon Johnson was willing to experiment with the bold new power over nature that Pentagon scientists (and powerful outsiders, such as Teller) sought to put at his disposal for both humanitarian and war-fighting purposes. His actions, particularly their moral and ethical dimensions, need to be further explored in the frame of the twentieth-century fascination with the control of nature.

The Politics of Noncooperation: The Boycott of the International Centre for Theoretical Physics

By Alexis De Greiff[*]

ABSTRACT

In 1974, the General Conference of UNESCO approved three resolutions condemning Israel. In retaliation, a group of physicists promoted a boycott of the International Centre for Theoretical Physics (ICTP), an institute created to foster collaboration in theoretical physics between industrialized and third world countries and partly supported by UNESCO. This political action against the "politicization" of UNESCO was led by American and Israeli scientists. I show that the position toward the boycott was very different among European scientists. I shall argue that the boycott of the ICTP was motivated as much by the formal connection between UNESCO and ICTP as by the identification of the ICTP with the third world, which was blamed for the "exclusion" of Israel from UNESCO. The episode reflects the contradictions and workings of scientific noncooperation. It also reveals the limits of scientific internationalism in the second half of the twentieth century. In this context, I investigate the meaning ascribed by the actors to the term the "politicization of science."

INTRODUCTION

Soon after its creation in 1964, the International Centre for Theoretical Physics (ICTP) became the best-known institution in which third world physicists came to have access to the latest developments in their field and had the chance to do research. Between 1964 and 1980, more than 6,000 scientists from the developing countries (and a sim-

[*] Departamento de Sociología/Centro de Estudios Sociales, Universidad Nacional de Colombia, Bogotá, Colombia; ahdegreiffa@unal.edu.co.

I wish to thank the staff of the National Catalogue Unit Archives of Contemporary Scientists, Bath University; Jean Marie Deken from the SLAC Archives and History Office, at Stanford University; and the ICTP Archives for their crucial help. I am deeply grateful to John Krige and David Edgerton for their careful reading and thoughtful criticisms and suggestions; this chapter owes much to their encouragement. My colleagues and students of the Seminario de Estudios Sociales de la Ciencia, la Tecnología y la Medicina of the Universidad Nacional de Colombia, Bogotá, made insightful comments. Jimena Canales, Paul Forman, Stefania Gallini, Francisco Ortega, Angélika Rettberg, Juan Federico Vélez, Fernando Viviescas, and Andy Warwick, as well as the participants of the Seminar of the Centre for the History of Science, Technology and Medicine at Imperial College, London, offered valuable critical suggestions. I thank also the participants of the workshop this volume is based on, especially Kai-Henrik Barth and Nikolai Krementsov, as well as two anonymous referees. For their support at various stages of this work, I thank the Universidad Nacional de Colombia, COLCIENCIAS, and the Bogliasco Foudation. I thank Professors Luciano Bertocchi, Haim Harari, Faheem Hussain, and John Ziman for sharing with me their memories on the history of the ICTP and UNESCO.

ilar number from industrialized countries) visited the ICTP. It was located in Trieste, and its deputy-director was a scientific diplomat from that city, Professor Paolo Budinich.[1] The central actor of the ICTP's early history was, however, Professor Abdus Salam, its first director. Born in the region of British India that would later become Pakistan, Salam read mathematics and physics at Cambridge. In 1958, he became the first professor of theoretical physics at Imperial College. In 1979, he was awarded the Nobel Prize for physics. Under his leadership, the ICTP became a reference point for scientists in developing countries as the model of international scientific collaboration for third world development, and Salam perhaps the most famous spokesperson of the "science for development" ideology amongst political and scientific milieus in both developing and industrialized countries.

In its early years, the ICTP operated under the auspices of the International Atomic Energy Agency (IAEA), a United Nations technical agency, and also had substantial financial support from the Italian government. In 1970, UNESCO joined the IAEA in the operation. The collaboration of the United Nations Educational, Scientific and Cultural Organization (UNESCO) proved vital to the survival of the center, whose financial condition was always precarious. Although Salam did not like UNESCO's tutelage in the ICTP's early years, by 1970, there were few alternatives for increasing the center's finances. Even then, the sense of instability continued, for although the agreement with UNESCO brought more resources, its future plans continued to be subjected to periodic approvals.[2] What is more, the center became ensnared in a major political confrontation inside UNESCO that almost destroyed it.

In the mid-1970s, tension in the Middle East put international cooperation with the third world in serious jeopardy. Three "anti-Israeli" resolutions approved in the 1974 UNESCO General Conference sparked a massive boycott against the international organization and, as a result of its association, against the ICTP as well.[3] This was led by American and Israeli scientists who believed that UNESCO and, by ricochet, Salam's institute were failing to respect their international callings and had become politicized and beholden to radical groups determined to attack Israel wherever and however they could.

"A boycott is a particular form of sanction against a country or a group in order to

[1] Alexis De Greiff, "The Tale of Two Peripheries: The Creation of the International Centre for Theoretical Physics in Trieste," *Historical Studies in the Physical and Biological Sciences* 33 (2002): 33–59; idem, "The International Centre for Theoretical Physics, 1960–1979: Ideology and Practice in a United Nations Institution for Scientific Co-Operation and Third World Development" (Ph.D. diss., Univ. of London, 2002).

[2] In November 1974, a committee presided over by CERN theoretical physicist Leon Van Hove stated that "a feeling of uncertainty concerning the future of the Centre is rampant, both among its scientific leaders and its administrative staff" and recommended the allocation of *stable sources* of income because "the staff should be given a highest security of position." IAEA, "Report of the Ad Hoc Consultative Committee of the International Centre for Theoretical Physics, Trieste, to the Director General of the International Atomic Energy Agency and the United Nations Educational, Scientific and Cultural Organisation" (Paris, 1975).

[3] UNESCO was one of the most interesting and active international organizations aimed at promoting cultural and scientific cooperation. However, critical works on its political, cultural, and scientific activities are scarce. (See Aant Elzinga, "Introduction: Modes of Internationalism," and idem, "Unesco and the Politics of Scientific Internationalism," in *Internationalism and Science,* ed. Aant Elzinga and Catharina Landström [London, 1996], 3–20 and 89–131.) In spite of its importance for the history of UNESCO, there is only one scholarly study devoted to the so-called Israel Resolutions and the boycott that followed, written from the international relations perspective: Clare Wells, *The UN, UNESCO, and the Politics of Knowledge* (London, 1987). Histories of UNESCO have systematically ignored the issue. Fernando Valderrama. *A History of UNESCO* (Paris, 1995).

pressure some change." It is based on a theory of action that identifies a relationship between the isolation of the target and the desired goal.[4] This phenomenon, as part of international politics, is relatively new. In the history of science, the best-known case concerns the isolation of German and Austrian intellectuals after the Great War. The chauvinistic stances adopted by scientific intellectuals on both sides of the trenches broke down the kind of scientific internationalism that had prevailed since circa 1870. As Forman has shown, scientific internationalism—like nationalism—is always a political stance.[5] Boycotts act as a negation of scientific internationalism. Yet despite potential interest in the international relations of science, the literature on scientific boycotts is practically nonexistent, especially in the period after World War II.[6]

This chapter tackles three related issues regarding a scientific boycott. The first concerns the organization of the boycott as well as the criteria and reasons scientists have to boycott certain institutions as opposed to others. I shall argue that the motivations for the boycott concerned essentially the image of the ICTP as a center for third world development rather than as a research institution. Such motivations never, or almost never, arose in an explicit way. Thus the boycott will allow us to investigate how the scientific community reacted when the center was trapped by a major crisis in UN politics and to scrutinize the workings of scientific internationalism in the second half of the twentieth century.

The second issue concerns the manner in which the boycotting scientists moved the boundary between science and politics to suit their interests and ideologies. Such boundary work served as a basis to defend Mertonian norms such as the need to keep science free from politics and the identification of science with the values of Western democracies.[7] I suggest that those opposing the "politicization of science" not only sought to keep international scientific institutions politically neutral but also purported to define "ideologically correct" science in the context of international exchange of scientific knowledge.[8]

The third issue concerns the manner in which the center handled the boycott, ex-

[4] Lorraine J. Haricombe and F. W. Lancaster, *Out in the Cold: Academic Boycotts and Isolation of South Africa* (Arlington, Va., 1995), 1–2.

[5] Paul Forman, "Scientific Internationalism and the Weimar Physicists: The Ideology and Its Manipulation in Germany after World War I," *Isis* 64 (1973): 151–80. See also Catharina Landström, "Internationalism between Two Wars," in Elzinga and Landström, *Internationalism and Science* (cit. n. 3), 54; and Daniel Kevles, "'Into Hostile Political Camps': The Reorganization of International Science in World War I," *Isis* 62 (1971): 47–60.

[6] Perhaps the only exception is the sanctions against the racist regime in South Africa (Haricombe and Lancaster, *Out in the Cold* [cit. n. 4]). The effects of the boycott upon South Africa are a source of major debates. Some authors have studied the activities of the boycotting activists, concluding that sanctions alone do not guarantee the desired effects. (P. Wallensteen, "Characteristics of Economic Sanctions," in *A Multi-Method Introduction to International Politics,* ed. W.D. Coplin and C. W. Kegley, [Chicago, 1971], 128–54; M. P. Doxey, *International Sanctions and International Enforcement* [New York, 1980]). Others criticize scientific collaboration because science and scientists are pivotal for the prolongation of the racist government (Yngve Nordvelle, "The Academic Boycott of South Africa Debate: Science and Social Practice," *Studies in Higher Education* 15 [1990]: 253–72).

[7] See Robert K. Merton, "Science and Technology in a Democratic Order," *Journal of Legal and Political Sociology* 1 (1942): 115–26. On the relation of boundary work to Merton's norms, see Thomas F. Gieryn, "Boundaries of Science," in *Handbook of Science and Technology Studies,* ed. Sheila Jasanoff, James C. Petersen, Trevor Pinch et al. (Thousand Oaks, Calif., 1995), 393–443. See also idem, *Cultural Boudaries of Science: Credibility on the Line* (Chicago, 1999).

[8] Michael Gordin, Walter Grunden, Mark Walker, and ZouYue Wang, "'Ideologically Correct' Science," in *Science and Ideology: A Comparative History,* ed. Mark Walker (London, 2003), 35–65.

ploring Salam's strategies to use the boycott to stabilize the center financially. As we shall see, despite the disruption it caused, the boycott itself provided Salam with an instrument to negotiate successfully with UNESCO and to ensure financial support from the organization.

THE UNESCO RESOLUTIONS: THE "NOVEMBER DIPLOMATIC REVOLUTION"

In the years between 1950 and 1970, decolonization gave the United Nations (UN) and its technical agencies a new composition. The emergence of a new majority, and the controversial character of some of the issues it raised, became a source of high tension within the system. The UN was transformed into a forum for the confrontation of the radical regimes in the third world and the former colonial countries.[9] By the end of the 1960s, the third world commanded more than two-thirds of the votes in the UN General Assembly, UNESCO, the Food and Agriculture Organization, and the World Health Organization.

The new states gained assertiveness in international scenarios under the leadership of young, charismatic revolutionaries.[10] For the first time, bloc positions and initiatives challenged the "existing institutional order."[11] The strategic alliance between some Arab states and the Pan-African movement crystallized the dreams of the Bandung Conference (1955): to unite the third world against colonizers and neoimperialists, and repressive regimes backed by the latter.

The confrontation in the UN reached a peak in 1974, after the Yom Kippur/Ramadan War, when the General Assembly and UNESCO's General Conference approved a number of resolutions against Israel and South Africa. Most of the African states developed angry anti-Israel sentiments because of Israel's close relationship with the white-dominated regimes of South Africa and Rhodesia, on the one hand, and Portugal and the United States, on the other. As an analyst pointed out in 1975, "Israel was considered too much part of the Western world. . . . [It] appeared to be virtually the fifty-first state of the United States . . . In that respect Israel seemed a piece of the West deposited in the heart of the third world."[12] This revolt against the traditional order in the UN institutional structure I call the Diplomatic November Revolution.[13]

The tone was set by the UN in New York. The General Assembly requested that the secretary general establish contacts with the Palestinian Liberation Organization "on all matters concerning the 'Question of Israel.'" Yasir Arafat was invited to address

[9] Between 1954 and 1974, the number of UNESCO member states increased from 70 to almost 130. In 1960 alone, 17 new states were admitted.

[10] Just to mention a few of the names: Egyptian Gamal Abdel Nasser, Patrice Lumumba from Congo, and Fidel Castro from Cuba.

[11] In the context of UNESCO, Wells defines it as: "matters of political representation and legitimacy," closely related to "the challenge to established patterns of resources allocation, that is, explicit questioning of the political ends which may be served by extensibly technical activity." Wells, *Politics of Knowledge* (cit. n. 3), 5.

[12] Ali A. Mazui, "Black Africa and the Arabs," *Foreign Affairs* 53 (1975): 725–42. See also Arthur Goldschmidt Jr., *A Concise History of the Middle East,* 4th ed. (Boulder, Colo., 1991).

[13] The U.S. and U.K. withdrawals from UNESCO were motivated by this kind of manifestation of anti-West movements. The "free flow of information" debate exacerbated this confrontation (see Elzinga, "Unesco and the Politics of Scientific Internationalism," [cit. n. 3], 114–5). Another instance was the confrontation around the Rio Declaration on the Establishment of a New International Economic Order (1974).

the Assembly, and a year later, in 1975, the Assembly passed a resolution condemning Zionism "as a form of racism and racial discrimination."[14] Furthermore, the PLO was admitted to the International Labour Organization (ILO), another UN agency that was increasingly opposed by the United States. Concomitantly, the UN suspended South Africa from its Assembly because of its racial policies, as a token of the quid pro quo alliance.[15]

Similarly, the 1974 UNESCO General Conference held in Paris produced two major outcomes. First, it saw the election of Amadou Mahtar M'Bow of Senegal as director-general. Second, three resolutions attacking Israel were approved. It is worth describing at least their most general points. The first Israel resolution "invite[d] the Director General to withhold assistance from Israel in the field of education, science and culture until such time as it scrupulously respects the resolution and decisions of the Executive Board and the General Conference." This referred to decisions regarding the archaeological excavations carried out by Israel at Muslim sites in Jerusalem, in violation of 1967 UN and UNESCO resolutions. The second Israel resolution was the conference's condemnation of Israel for violating the rights of the population of the occupied Arab territories to "national and cultural life." However, the bitter attacks against UNESCO originated when the United States, Canada, and Israel introduced a draft resolution asking that they be included "in the list of countries entitled to participate in the European regional activities in which the representative character of States is an important factor." On November 20, the United States and Canada were admitted, while Israel was turned down.[16] The rejection (the third Israel resolution) was interpreted as an effective exclusion of Israel from UNESCO.

The United States Department of State, through its secretary of state, Henry Kissinger, objected at once, stressing that these moves meant the "politicization of UNESCO." The *New York Times* echoed declarations of U.S. and Israeli diplomats about "the tyranny of the majority" and the UN as the "World center for Anti-Semitism," while the editorial pages denounced the way in which the Arab bloc "and its allies, on behalf of the PLO, amassed votes of vengeance against Israel."[17] Virtually without exception, the discourse of the American foreign affairs top officers and the mass media led to the identification of the "politicization of UNESCO" with the imposition of the majority, that is, the third world countries and "its allies"—the communists and the terrorist organization PLO. As Robert Jordan, an American UN research director, observed some years later, "[F]or the United States to bemoan the 'politicization' of

[14] A similar resolution had been adopted in Kampala by the Assembly of Heads of State and Government of the Organization of African Unity. For a complete version of the resolutions, see UN Doc. A/RES/3236 (XXIX), Nov. 22, 1974, and UN Doc. A/RES/3379 (XXX), Nov. 10, 1975, in J. N. Moore, *The Arab-Israeli Conflict: Readings and Documents,* abr. and rev. ed. (Princeton, 1977).

[15] The December 1975 issue of the *UNESCO Courier* advertised two studies sponsored by that agency: one on "Racism and Apartheid in Southern Africa" and another on "South Africa and Namibia and Portuguese Colonialism in Africa: The End of an Era."

[16] The draft resolution was voted in the Commission for Social Sciences, Humanities and Culture with the following results: 85 against; 2 in favor (Israel and Paraguay); 11 abstentions (Australia, Austria, Chile, China, Finland, France, Honduras, Japan, Nepal, Switzerland, Uruguay). The socialist countries voted against, and many European countries were not present in the room at the time of the vote; see Moore, *The Arab-Israeli Conflict* (cit. n. 14).

[17] "Kissinger Role on U.N. Force Related," *New York Times,* Dec. 1, 1974, 14. The following day, the editorial reported that "the Arab bloc and its communist and African allies ha[d] succeeded in politicizing heretofore non-political UNESCO" (Editorial, "P.L.O. vs. UNESCO," *New York Times,* Nov. 23, 1974, 30).

UNESCO (or ILO) is merely a way of saying that U.S. influence has been on the wane."[18] The United States declared that it would stop paying its contribution to UNESCO if the resolutions were not lifted.[19] France and Switzerland, in spite of their ambiguous positions during the General Conference, stated similar intentions.[20] Director-General M'Bow replied to the attacks by pointing out that Israel had not been "excluded from UNESCO," as might be inferred from the presentation of the resolutions in the mass media. Its exclusion was from "the list of countries entitled to participate in activities in which the representative character of states is an important factor."[21] The statement was also published in the *New York Times,* but it passed unheeded: even the distinction was significant for it still allowed for the exclusion of Israel from some UNESCO activities.

Jewish intellectuals had been mobilizing support against an Arab boycott ever since the Yom Kippur/Ramadan war. A number of voices within the Israeli political sector began a campaign to push the governments of Israel and the United States—as well as those of Canada, France, and the United Kingdom—to adopt appropriate countermeasures. In the words of Danny Halperin, founder in 1975 of the Israeli Economic Warfare Authority, the "philosophy" was "not to act, but to activate." Years later, he outlined the effectiveness of this strategy:

> I think it would be true to say that before 1973 people in Israel looked at the boycott [of Israel] as a nuisance. Something one could use to badmouth those applying it, but nobody was involved in a real struggle against the boycott . . . But after 1973, we all realized that the boycott is not only a problem but a danger as well.[22]

A central issue was the mobilization of the public in the United States and Europe. As Susan Rolef eloquently put it: "The logic behind this approach was that *the more noise one made around the issue, both in Israel and abroad [the better]* . . . With regard to North America and Europe this was part of a broader approach which sought to convince the public opinion that the Arabs were up to no good, and that the West could and should stand up to them."[23] Although the initiative involved the three major American Jewish organizations, it was led by "panic-stricken persons" from outside the Israeli government and "in cooperation with well-wishers from abroad to make greater effort than ever before to face up to the Arab-boycott on the legal, practical and moral levels."[24] In addition, since 1974, Jewish associations in the United States, especially the American Jewish Congress, had reacted definitively against the Diplomatic November Revolution. Several demonstrations were organized in protest against the UN's invitation to the PLO to address the General Assembly. Articles and reports about the

[18] Robert S. Jordan, "Boycott Diplomacy: The US, the UN, and UNESCO," *Public Administration Review* (July-August, 1984): 283–91, 283. In a speech at the Institute of World Affairs, University of Wisconsin, in 1975, Kissinger insisted that the "Third World sanctions against Israel" were the expression of the "heavy politicization" of UNESCO.

[19] "Kissinger Warns Majority in U.N. on U.S. Support," *New York Times,* July 15, 1975, 1, 4–5. M'Bow vehemently criticized the United States (Paul Hoffmann, "UNESCO Prodding U.S. on Payments," *New York Times,* Oct. 21, 1975, 9).

[20] "U.S. Threat to UNESCO Budget," *Times Educational Supplement,* Dec. 6, 1974, 16.

[21] Amadou M. M'Bow, "A Statement on Israel," *UNESCO Courier,* Jan. 1975.

[22] Danny Halperin, "Combatting the Arab Boycott—An Historical Survey," in *Freedom of Trade and the Arab Boycott,* ed. Susan H. Rolef (Jerusalem, 1985).

[23] Susan H. Rolef, *Israel's Anti-Boycott Policy* (Jerusalem, 1989), 36 (my italics).

[24] Ibid., 45.

Arab boycott and the "new Arab strategy" in the UN appeared every month on the pages of the *Congress Monthly.*[25] However, the conservatives were not the only ones who condemned the UN and UNESCO; liberal writers—"the New York intellectuals"[26]—also deplored the General Assembly's decision. Collaboration with UNESCO was presented by intellectuals and scientists in Israel as an endorsement of its policy toward Israel: "Israelis are disappointed at the lack of reaction from the scientific community abroad but grimly resigned to their increasing isolation," reported two observers.[27]

The UNESCO resolutions were also widely repudiated by intellectuals across the political spectrum on both sides of the ocean. The day after the three "Israel resolutions" were voted, the *New York Times* published a one-page advertisement with the heading "WE PROTEST," condemning UNESCO "in view of the increasing open and blatant anti-Israel bias shown by the recent decisions."[28] More than 100 intellectuals signed it, including scientists such as Hans Bethe, Owen Chamberlain, Robert Hofstadter, Isidor Rabi, Edward Teller, and Eugene Wigner.[29] It was followed by another statement with a similar text signed in Paris by European intellectuals of all political affiliations, from Raymond Aron to Jean-Paul Sartre. In early 1975, an ad hoc committee, convened by Nobel laureate André Wolf and including intellectuals such as Kenneth Arrow and Julian Huxley and writers Ernesto Sabato and Ignazio Silone, was set up to "look for the means and ways to bring UNESCO back to its vocation."[30]

Somewhat ironically, it was the ICTP itself that, at just this time, drew attention to its financial links with UNESCO. Salam had always used the scientific journals and magazines to call for further support for his institute. Just a few months before the 1974 General Conference, a long article about the ICTP appeared in the pages of *Nature.* "Financial support is shared in about equal proportions by the IAEA, UNESCO and the Italian government," the author wrote.[31] On November 8, just twelve days before the UNESCO scandal, another article in the same journal detailed the finances of the ICTP: "UNESCO support, although modest at first, is, at least formally, ten years old and UNESCO pursues a policy of regarding its financial aid as no more than seed money to get an institution ongoing." The article also pointed out the fragile situation in which the center found itself due to its financial instability and the hard line adopted by some delegations at UNESCO.[32] Therefore, when the Israel resolutions were approved, the scientists knew that the ICTP, an institute well known for its concern for

[25] The journal of the American-Jewish Congress. Joseph B. Shatta, "The New Arab Strategy," *Congress Monthly* 41 (Nov. 1974): 8. See also Richard Cohen, "The American Jewish Congress vs. the Arab Boycott," *Congress Monthly* 43 (Dec. 1975): 9–11; "Fighting the Arab Boycott," *Congress Monthly* 42 (Oct. 1975): 2; "UNESCO Assailed," *Congress Monthly* 41 (Dec. 1974): 5.

[26] David A. Hollinger, *Science, Jews, and Secular Culture: Studies in Mid-Twentieth-Century American Intellectual History* (Princeton, 1996), 8.

[27] John Hall and Peter Newmark, "Problems in Israel," *Nature,* Dec. 20, 1974, 626–7.

[28] "We Protest," *New York Times,* Nov. 21, 1974. The advertisement, with more signatures, reappeared in the following weeks.

[29] "UNESCO Adopts Resolution to Deny Israel Cultural Aid," *New York Times,* Nov. 21, 1974.

[30] "Rencontre pour l'Universalité de l'UNESCO," undated, mimeo, Abdus Salam Papers. Cataloged by the National Cataloguing Unit for the Archives of Contemporary Scientists; catalog no. 99/4/1. Library of the Abdus Salam International Centre for Theoretical Physics Archives, Miramare-Trieste, Italy (hereafter cited as ASP).

[31] "Centre for Practice of Theory," *Nature,* March 22, 1974, 270–1.

[32] "Support for Trieste," *Nature,* Nov. 8, 1974, 87.

the third world, depended on UNESCO. In Israel, this knowledge produced among the scientific community the decision to boycott the ICTP.[33]

THE BOYCOTT OF THE ICTP'S PROGRAMS

In June 1975, science writer John Maddox reported: "I heard that a conference due to be held at the ICTP during July has to be moved to Venice, simply because there are limits to the freedom of international scientific centres, such as that at Trieste, which are supported by UNESCO, to sponsor conferences at which Israelis may or may not attend."[34] Maddox was talking about the sixtieth birthday celebration Salam had arranged for Fred Hoyle. Soon after the announcement of the event, Israeli scientists made it clear that they would not visit the ICTP because of its links with UNESCO, which had "excluded" Israel. The Israeli scientists were led by Salam's former pupil Yuval Ne'eman, then president of Tel Aviv University. Earlier in 1975, Ne'eman had been elected "corresponding member" of the ICTP, presumably as a maneuver to demonstrate that the center wanted to stay away from the political feud at UNESCO. Ne'eman had replied that he would not be willing to visit or accept any honor from the ICTP.[35] After a few months, the center had been flooded with letters from Israeli physicists who followed the line taken by Ne'eman. Thus Hoyle's birthday conference had to be moved to Venice. This, however, was only one of the problems in a difficult year for the ICTP.

The ICTP had scheduled sessions in complex analysis, solid-state physics, nuclear physics, plasma physics, and high-energy physics for 1975. By mid-December 1974, however, a number of American and Israeli physicists and mathematicians had resigned as organizers of the ICTP courses and refused to attend any activity there. The boycott would badly affect virtually all the 1975 programs and activities.

Lipman Bers was invited to participate in the complex analysis course. Bers, a world authority in the field from Columbia University, took an openly hostile position. He sent a strong letter of resignation, with copies to a large number of his colleagues.[36] He convinced his good friend, the eminent mathematician Lars Valerian Ahlfors, to follow in his footsteps. In his own resignation letter to Salam, Ahlfors explained: "the fact that some of my closest friends are staying away makes my participation quite unattractive."[37] Wolfgang Fuchs, professor at Cornell and the organizer of the same course, did likewise: "the only way of protesting that [went] beyond empty words [was] resigning."[38] More letters followed, and the trickle of resignations became a flood from American mathematicians.

The Solid State Winter Course followed a similar pattern. In late April 1975, Walter Kohn and Norton Lang, from the University of California at San Diego, both central figures to the course, added their names to the list of boycotters. An attempt was made to reorganize the course by replacing the Americans with lecturers from Spain and Latin America and requesting that remaining speakers assume additional sessions.

[33] Luciano Bertocchi to Salam, internal memo, July 26, 1975, G.118, ASP.
[34] John Maddox, "Affront to Freedom in Science," *Times Educational Supplement,* June 1975.
[35] Yuval Ne'eman to Salam, May 13, 1975, G.119, ASP.
[36] Lipman Bers to Salam, Jan. 20, 1975, G.118, ASP.
[37] Lars Ahlfors to Salam, Jan. 27, 1975, G.118, ASP.
[38] Wolfgang Fuchs to Salam, Dec. 12, 1974, G.116, ASP.

The situation became critical in December 1975, when, a matter of weeks before the course was due to start, Leo Falicov, an Argentinean solid-state physicist who worked in the United States, resigned. Falicov was the leading speaker, in charge of fifteen lectures. John Ziman, the organizer of the course, had to ask two Spanish and Latin American speakers to fill the breach.

The case of the nuclear physics course, which was scheduled for summer 1975, mirrored those of the mathematics and solid-state courses.

Very few scientists apart from the Americans and the Israelis joined the boycott. However, the absence of researchers from leading institutions in North America and from the Weizmann Institute badly disrupted the courses. The ICTP was precisely a space where third world scientists could meet colleagues from leading centers in the West. Young European scientists were attracted to the ICTP meetings mainly because of the presence of leading physicists, most of them working in American universities. The resignation of those physicists was a terrible blow to young scientists' expectations.

The boycott against the ICTP took place while a big revolution in high-energy physics was under way. High-energy physics was the main field of research and activities at the ICTP because of Salam's group. The 1974–1976 events have been analyzed by various authors, including insiders, largely concerned with the intellectual development of particle physics and the relations between theory, experiments and machines.[39] On November 11, 1974, two American laboratories investigating the e^+e^- (positron-electron) annihilation detected an enormous resonance around 3.1 GeV. Burton Richter's group at the Stanford Linear Accelerator Center (SLAC) called the new particle psi (ψ), and the Brookhaven National Laboratory–MIT collaboration, led by Samuel Ting, called it J. A few days later, the Italian laboratory at Frascati confirmed the discovery of the J/ψ particle. The outcome was the establishment of charmed quark and quantum chromodynamics, the late-twentieth-century model of particle interactions. It was the community of particle physicists who coined the term the "November Revolution" in high-energy physics to refer to this period.

A few weeks after the J/ψ discovery, Salam and Jogesh Pati, his Indian collaborator, offered a particle spectroscopy alternative to the one predicted by the charm model. They claimed that it was necessary to begin a search in the energy regions in which their model predicted the existence of new particles. Salam hoped the color gluons would be detected as well as charm. Salam proposed holding a meeting in Trieste during the summer of 1975. Its title would be *Phenomenology in High Energy Physics and the Missing Particles,* referring to the possible companions of the J/ψs. Leading theoreticians and experimentalists were invited, including Richter, but the Americans and the Israelis refused to participate. Some influential theorists in Israel, such as Ne'eman and David Horn, had already rejected any collaboration with the ICTP, as did Haim Harari. Refusing any collaboration with the center, he told Salam that "the only possible reaction of the civilized world must be to reject any participation of UNESCO in any . . . event." He also advised Salam that he intended to circulate the letter in which he had discouraged the participation of scientists in ICTP activities.[40]

[39] On the November Revolution, see Andrew Pickering, *Constructing Quarks: A Sociological History of Particle Physics* (Chicago, 1984), 180–8, 213–28, and 253–81; Lillian Hoddeson, Laurie Brown, Michael Riordan, and Max Dresden, eds., *The Rise of the Standard Model: Particle Physics in the 1960s and 1970s* (Cambridge, 1997).

[40] Haim Harari to Salam, Feb. 4, 1975, B.246, ASP.

Salam tried to persuade him and the others to stop the boycott, pointing out that it would damage the center, not UNESCO.

The new particles were discussed in August at a major SLAC conference. Both Salam and Pati felt excluded from the debates being held in the United States, not because their model was rejected, but because it was being utterly ignored and the chance to explain it in Trieste denied. Harari gave a talk in which he referred to "[t]he many versions of the Han-Nambu color," stressing that "[a]ll such models suffer from common difficulties." Thus the only tacit reference to the Pati-Salam model was in a reference to ten models that had to be discarded. After a short comment, Harari concluded: "The rejection of the possibility that the J/ψ particles are colored *returns us to the conventional theoretical framework* of hadron physics."[41] Indeed, Salam knew that alternative theories needed advocates and that direct access to experimentalists was crucial. He learned about SLAC's official stance through an internal memo signed at the laboratory by fifteen physicists stating that "no experimental results obtained at SLAC could be exhibited at a UNESCO institute like Trieste."[42] This was the coup de grace to the Trieste meeting, and the only ICTP activity that was actually canceled as a result of the boycott.

In short, the boycott seriously disrupted all the activities held at the center in 1975.

THE PROPONENTS OF THE BOYCOTT

Although there was no explicit coordination of the boycott of the ICTP, there was a clear national pattern: it was spearheaded by Israeli physicists who asked their American colleagues to join them.[43] Furthermore, in each subdiscipline there was at least one promoter of the boycott. Kohn (solid-state physics) circulated letters urging his colleagues, including Falicov, not to visit the ICTP. Bers (complex analysis), Harari (high-energy physics), and T. E. O. Ericson (nuclear physics)—the latter from CERN and one of the few physicists in Europe who boycotted the center—sent similar letters urging their colleagues to follow their examples.[44] There were no contacts between boycotting scientists from different subdisciplines; in this sense, the phenomenon was local.

The motivations of those promoting the boycott were very different. Lipman Bers was born in Latvia, where he was politically active. In 1940, because of his Jewish background, he immigrated to the United States, where he was supported by a Yiddish research organization. His contributions in complex analysis made him one of the leading mathematicians of his day. A Fellow of the American Association for the Advancement of Science since 1965, he was appointed Davies Professor of Mathematics at Columbia University in 1972. Bers led the creation of the Human Rights Committee of the National Academy of Sciences. He was a left-wing liberal and widely respected

[41] Haim Harari, "Theoretical Implications of the New Particles" (paper presented at the International Symposium on Lepton and Photon Interactions at the High Energies, Stanford Univ., Stanford, Aug. 1975).

[42] Salam to J. Harrison, July 28, 1975, G.118, ASP.

[43] The idea of a conspiracy did arise in the minds of those in charge of the ICTP's activities. The most explicit reference to someone pulling strings in order to damage the center came from Stig Lundqvist. He thought that Walter Kohn was behind the boycott (Lundqvist to Budinich, May 27, 1975, G.121, ASP).

[44] T. E. O. Ericson to Salam, Dec. 18, 1974; and Bers to Salam, Jan. 20, 1975, G.118, ASP. Harari does not recall what his personal involvement in the affair was, but he remembers sending copies of his "very strong letter to Salam" to other speakers (Haim Harari, email to author, June 11, 1999).

among his American colleagues as a mathematician and as someone committed to humanitarian causes.[45] His friendship with Ahlfors dated from 1951.

Wolfgang Fuchs was another émigré from the Nazi government. After studying in Cambridge, he went to the United States, where he pursued a brilliant career at Cornell University. He was well known for his political activism. In Ithaca, he was a member of the local chapter of Amnesty International. More significantly, during the cold war he promoted contacts between American, Russian, and Chinese mathematicians.[46] Neither Bers nor Fuchs was a Zionist. Fuchs, in fact, had signed letters against the violation of human rights by Israel. Therefore, their goal was not necessarily to support Israel's policy through the boycott but certainly to oppose the decision made at UNESCO. Excluding a country from the United Nations violated the most elementary principles defended by human rights activists.

Walter Kohn's case was different. Born in Austria to an orthodox Jewish family, he studied in a Jewish school in Austria in the 1930s. When the Nazis occupied the country, he immigrated to England; his parents died in Auschwitz. After the war, he studied first in Canada and later in the United States, at Harvard. His identity was always a matter of permanent reflection for Kohn: "In terms of my identity, I see myself as an American, a world citizen, a Jew and a former Austrian." Yet his strongest ties were with the Jewish culture and community. In San Diego, he worked on several Jewish projects. In Israel, where he "had some of [his] closest friends," Kohn had a reference point.[47]

Despite the differences, we should notice a common factor. Bers, Fuchs, and Kohn were émigrés. They had learned firsthand about the politicization of the German academic world. Furthermore, they not only strove against the depoliticization of academia but also, as scientific émigrés, struggled for the secularization of academic life to confront the anti-Semitism that prevailed in American universities in the 1930s and 1940s. The importance of keeping scientific institutions "politically neutral" had been crucial for their survival. Those who presented the Israel Resolutions as the "politicization of UNESCO" thus brought back bitter memories to the minds of several Jewish scholars. The way the resolutions were presented to these scientists was very effective in gaining their unquestioning support.

In high-energy physics, Haim Harari overtly promoted the boycott. Privately, Ne'eman also campaigned against the ICTP, but he preferred keeping a low profile; Harari took the mission of mobilizing high-energy physics in the United States. He had finished his Ph.D. in Israel in 1965 and visited the ICTP that same year. He became an associate member in 1967, although he only visited the center to attend a few conferences, not to do research.[48] In 1974–1975, he was a visiting theoretician at SLAC, and it was from there that he energetically campaigned against UNESCO and the ICTP. Wolfgang Panofsky, one of the most prominent Jewish physicists on the West

[45] William Abikoff, "Lipman Bers," and Carol Corillon and Irwin Kra, "On the Social Activism of Lipman Bers," in *Notices of the American Mathematical Society* 42 (Jan. 1995): 8–18 and 18–21.

[46] J. Milne Anderson, David Drasin, and Linda R. Sons, "Wolfgang Heinrich Johannes Fuchs (1915–1997)," *Notices of the AMS* 45 (Dec. 1998): 1472–8.

[47] From *Les Prix Nobel,* ed. Tore Frängsmyr (Stockholm, 1999), available at http://www.nobel.se/chemistry/laureates/1998/kohn-autobio.html [in 2004]. Karin Hanta, "From Exile to Excellence," *Austria Culture* 9 (Jan./Feb. 1999).

[48] ICTP, *ICTP Annual Reports,* 1967, 1968, ICTP Archives, Trieste.

Coast, was director of the laboratory. Its deputy-director, Sidney D. Drell, joined other scientists who signed a "Statement on UNESCO," which stated that they would "not participate in, cooperate with, support, or contribute to any UNESCO programs or Activities" unless the decision to exclude Israel was reversed.[49] As we saw, several physicists from that laboratory refused to present their results at the Trieste center. SLAC did not oppose that decision, which can be interpreted as tacit support by its directorship.

American scientific academies were also embroiled in the issue. In early 1975, a joint committee of the American Academy of Arts and Sciences and the National Academy of Sciences was appointed to recommend what action should be taken regarding the political "misuse of UNESCO." Harvey Brooks, president of the American academy, reported to Salam about the angry feelings among his American colleagues "due to the politicization of UNESCO."[50] He asserted that although he would try to help the center, "this new turn place[d] grave obstacles." The ICTP should be prepared for a massive boycott.[51] Most of the boycotting scientists of the center belonged to at least one of these academies. The "exclusion" of Israel meant, in their view, an occupation of the free cultural field by the third world and the Communists. Jewish American scientists found a chance to contribute to the defense of Israel on their own battlefield. Scientists aligned themselves with other members of the Jewish Diaspora: they were also part of the machinery that was "activated" against the "Arab boycott."

SALAM'S SUPPORTERS IN EUROPE

Most members of national communities outside the United States and Israel refused to join the boycott. The Europeans were critical about the resolutions but moderate with regard to the idea of taking action against UNESCO programs. For several years, the solid-state physics and the mathematics courses had been coordinated by Europeans. John Ziman, a British fellow of the Royal Society, and the Swede Stig Lunqvist had been collaborating with the ICTP since the end of the 1960s. Both had experience in programs sponsored by UNESCO; for instance, UNESCO had sent Ziman to Cuba as a scientific expert. They considered the Israel resolutions to be a big mistake by the third world representatives and to be grave obstacles for the development of the ICTP. Throughout the two years of the boycott, Ziman and Lunqvist strove to explain to their colleagues that the UNESCO resolution had no "practical effects" whatsoever for the organizations under its tutelage and that there were no "strong administrative links between" ICTP and UNESCO.[52] Furthermore, both scientists were convinced that

[49] *A Statement on UNESCO,* papers of Professor Sidney Drell, SLAC Archives, Stanford Univ., Stanford.

[50] Harvey Brooks, letter read at the Annual Meeting of the American Academy on May 14, 1975, G.116, ASP.

[51] Harvey Brooks to Salam, Dec. 2, 1974, G.116, ASP. See also Harvey Brooks, American Academy annual meeting letter (cit. n. 50). The joint committee considered the possibility of having a meeting between scholars and scientists from both industrialized and developing countries, "under the impartial aegis of the Swedish Academy, but the idea of a conference was finally dropped in favor of a more thorough long range approach through the carefully planned commission, discussion, and publication of scholarly papers." S. R. Davis, "Report of the Special Committee on UNESCO," in *Records of the Academy, 1975–1976* (Cambridge, 1976).

[52] Ziman to Henry Ehrenreich, Jan. 15, 1976, G.118, ASP.

Walter Kohn was behind the boycott of their course. Hence Ziman's reaction to the resignation of Norton Lang, who had been Kohn's pupil, was severe:

> If you are seriously concerned about the general policies of all the various organisations, corporations, institutions, governments etc. that happen to give support to your scientific work, you should investigate these thoroughly and decide where your moral allegiances really lie.[53]

Falicov's reasons for boycotting the ICTP added another ingredient to the anti-Israel affair, namely the exclusion of a nation such as Taiwan from the center's activities. Falicov, however, considered this a violation of the "principle of universality" that should prevail in science.[54] Ziman perceived the danger of this argument and warned Salam: "I wrote back immediately reassuring him as far as I possibly could and putting him right over Israel but the linking with Taiwan is a very serious danger, since we have, indeed, excluded people from there."[55] In line with the replacement of Taiwan by China in all UN agencies in 1971, the ICTP had to follow UN rules. Paolo Budinich, deputy-director of the ICTP, too, replied to Falicov, explaining the legal reasons why the center could not invite non-UN members.[56] Ziman knew that this argument could open a new front in an already difficult battle, igniting inconvenient debates about the real nature of scientific internationalism. The ICTP was unwilling to open a discussion regarding which countries could and could not participate in the center's activities.

Other scientists in Europe preferred to maintain UNESCO programs, like those sponsored at the ICTP, independent of the political feud ignited by the General Conference. While Harari and the SLAC group boycotted the ICTP, Leon Van Hove, former director of the CERN Theory Division, disapproved of the episode at UNESCO "as a personal position," but he "did not feel that [his] contacts and doings with the Centre should be affected."[57] Similar reactions came from mathematicians in England. J. Eells, coordinator of the mathematics courses at the ICTP and a professor in London, informed Salam that the general position within the American Mathematics Society was to boycott any UNESCO-affiliated organization, which he deplored.[58] M. J. Field, a colleague of Eells's, disagreed with the outcome of the 1974 General Conference but was very upset with Fuchs's resignation.[59] The sharp contrast between Americans and Europeans was mirrored in music: the public discussion between virtuoso Yehudi Menuhin, on the one hand, and his colleagues in America—Isaac Stern, Arthur Rubinstein, and Leonard Bernstein—on the other, is a case in point. Menuhin, then president of the Music Council of UNESCO, refused to resign, as his American colleagues urged him to do.[60]

An initiative taken by Victor Weisskopf (in the United States), Aage Bohr (in Copen-

[53] Ziman to Norton Lang, May 20, 1975, G.118, ASP.

[54] This "principle of universality," as a token of an idealized scientific internationalism, was invoked by a number of scientists, including those against the boycott.

[55] Ziman to Salam, Dec. 2, 1975, G.120, ASP.

[56] Budinich to Leo M. Falicov, Feb. 4, 1976, G.122, ASP.

[57] Leon Van Hove to Salam, Feb. 3, 1975, G.120, ASP.

[58] Eells to Salam, Jan. 18, 1974, G.120, ASP.

[59] Field to Salam, Jan. 2, 1975, G.120, ASP.

[60] The exchange of open letters in the pages of the *New York Times* spanned three months in early 1975; see, e.g., the following dates: Jan. 19, 18; Feb. 11, 2; Feb. 14, 5; Feb. 23, 12.

hagen), Alfred Kastler (in France), and John Ziman demonstrates their concern about the future of scientific internationalism as a consequence of the boycott. Weisskopf, an Austrian émigré who had built close ties with postwar European physicists as director of CERN in the early 1960s, decided to write a letter to *Physics Today* on the grave consequences of such a "division of the scientific community."[61] After long discussions about its terms, he, Kastler, Bohr, and Ziman published the letter in June 1976. It was the only public statement in favor of the ICTP addressed to the scientific community. The message, however, went beyond support to the Centre; the authors were concerned with the division not only between first and third world scientists, but also between the scientific communities in Europe and the United States.

For intellectuals in Europe, UNESCO had been important in offering a new space for international cooperation within Europe. In particular, UNESCO had been involved in the negotiations to create CERN and therefore in the reconstruction of European science. For scientists such as Van Hove and Kastler the fact that UNESCO was "politicized" was not new. The point was whether politics *really* obstructed international scientific exchange. American scientists were more skeptical about multinational endeavors in general, and UNESCO in particular. A study of sales of UNESCO publications for 1968 indicated that the most sizable readership was in Europe, while attentiveness was greater in Latin America than in North America.[62] Europeans felt that the American and Israeli scientists were politicizing the issue, while the Americans and Israelis argued that it was Arabs who had politicized UNESCO. All were violently against the "November Diplomatic Revolution," but their different experiences and interests with regard to UNESCO led to sharply different attitudes toward the ICTP.

THE ANATOMY OF THE BOYCOTT

The common argument calling for boycott of the ICTP was UNESCO's sponsorship. To enter into the anatomy of the boycott, we need to investigate what other scientific initiatives, in areas similar to those pursued by the center, were sponsored by UNESCO.

Since the early 1950s, UNESCO had sponsored international activities related to the exchange of scientific information and to the establishment of regional centers for the promotion of scientific research. Since 1946, the International Council of Scientific Union (ICSU) had been UNESCO's consultant on international cooperation in science. Although the ICTP's financial dependence on UNESCO had decreased since then, for the 1975–76 fiscal year UNESCO contributed US$560,000 to ICSU for advisory services and specific activities, which was twice UNESCO's contribution to the ICTP.[63]

The International Union of Pure and Applied Physics (IUPAP) was the foremost international association of physicists. ICSU channeled funds for physics events through IUPAP. UNESCO's official documents explicitly stated this collaboration, extending its commitment to support ICSU's partners: "Further financial support will be provided, as appropriate, to the unions, associations and other organs of ICSU for the execution

[61] Kastler, in France, went even further, suggesting a letter urging Victor Weisskopf not only not to boycott organizations affiliated with UNESCO but also "not to boycott UNESCO itself" (Kastler to Weisskopf, Dec. 1975, G.120, ASP).

[62] Elzinga, "Unesco and the Politics of Scientific Internationalism," (cit. n. 3), 114.

[63] UNESCO, *Approved Programme and Budget 1975–1976* (Paris, 1975), 186, par. 2113.

of specific activities."[64] IUPAP's reports also established the links between the union and UNESCO, and although it concentrated on educational programs, the agreement gave no limit to subject areas.[65]

A significant, though not surprising, feature of IUPAP was its Western constituency and image. In 1972, ten out of the thirty-nine national committees of IUPAP were in third world countries, and of the total number of votes, allocated according to the number of shares belonging to each country, the third world had only 15 out of 102. By July 1975, Salam had decided that it was worth trying to expose the potential danger faced by ICSU and IUPAP if the boycott widened. He wrote to his friend Richard Dalitz, "If the boycott of all UNESCO-sponsored institutions continues, ICSU is going to have a very difficult time soon."[66] Assuming political and moral consistency on behalf of the boycotting scientists, the equation was simple: if ICSU was worth protecting, the scientific community should lift the boycott against the ICTP. The assumption proved to be wrong.

The only IUPAP *Book of Nomination Forms* Salam kept in his personal library in Trieste was the 1975 issue, the contents of which are revealing.[67] At least three nominations from Israel's IUPAP National Committee were presented for consideration by the Fifteenth General Assembly, due to be held in Munich in 1975: A. Muny, for a post at the Commission on Superconductors; W. Low, for a post at the Commission on Magnetism; and Haim Harari, who was nominated for membership at the Commission on Particles and Fields. The form was presumably submitted in May 1975, after Harari started his campaign against the ICTP. Another member of IUPAP throughout these years was Yuval Ne'eman.[68] It is remarkable that Salam did not expose this apparent contradiction by some the boycotting scientists. He recognized that, to retain a foot in both fields, third world development and first world science, he had to ensure that certain boundaries were not transgressed.

This was not the only incongruity between the adduced reasons for boycotting the ICTP and thc attitude toward other UNESCO-sponsored scientific initiatives. The August 1975 SLAC (high-energy physics) conference was partly sponsored by IUPAP. Salam built his hopes on the SLAC-IUPAP-UNESCO connection and wrote to the assistant director-general for science, Canadian J. M. Harrison: "If SLAC knows of IUPAP's relation with UNESCO, they must have withdrawn hostility to UNESCO. This is good news for our programmes next year."[69]

A few months later, Salam learned that the boycott was provoked not just by the connection with UNESCO but also by something deeper. Immediately after the SLAC conference, Salam received a letter from an attendee, Tai Tsun Wu, a Harvard researcher. He had visited the center only a couple of times, and his contacts with Salam were sporadic,[70] but "strange conversations" among some participants on "what to do about the Trieste Conference" elicited the letter. The official position was to boycott the conference because it was sponsored by UNESCO, Wu told Salam. But he added:

[64] Ibid.
[65] IUPAP, "Report on the 14th General Assembly" (Washington, D.C., 1973), 5.
[66] Salam to Dalitz, July 1, 1975, G.119, ASP.
[67] In Salam Personal Library, held in the Abdus Salam ICTP Library (AS 341.16 IUPAP).
[68] In 1971, Ne'eman organized an international conference in Tel Aviv partially supported by the union through a grant of US$1000. IUPAP, "Report on the 14th General Assembly" (cit. n. 65).
[69] Salam to Harrison, July 28, 1975, G.120, ASP.
[70] Tai Tsum Wu, email to author, June 20, 1999.

> Somebody then mentioned that the SLAC conference was also partially supported by UNESCO, and that Harari was well aware of this. I could not judge the accuracy of this statement, but it was in any case not challenged. After some further discussions, *a pro-Israel physicist finally admitted that the real reason for the boycott was not against UNESCO, but because of your close tie to the developing countries, who were responsible for kicking Israel out of UNESCO.*[71]

Salam replied to Wu: "It had always puzzled me why Harari had taken such an initiative against us . . . Your letter seems to make the issues a little clearer."[72]

This letter deserves careful examination. The "politicization" of an international cultural institution such as UNESCO had been linked in the United States to the "tyranny of the majority." Accordingly, the third world "and its allies" put in jeopardy the normal course of international cooperation. Wu's letter allows us to learn how the boycotting scientists translated such a link and defined "normality" in the scientific field. I should like to extend Jessica Wang's thesis to argue that their aim was to call for an "anti–third world" science as a discursive strategy to define a "politically correct" science.[73] For the promoters of the boycott, a "politically correct" science was, thus, keeping scientific institutions neutral regarding any political conflict, meaning science should not be used to upset existing power relations in international politics. The exclusion from international scientific exchange of those who threatened the status quo was a corollary of that norm.

What was Salam's link to third world regimes interested in "politicizing" UNESCO and other international organizations? Salam was indeed the leader of the third world cause in the Western physics community, and the ICTP embodied such a crusade to modernize the developing countries through science. However, it is worth considering another facet of Salam's life and his links to the third world. Salam's ties to Pakistan were at the highest level; for more than fourteen years, he had been scientific adviser to three different presidents. In early 1972, Pakistan left the commonwealth, and by November it had left the Southeast Asia Treaty Organization (SEATO) security pact. It was hoped that cutting economic and military ties with Britain and the United States would pave the way to a leading position among Arab nations. After being appointed as prime minister in 1973, Zulfiqar Ali Bhutto made a radical turn toward the Middle East and North Africa, hoping to impede an eventual recognition of Bangladesh and to finance a Pakistani nuclear bomb.[74] In February 1974, the Islamic Summit gathered at Lahore. Radical leaders of the Arab world convened: Arafat, King Faisal of Saudi Arabia, Colonel Muammar al-Qaddafi, and Presidents Hafez al-Assad, Anwar el-Sadat, and Houari Boumedienne.[75]

Salam in fact disliked Bhutto's anti-Western views, rhetoric, and actions, but his position was unclear to those outside his immediate circle—after all, he continued being chief scientific adviser to the president of Pakistan. Ironically, Salam was linked by

[71] Wu to Salam, Sept. 22, 1975, G.119, ASP (my italics).

[72] Salam to Wu, Oct. 20, 1975, G.119, ASP.

[73] Jessica Wang, *American Science in an Age of Anxiety: Scientists, Anticommunism, and the Cold War* (Cambridge, Mass., 1999).

[74] Ashok Kapur, *Pakistan's Nuclear Development* (London, 1987), 74, 150; Ian Talbot, *Pakistan: A Modern History* (London, 1998), 238.

[75] In addition, the young Libyan leader, in a packed stadium, spoke before the crowd regarding Pakistan as the "citadel of Islam in Asia" and stating that "our resources are your resources." Talbot, *Pakistan* (cit. n. 74), 237.

his fellows in the West with a movement in his native country with which he disagreed. In mid-1974, Salam resigned his position as presidential adviser, and a few months later he resigned his membership on the National Science Council—a body of which he had been part since 1963. His discrepancies with Bhutto's foreign policy were not the cause, but rather his concern about the domestic policy of the new president. Salam belonged to an Islamic heterodox sect called the Ahmadiyya Jamaat. In 1974, an eight-party coalition of the ulama launched a campaign against the Ahmadiyyas, and as a consequence the sect was legally expelled from Islam.[76] Two of his lives clashed, leaving Salam an easy target for attacks from both extremes, the populist Muslims in Pakistan and the "pro-Israel" physicists in the scientific community. The ICTP was scapegoated for Salam's conflicting and ambiguous links with both a third world country striving to strengthen its ties with the Islamic world and a third world center seeking Western support. The situation was ambiguous as well because, ignoring the callings of various colleagues from both sides of the dispute, Salam never clarified publicly his position about the Israel resolutions.

Yet Salam's association with Islam was not enough to warrant the boycott; during its first fifteen years, the ICTP did not achieve the status of a mainstream research school. It was perceived as a center for fostering third world development rather than a mainstream research institution. As a result, boycotting the ICTP had very different professional consequences from boycotting an elite institution such as Stanford. While the former would not harm the professional situation of a boycotter, the latter would have resulted in scientific suicide. Harari described his first contact with the center in a letter as follows: "It was an unbelievable opportunity for me, as a young scientist, to meet just about every prominent theoretical particle physicist in the world (including USSR) in one place before I even become a postdoc."[77] By the mid-1970s, Harari had become a player in the big leagues, a regular visitor to SLAC and a professor at the Weizmann Institute. For most Israelis, the prospect of strengthening their links with a prestigious laboratory in the United States was certainly more attractive than an association with a center identified with third world development. We can, then, begin to understand why the ICTP was an obvious scapegoat; its low scientific reputation, its identification with third world aspirations, and its unstable financial situation made it a soft target for an attack against UNESCO.

NEGOTIATING WITH UNESCO

To understand Salam and Budinich's strategy in handling the boycott, it is necessary to recall the financial situation of the ICTP by the mid-1970s. Between 1972 and 1977, the combined contributions from "unstable sources," especially the Ford Foundation, the United Nations Development Program (UNDP), and the Swedish Interna-

[76] As Talbot explains, Bhutto "found that while closer ties with the Islamic world were all well and good for strengthening Pakistan's diplomatic position, money from the oil-rich Middle East also flowed freely into the coffers of his would-be opponents" (Talbot, *Pakistan* [cit. n. 74], 238). The 1973 constitution declared that Islam was the state religion. Hence, the 1974 expulsion had not only religious consequences that led to further persecution but also political and legal ones. Indeed, since 1974, the harassment and human rights violations against the Ahmadiyya community escalated with the tacit complicity of the state, as organizations such as Amnesty International and others have made public. A. R. Gualtieri, *Conscience and Coercion: Ahmadi Muslims and Orthodox in Pakistan* (Montreal, 1989).

[77] Haim Harari, email to the author, June 11, 1999.

tional Development Cooperation Agency was greater than UNESCO's or the IAEA's. To make matters worse, those contributions would end in 1977–78, placing the continuation of such programs in jeopardy. Salam spelled out the situation in an eloquent letter to Sigvard Eklund, IAEA's director, concluding that "under the actual conditions [the ICTP was] not viable."[78] This was a dramatic touch to a longtime "plea for our parent organizations to take charge of the funding."[79]

Hans Bethe then suggested a solution: "In the meantime, and until the UNESCO Board rescinds its decision, could you not ask the government of Iran to support you for the intervening years directly? In that case you could renounce UNESCO support for the time until the political decision is reversed."[80] In fact, Salam had already taken the initiative to approach Iran, motivated by its government's pro-Western leanings and public discourse about the importance of promoting science and technology for development. In the winter of 1974 he traveled to Teheran seeking funds for the center. On returning home, he was optimistic, but the offer did not materialize.

During 1975–1976, Salam and Budinich negotiated with UNESCO and the IAEA on two points: how to bring financial stability to the center, and how to survive the boycott. On the financial side, they pointed to the ICTP's chronic deficit and its new programs, the most vulnerable of which were those of most value to UNESCO. They lobbied UNESCO through their Italian contact, Dr. A. Forti, to get support for courses on "appliable" subjects, such as oceanography and applied mathematics. This trend of aligning the ICTP scientific program with UNESCO science policies had started before this date. However, given M'Bow's instrumentalist view of science, it was crucial to emphasize the ICTP's commitment with something other than theoretical physics. On April 21, 1975, the director-generals of the IAEA and UNESCO arranged a lunch meeting in Paris, to which Salam was invited, to discuss the future of the ICTP. Salam requested that the physicists Leon Van Hove (CERN) and Alfred Kastler (Strasbourg), both strong supporters of the ICTP, be invited. These men had been members of the 1974 ad hoc committee, whose recommendations served as a reference point to prepare the agenda of the meeting.[81] The agenda did not explicitly refer to the boycott, but it was the political background to the meeting. In that meeting, the extension of the agreement was arranged, admitting the necessity of increasing the "stable" contributions, and on July 3, 1975, the formal extension, valid until 1978, was signed by UNESCO and the IAEA.

Did Salam invoke the boycott during the negotiations? He certainly played with an elementary feature of a patronage relationship: that at its most elemental level it entails an exchange of loyalty for material support. In January 1975, Salam wrote to M'Bow thanking him for his note to Eklund recommending an extension of the IAEA-UNESCO agreement. He also briefed the director-general about the risk of a massive boycott from the United States and Israel. Salam sent copies of his letter to M'Bow and Harrison, emphasizing that, in view of the boycott and of the critical financial situation, he "would deeply appreciate guidance."[82] The word "guidance" appears in both his letter to M'Bow and the annexed note to Harrison, but nowhere else

[78] See Salam to Eklund, Jan. 19, 1976, D.169, ASP; Salam to Eklund, July 11, 1975; and Salam to Eklund, Sept. 12, 1975, D.170, ASP.
[79] Salam to Eklund, July 11, 1975, D.170, ASP.
[80] Bethe to Salam, Dec. 27, 1974, G.119, ASP.
[81] Budinich to C. R. O'Neil, March 16, 1975, D.169, ASP.
[82] Salam to M'Bow and Harrison, Jan. 13, 1975, G.118, ASP.

in his correspondence concerning the ICTP's management did Salam make such a request. He was a master at seeking help for the center, but "guidance" also meant advice, and such a plea for instructions was unusual. Salam continued to consult UNESCO about the best steps to take and to brief it about the boycott. He wanted to transmit a clear message: although he disliked the Israel resolutions for the damage they could cause to the UNESCO programs, he would remain loyal to the organization. Harrison insisted: "you [Salam] are in a unique situation to make a statement . . . to show the western scientists that their action in denying support to the UNESCO secretariat will do precisely what western science does not want to happen: impede the implantation of science and technology in the developing world."[83] Salam never made a public statement for or against the resolutions and its consequences. The key, he believed, was keeping a prudent distance from the political debate and, in private, displaying the art of opportunism in the political context. The ICTP should not state an official position because that might seriously jeopardize its relations with UNESCO or with the Western scientific community, both of which were crucial.

This survey of their actions helps one appreciate Salam and Budinich's strategy. Salam's punctual reports of the effects of the 1974 General Conference upon the center's activities; the desperate appeal for "guidance" just when the boycott was starting; and the combined efforts with the Italians to make sure that UNESCO became aware of their efforts to widen the ICTP's programs—in these ways Salam and Budinich delivered a message of solidarity and loyalty during a crisis *generated by* and *within* UNESCO. However, if the center had to suffer the consequences of the confrontation between blocs in the General Conference, then UNESCO had to show its commitment in supporting the ICTP's demands for more funds. The center, with its halo of neutrality, would try to clarify the situation among the scientific community. In this sense, the ICTP could be instrumental to the purposes of UNESCO. If the center disappeared, UNESCO would lose an important ally within the scientific community.

In 1976, the ICTP achieved an unprecedented increase in the IAEA's and UNESCO's appropriations. Shortly before the beginning of both General Conferences in Vienna and Paris, Salam mobilized all his allies, in the third world and in the industrialized countries, to "exercise their influence" upon their national delegations in support of the director-generals' recommendations. For fiscal year 1977–78, they recommended raising IAEA's contribution from US$230,000 to US$450,000, and UNESCO's from US$225,000 to US$300,000.[84] The recommendations were passed by both conferences. After the UNESCO conference, M'Bow allocated an extra US$100,000 grant, a gesture that one could interpret as compensation for a difficult year. Overall, it was the largest increase by both agencies during the 1964–1979 period. (See Table 1.) In current dollars, compared with the year before, the IAEA and UNESCO allocations had increased by 85 percent and 38 percent, respectively.[85] In constant dollars, the effective combined contribution had increased by 20 percent. Finally, the negotiation

[83] Harrison to Salam, Jan. 24, 1975, G.118, ASP.

[84] Salam's letter to a network of scientific-bureaucratic allies was dated March 11, 1976. Among them: V. Latorre (Peru); S. Mascarenhas (Brazil); J. J. Giambiagi (Argentina); A. Kastler (France); V. Weisskopf (United States); B. D. Nag Chaudhury (India), M. Menon (India); A. Baiquni (Indonesia); H. B. G. Casimir (Netherlands); F. García-Moliner (Spain); Edmundo de Alba (Mexico); F. Claro (Chile); D. A. Akyeampong (Ghana) (D.171, ASP). For a discussion on the ICTP's associates network, see De Greiff, "Intenational Centre for Theoretical Physics" (cit. n. 1), chap. 5.

[85] The 1970 figure might be misleading without recalling that in that year UNESCO joined the IAEA in the operation of the ICTP.

Table 1. Increment of the Contributions to the ICTP

	1965	1966	1967	1968	1969	1970	1971	1972	1973	1974	1975	1976	1977	1978	1979	1980	1981	1982
Italy	–1.77	–2.54	–1.73	–8.2	–2.3	–2.3	–2.1	–1.7	–1.2	2.98	1.82	–1.5	–0.3	5.42	11.1	2.67	26.2	10.353
IAEA	–0.35	14.49	8.742	–1.58	–0.5	–1.4	–1.3	–1.7	0.22	2.73	15.4	–7	**16.3**	–5.9	–0.1	7.32	–1.7	6.235
UNESCO	–0.14	–0.2	–0.14	–0.23	–0.2	23.9	–1.3	–1	–0.3	2.26	1.46	–1	**4.86**	–1.4	–1	–1.5	1.52	–0.84
Other	–2.26	11.75	6.88	–10	–3	20.2	–4.6	–4.4	–1.3	7.97	18.7	–9.5	20.8	–1.9	9.92	8.51	26	15.748

* Comparison between two successive years presented taking into account inflation, using constant dollars.

with both agencies was crucial to increasing the Italian component; as mentioned before, raising UNESCO and IAEA appropriations would ease negotiations with the Italian government. As can be seen from Table 1, in 1978 and 1979 the Italian contribution also increased significantly.

I should stress that in 1976 UNESCO itself was boycotted financially by the United States, France, and Switzerland. Thus it is remarkable that, under such financial pressure, added to the world financial crisis and inflation, UNESCO's secretariat decided to allocate more funds to the ICTP.[86] Salam described the improved relationship in a letter to his ally on the other side of the Atlantic, Victor Weisskopf. He described the only consolation in the critical situation created by UNDP's withdrawal and the boycott: "Fortunately, we have now at the UNESCO secretariat, some real friends. Both the new Director-General, M'Bow and the Head of the Scientific Division, Prof. A. Kaddoura—a nuclear physicist—are good and courageous friends who try to help us in every possible way . . . They will propose in the next meeting of Executive Board (Apr-May) for an increase of 28% of ICTP." Salam thus suggested to Weisskopf that he say "a word to the US delegation at UNESCO" in favor of their initiative.[87]

NEGOTIATING WITH THE BOYCOTTING SCIENTISTS

So far, I have not discussed Salam's strategy for dealing with the boycotting scientists. In order to analyse the director's modus operandi among the American scientists, it is necessary to look back to the summer of 1975. Salam suggested to Luciano Bertocchi, professor at the Institute of Theoretical Physics of Trieste University, that he contact some Israeli physicists during his trip to the United States to attend conferences. Salam thought that this would be a good occasion to investigate the magnitude of the attack without exposing the center or himself directly. Bertocchi could do what Salam could not: explain that the ICTP did not approve of the UNESCO resolutions either. Bertocchi provided Salam with a detailed report titled "Report of His Visit to US" explaining that the Israelis wanted the ICTP to issue a public statement against UNESCO. That would be enough, at least for some Israelis to lift the boycott. However, Salam had no doubts that signing any statement against UNESCO would jeopardize his negotiations to secure the financial stability of the center. Bertocchi, however, was in a position to sign such a letter. A few months later, he sent a letter in which the boundary between his personal opinion and the ICTP's official position was left deliberately unclear.[88] That was not enough for the Israelis. Although he did not recall the episode, in an interview with the author he eloquently speculated about the case:

> [I]t was clear that for certain things Salam, even if he wanted, probably preferred not to appear in first person. Because, after all, he was a United Nations officer . . . I was just a university professor. I had a sort of contract with the center as an adviser, but I was not directly responsible for it. Therefore, I could take a position that was a bit different. And, of course, knowing Salam, and remembering the general situation, this letter, if it was written, was written not *against* Salam's will, but I would say, instead, on Salam's suggestion.[89]

[86] Salam to M'Bow, Dec. 3, 1977, G.118, ASP.
[87] Salam to Weisskopf, March 3, 1976, G.120, ASP.
[88] See Adam Schwimmer, Avraham Rinat, and Julius Davis to Luciano Bertocchi, Jan. 29, 1976, G.122, ASP.
[89] Luciano Bertocchi, interviewed by author, Trieste, July 5, 1998 (my translation).

Avoiding direct public confrontation on political matters would preserve the boundary between science and politics. Salam was definitively not allowed to mix them. He must demonstrate that his political actions were completely disconnected from his politics. While the balance of power allowed the boycotting scientists to "politicize science," Salam could not. Furthermore, he was fully aware that his political strength depended on preserving an image of a nonpoliticized scientist. Indeed, Salam appealed to his allies to confront the political controversy.

Wu's letter to Salam following the conference provided him with an instrument to mobilize his allies. Wu's testimony showed that the boycott was motivated by ill feelings toward some members of the scientific community. Aware of its power, Salam forwarded the letter to several scientists in the third world as well as to some of his allies in the United States and Europe, including Aage Bohr, Kastler, and Weisskopf.[90] The decision and terms of their letter to *Physics Today,* published the following year, was triggered by their indignation after reading Wu's revelations:

> It would appear that the boycott is in itself an attempt to use a *bona fide* international scientific activity as an instrument in the political conflict. It was thus at variance with the very principle that provides the basis for the criticism of the developments in UNESCO.[91]

In the meantime, the development of the high-energy physics November Revolution was unfolding quickly. Pati and Salam knew that their theory would not stand on its epistemological merits alone; to survive it had to be circulated within the appropriate social circles. As I pointed out, the aim of the conference at Trieste was to have access to the experimenters. Salam, however, never had this chance at the ICTP, although, in the summer 1976 the center held a conference on the topic with the altered title "Lepton Interactions and *New Particles.*" Salam approached Wolfgang Panofsky, the director of SLAC, asking him to suggest names of participants. Panofsky replied dryly that arrangements should be made on a "personal basis" and that he could "not guarantee that some of the problems which beset your last conference may not arise again."[92] The 1976 conference was a failure for one reason: by then, as Pickering points out, "the critical phase of the November Revolution was over."[93] There was little space to convince the experimentalists to start a search for alternatives to charm and confinement. Apparently, Salam did not even go to the meeting; he had realized that the ICTP had been isolated from the high-energy physics revolution. A new "established tradition" was being created, and the ICTP could only learn what was happening elsewhere.

CODA

The boycott was the instrument some Israeli and pro-Israeli scientists deployed to sensitize and unify the American scientific community against the Arab boycott and, more generally, to favor any eventual Israeli countermeasure. Whatever the resolutions said, UNESCO represented an ideal opportunity to mobilize American scientific

[90] Kastler to Weisskopf, Dec. 1, 1975, G.120, ASP.
[91] Hans Bethe, Aage Bohr, Ben Mottelson, Victor Weisskopf, and John Ziman, "No Boycott for Trieste," *Physics Today* (June 1976): 9.
[92] Wolfgang K. H. Panofsky to Salam, Feb. 17, 1976, G.121, ASP.
[93] Pickering, *Constructing Quarks* (cit. n. 39), 268.

intellectuals, with the ICTP serving as a scapegoat. Scientists who boycotted the center had different motivations. However, their individual actions produced a global effect: a serious disruption in the ICTP activities and a debate about it.

What were the effects of the boycott? Boycotts are intended to produce political changes in the target. The boycotting scientists thought that denying scientific collaboration with the third world through the ICTP would pressure the delegations to reverse their decision about Israel. At the outbreak of the boycott, Leon Van Hove was confident that "under [Salam's] direction the Centre [would] avoid any form of political prejudice" and estimated that the crisis would last only a "few years."[94] Both judgments proved to be correct. The time factor is, in fact, crucial in a boycott; this form of sanction requires a sustained action during extended periods of time because of the complexity of the networks involved in the academic field. By 1977, after the UNESCO General Conference lifted the sanctions on Israel, independently from the boycott of the ICTP, the Trieste center was again running normally.

This episode shows the tensions and contradictions of international science.[95] In spite of being aware that a clear boundary between science and politics was essential for the public image of the ICTP, Salam carefully strengthened the link between the center and UNESCO during the negotiations and kept out of the public debate. He quickly realized that while the boycott was contingent, the shortage of funds was the real obstacle to the ICTP's consolidation as a research center. It was essential to avoid any direct confrontation because, otherwise, he would be charged with "politicizing" science. Accusing someone of "politicizing" science, as American and Israeli scientists did, is a political maneuver to discredit the opponent by showing that he or she is violating the supposedly neutral character of the scientific ethos. The ideological force of the scientists' own rhetoric about scientific internationalism lies in its power to mobilize allies, even though countermeasures, such as participating in a boycott, demonstrate the political character of science. Such behavior provides, perhaps, an ideal crucible in which to explore the contradictions of the practice of scientific internationalism.

In relation to the politics of international scientific cooperation, Elzinga points out that the idea of UNESCO as an intergovernmental organization, in which scientific actions were supposed to promote political consensus, instead of a nongovernmental organization, concerned with scientific knowledge, was the outcome of the Anglo-American "populism" during the initial negotiations. It was opposed by the French, who wanted a more "intellectual" UNESCO. One could add thus that, since its inception, the politicization issue has been part of the organization's history.[96] The Israel resolutions were manifestations of the process of "detechnicization" that M'Bow's administration represented in the history of that institution. It was not the politicization of the institution but an effort to bring it back to the "original more activist spirit of its Constitution," as conceived by its Anglo-American fathers-founders. Hence, rather than the politicization it was the repoliticization of UNESCO. Thirty years after the establishment of UNESCO, the positions about the politicization of

[94] L. Van Hove to Salam, Feb. 3, 1975, B.286, ASP.

[95] Ron Doel, "Scientists as Policy Makers, Advisors, and Intelligence Agents: Linking Contemporary Diplomatic History with the History of Contemporary Science," in *The Historiography on Contemporary Science and Technology,* ed. Thomas Söderqvist (Amsterdam, 1997), 215–44, 216.

[96] Elzinga, "Unesco and the Politics of Scientific Internationalism" (cit. n. 3), 90–1.

that institution were inverted. In the 1970s, Americans and Israelis presented such re-politicization as a misrepresentation of its "original" function. Conversely, the Europeans and third world nations saw UNESCO as a political institution. Indeed, politicization of science and scientific institutions is a free-floating boundary between knowledge and power that, during a controversy, every actor draws upon according to his or her own interests.

Exporting MIT:
Science, Technology, and Nation-Building in India and Iran

By Stuart W. Leslie and Robert Kargon*

ABSTRACT

Massachusetts Institute of Technology (MIT) emerged from World War II with an impressive, worldwide reputation in basic and applied science and engineering. After redefining its own engineering education in the 1950s, MIT responded to the challenge of U.S. policy makers and foundation officials and its own sense of mission in engineering research, teaching, and practice by assisting in establishing new technical institutions of higher education around the world. This paper focuses on MIT's participation in the creation of such institutions in India and in Iran. Three case studies explore the Indian Institute of Technology Kanpur, the Birla Institute of Technology and Science, and the Aryamehr University of Technology. The aim of establishing an international system of expertise with MIT at its apex reveals both the strengths and the limitations of the "export" effort.

INTRODUCTION

The Massachusetts Institute of Technology (MIT) redefined engineering education in the 1950s, then became a model and mentor for the rest of the world in the 1960s and 1970s. Responding to the challenge of U.S. policy makers and foundation officials and driven by its own sense of mission as the center of an international network of engineering research, teaching, and practice, MIT assisted in the establishment of two new technical institutes in India and one in Iran. The sponsors and supporters of these efforts, both in the United States and in the host countries, expected these junior MITs to provide the engineering expertise and leadership considered essential for economic and political modernization. While acknowledging that the "MIT idea" might be difficult to define precisely, and even more difficult to emulate, its proponents agreed that they could "identify the major characteristic of MIT which has made it different from other institutions of technology, and . . . that this characteristic is an exportable quantity."[1]

* Stuart W. Leslie, Department of History of Science and Technology, The Johns Hopkins University, 3505 North Charles St., Baltimore, MD 21218; swleslie@jhu.edu. Robert Kargon, Department of History of Science and Technology, The Johns Hopkins University, 3505 North Charles St., Baltimore, MD 21218; kargon@jhu.edu.

[1] Gordon Brown, Norman Dahl, C. H. Norris et al. to J. A. Stratton, Oct. 27, 1960, Box 5 f.218, Gordon S. Brown Papers, Institute Archives and Special Collections, Massachusetts Institute of Technology, Cambridge, Mass. (hereafter cited as MIT MC 24).

Predictably, given MIT's long tradition of relative autonomy among schools and departments, the "MIT idea" could be interpreted any number of ways. Some faculty and administrators looked back to prewar MIT, when an emphasis on engineering practice and cooperative education set the pace. Others looked to postwar MIT, when "engineering science" and a closer coupling of the basic sciences and engineering throughout the curriculum and through interdepartmental laboratories prevailed. Postwar MIT, they recognized, encouraged a new entrepreneurial spirit most visible in the startup companies that turned Route 128 from "the road to nowhere" into the main street of high-technology industry.[2] Still others looked ahead to a future MIT, when interdisciplinary centers would reorganize research and teaching around sets of problems rather than by conventional departments. Each version of MIT would have its champions, and its opportunity, as an appropriate model for engineering education in the developing world.

Gordon Brown, a key figure in all three technical assistance programs, embodied the "MIT idea"—past, present, and future. As an MIT undergraduate and graduate student in the 1930s, Brown studied in an electrical engineering department still dominated by power systems and analog computing. During the war, as an ambitious young professor, he founded the Servomechanisms Laboratory, which pioneered digital computing and numerical control for machine tools. Named head of the electrical engineering department in 1952, Brown overhauled the curriculum for the electronics age, with a solid foundation in advanced mathematics and fundamental science.[3] As dean of engineering, beginning in 1959, Brown extended his ideas about "the engineering of science" to the entire school, backed by a $9 million grant from the Ford Foundation for "the development of a science-based engineering curriculum."[4] Engineering, for Brown, would be more theoretically rigorous but no less practical: "The tough part of the program that we now envision at MIT will be to help students acquire the purposefulness, the creativity and the sound judgment found in the brilliant engineering of science—and become men who get things done."[5] Brown called his vision a "university polarized around science," a place where the basic sciences encompassed and contributed to interdisciplinary centers, constituent departments, and education at all levels.

Whatever else the MIT idea may have implied, for Brown and his colleagues it meant national, indeed international, leadership. MIT considered itself a national resource, never more so than in the 1960s when its laboratories constituted America's "first line of defense,"[6] and its faculty and administrators served as prominent policy advisers to the White House. Was any other university better positioned to make good

[2] AnnaLee Saxenian, *Regional Advantage: Culture and Competition in Silicon Valley and Route 128* (Cambridge, Mass., 1994); Susan Rosegrant and David Lampe, *Route 128: Lessons from Boston's High Tech Community* (New York, 1993); and Alan Earls, *Route 128 and the Birth of the Age of High Tech* (Charleston, S.C., 2002).

[3] For the postwar reform of the MIT curriculum, see Karl Wildes and Nilo Lindgren, *A Century of Electrical Engineering and Computer Science at MIT, 1882–1982* (Cambridge, Mass., 1985). For the spread of engineering science across the United States, see Bruce Seely, "Research, Engineering, and Science in American Engineering Colleges, 1900–1960," *Technology and Culture* 34 (1993): 344–86.

[4] Gordon Brown to Carl Borgmann, July 15, 1959; Julius Stratton to Joseph McDaniel, Oct. 23, 1959, Box 23, f. "Ford Foundation," Julius A. Stratton Administrative, Records, 1957–1966, Institute Archives and Special Collections, MIT (hereafter cited as MIT AC 134).

[5] Gordon Brown, "The Engineering of Science," *Technology Review* 60 (July 1959): 19–22.

[6] Michael Dennis, "'Our First Line of Defense': Two University Laboratories in the Postwar American State," *Isis* 85 (1994): 427–55.

on the challenge, first laid down as the "fourth point" in President Truman's inaugural address of 1949, to "embark on a bold new program for making the benefits of our scientific advances and industrial progress available for the improvement and growth of underdeveloped areas"?[7]

Oddly enough, given the relative numbers and reputation of their faculties, the work of MIT's social scientists has overshadowed the arguably more enduring foreign policy legacy of its engineers, who believed that the institute itself could be a powerful model for economic development and nation building. MIT's Center for International Studies (CENIS), under the leadership of Walt Rostow and Max Millikan, certainly helped put modernization theory and nation building at the center of America's foreign policy agenda for the developing world. Rostow's influential *The Stages of Economic Growth,* provocatively subtitled "a Non-Communist Manifesto," provided a compelling vision for a postcolonial world and led to Rostow's appointment as a highly placed adviser on foreign policy for the Kennedy and Johnson presidencies.[8]

Though certainly aware of CENIS and modernization theory, MIT's engineers had another agenda: to train the future engineers and engineer-administrators capable of leading developing nations to modernization. Having spent a decade perfecting engineering education at home, they welcomed the opportunities offered by the Department of State, by the Ford Foundation, and by businessmen and political leaders in developing countries to share their hard-won success abroad. The MIT engineers recognized that the institute drew much of its strength from its relevance to the particular technological challenges facing the United States and that any foreign version of MIT would have to do the same within its national context. Still, they believed that the MIT idea could provide at least a road map for other countries. Much like Rostow's stages of economic growth, there might be regional variation but no serious alternative. Brown and his colleagues believed that modern engineering, like modern capitalism, was essentially global and linear. The less developed would advance by learning from, and emulating, the more developed.

Yet however committed in principle to modifying the MIT idea to accommodate local goals and resources, in practice the intellectual architects of these new MITs could never really let go of their original blueprints nor imagine genuine alternatives. Had they been able to understand how much the models of technical education they offered India and Iran embedded within them distinctly American experiences and expectations, they, and their sponsors, might have been less surprised when these new schools found themselves at odds with the political and economic realities of places with different histories, visions, and values.

[7] *Point Four: Cooperative Program for Aid in the Development of Economically Underdeveloped Areas* (Washington, D.C., 1949), 95; and Wilfred Malenbaum, "America's Role in Economic Development Abroad," *Department of State Bulletin,* Economic Cooperation series, 18–20 (1949): 1–6.

[8] Michael E. Latham, *Modernization as Ideology: American Social Science and "Nation Building" in the Kennedy Era* (Chapel Hill, N.C., 2000), charts the rise and fall of modernization theory through the Alliance for Progress, the Peace Corps, and the Strategic Hamlet Program in Vietnam. For an insider's look at CENIS, see Walt Whitman Rostow, *Eisenhower, Kennedy, and Foreign Aid* (Austin, 1985). For a more critical perspective, Nils Gilman, *Mandarins of the Future: Modernization Theory in Cold War America* (Baltimore, 2003); Allan A. Needell, "Project Troy and the Cold War Annexation of the Social Sciences," and Bruce Cumings, "Boundary Displacement: Area Studies and International Studies during and after the Cold War," in *Universities and Empire: Money and Politics in the Social Sciences during the Cold War,* ed. Christopher Simpson (New York, 1998), 3–38, 159–88.

IIT KANPUR

IIT Kanpur, which would be established in 1960, would take contemporary MIT as its model. Its supporters believed that postwar MIT, with its emphasis on engineering science and cutting-edge research in fields such as electronics, computers and aeronautics, would be appropriate for the developing world.

No nonaligned nation seemed more pivotal to U.S. interests in the late 1950s and early 1960s than India, and none more supportive of efforts to upgrade its science and engineering education.[9] John F. Kennedy, as senator and later as president, considered India a critical yardstick of democracy and economic development in the contest with China and so a major target for U.S. foreign aid.[10] The Ford Foundation likewise looked to India as a testing ground for initiatives in economic planning and development.[11]

In technical education, India had inherited from the British a system "geared only to produce overseers, surveyors and mechanics of various hues, just as literary education produced clerks and pleaders."[12] While India could boast some notable scientific institutions (the Indian Institute of Science) with some world-class talent (C. V. Raman, Homi J. Bhabha), engineering lagged far behind. Its few strengths lay in civil engineering, primarily for railroad and irrigation projects intended to sharpen Britain's "tools of empire."[13] Britain opened a half-dozen engineering colleges under the raj but kept the graduates clearly subordinate to their imperial supervisors.[14]

In planning for independence, Indian and British officials alike looked to MIT as the appropriate model for technical education in the national interest. Even before World War II, MIT had been the destination of choice for many aspiring Indian engineers, who considered its science and laboratory-based instruction a refreshing departure from an Indian educational system still dominated by lecture and recitation and the "affectation and snobbery often found at elite British universities."[15] Separate studies by British Nobel laureate A. V. Hill and by Ardeshir Dalal, the director of the Tata Iron and Steel Company (and member of the viceroy's Executive Council) concluded that an "Indian MIT," indeed several of them, would be critical in helping the country prepare itself for economic as well as political independence. With support at the highest levels of Indian industry and government, a blue-ribbon panel headed by N. R. Sarkar formally recommended "not less than four higher technical institutions," geographically dispersed throughout the country but sharing a curriculum

[9] Andrew Rotter, *Comrades at Odds: The United States and India, 1947–1964* (Ithaca, 2000); and M. Srinivas Chary, *The Eagle and the Peacock: U.S. Foreign Policy toward India Since Independence* (Westport, Conn., 1985).

[10] Arthur M. Schlesinger Jr., *A Thousand Days: John F. Kennedy in the White House* (Boston, 1965), 437–40. Rostow, *Eisenhower, Kennedy, and Foreign Aid* (cit. n. 8).

[11] George Rosen, *Western Economists and Eastern Societies: Agents of Change in South Asia, 1950–1970* (Baltimore, 1985).

[12] Deepak Kumar, *Science and the Raj, 1857–1905* (New Delhi, 1995), 143.

[13] Daniel Headrick, *The Tools of Empire: Technology and European Imperialism in the Nineteenth Century* (New York, 1981); idem, *The Tentacles of Progress: Technology Transfer in the Age of Imperialism, 1850–1940* (New York, 1988).

[14] Arun Kumar, "Colonial Requirements and Engineering Education: The Public Works Department, 1847–1947," in *Technology and the Raj,* ed. Roy MacLeod and Deepak Kumar (New Delhi, 1995), 216–32.

[15] L. M. Krishnan, "Memories of MIT, 1936–1939," *MIT Review* (April 1995): 24–5.

modeled on MIT's.[16] Prime Minister Jawaharlal Nehru, a strong advocate of science and technology in the service of the state, had personally laid the foundation stone for the first Indian Institute of Technology (IIT), at Kharagpur, near Calcutta, in 1951, calling it India's "future in the making."[17] IIT Kharagpur's founders envisioned it as the template for the IITs to come, with sufficient autonomy to ensure its standing as an "institution of national importance." India's faltering economy during the first five-year plan and an apparent surplus of engineers put the other IITs on hold for the moment.

Paradoxically, the first "Indian MIT" got no direct advice from MIT. In fact, despite assistance from the Americans (principally through the University of Illinois), the Soviets, and even the West Germans, IIT Kharagpur never received sufficient financial or intellectual resources to break the traditional mold of Indian higher education. While perhaps inspired by the Massachusetts Institute of Technology, IIT Kharagpur, as its first ten-year review concluded, was no MIT. Prime Minister Nehru had wished to balance influences of East and West and sought to diversify India's educational portfolio by establishing IITs based on several national models. Determined to push ahead, Nehru jump-started the IIT program by challenging the UN Educational, Scientific and Cultural Organization (UNESCO) and its members to support India as generously as it had developing nations elsewhere. He subsequently secured cooperative agreements for additional IITs in Bombay (in partnership with the Soviets), Madras (with the West Germans), and New Delhi (with the British).[18]

India clearly expected U.S. assistance for IIT Kanpur, already slated for a textile city southeast of Delhi. In 1958, the International Co-operative Alliance (ICA) invited MIT to send a team to India and help to prepare an initial blueprint for IIT Kanpur. When MIT begged off, citing a shortage of manpower, the United States sent the American Society for Engineering Education (ASEE) instead. Gordon Brown, for one, considered the ASEE's subsequent recommendations little more than a blueprint for "an institution similar to the engineering school one would find in a good, middle-western state university" and sought assurances that if MIT did get involved, IIT Kanpur would become "*the* graduate and research technological institute" of India.[19]

The Indian government did its best to make MIT an offer it could not refuse. Max Millikan, then in India for CENIS, reported to MIT president Julius Stratton that India's government adviser on science and engineering education, and the former head of the Indian Institute of Science, M. S. Thacker, had "underlined the willingness of the Indian government to meet almost any conditions to persuade M.I.T. to take on this task." Millikan added, "[W]e are unlikely to find any opportunity for institutional

[16] Kim Patrick Sebaly, "The Assistance of Four Nations in the Establishment of the Indian Institutes of Technology, 1945–1970" (Ph.D. diss., Univ. of Michigan, 1972), 12–28, provides the best early history of the IITs and the discussions surrounding MIT as the model for Indian technical education. Ross Bassett has begun a comprehensive study of the IITs. See Bassett, "Facing Two Ways: The Indian Institute of Technology Kanpur, American Technical Assistance, and the Indian Computing Community, 1961–1980," unpublished paper, Society for the History of Technology, Minneapolis, Nov. 5, 2005.

[17] Sudrid Sankar Chattopadhyay, "Kharagpur's Legend," *Frontline*, April 27–May 12, 2002, http://www.flonnet.com/fl1909/19090840.htm; Gyan Prakash, *Another Reason: Science and the Imagination of Modern India* (Princeton, 1999), suggests that Nehru's nationalist strategy of centralized planning and state-sponsored industrialization should be understood as a search for "a different modernity," learning from the West, as well as from the Soviets, but rooted in older, indeed ancient, indigenous traditions.

[18] Suvarna Rajguru and Ranjan Pant, *IIT: India's Intellectual Treasures* (Rockville, Md., 2003), 52–3.

[19] Brown et al. to Stratton (cit. n. 1).

assistance to science and engineering in the underdeveloped world more promising and more practicable than this one."[20] Ford Foundation president Henry Heald (who had just given MIT its largest single grant) appealed to MIT's sense of obligation—and to its vanity:

> MIT has such a splendid reputation throughout the world that it would be an excellent thing for it to sponsor an institution which could hope to have something like equal significance in the Asian area. If the proposed Indian Institute is intended to aspire to such a position of leadership, then MIT should help. On the other hand, if this is to be just another college of engineering then some other American institution would do as well.[21]

Bowing to the pressure, MIT appointed a three-man delegation, led by mechanical engineer Norman Dahl, to study the prospects for IIT Kanpur. Dahl and his colleagues learned what they could from catalogs and other sources, then spent January 1961 on a whirlwind tour of India that included meetings with government officials and visits to the other IITs, universities, national laboratories, and selected industries. The MIT team praised Indian undergraduate education—"They pray to the same gods we do!" one member commented[22]—and discovered that Kanpur was not entirely the industrial backwater they had imagined. The newly appointed head of IIT Kanpur, P. K. Kelkar, the former deputy director of IIT Bombay (which was established in 1958) genuinely impressed them as a person of intelligence, energy, and vision. He seemed to them to have "a philosophy of engineering education similar to our own and an eagerness to push ahead at Kanpur with an experiment along completely American lines," with American rather than British-style examinations, U.S. textbooks, and strong graduate and faculty research programs.[23] MIT agreed to organize and lead the Kanpur Indo-American Program (KIAP), to be funded by the U.S. Agency for International Development (USAID), and subsequently invited California Institute of Technology (Caltech), Carnegie-Mellon, Case Institute, Berkeley, Purdue, Ohio State, Michigan, and Princeton to join in advising and assisting IIT Kanpur.[24]

As Dahl read the landscape: "The primary engineering need there is for 'problem recognizing' and 'problem solving' graduates who will have the confidence, inclination, and training to do something about India's problems."[25] Given the limited numbers of potential U.S. faculty, KIAP's long-term goal would be recruiting and training a permanent Indian faculty. Top-quality Indian engineers could be found in abundance in U.S. universities and industries. How many, though, would be willing to relocate and remain in Kanpur? The Americans wondered if India was even ready for modern engineers. A future program director from MIT told Dahl after an initial visit:

> I have come to realize that the Indian culture is straining through a transition period and is in many ways only superficially receptive to the objective techniques of science and

[20] Max Millikan to Julius Stratton, Aug. 26, 1960, and Henry Heald to Julius Stratton, Aug. 19, 1960, MIT AC 134, 23/8 4.

[21] Henry T. Heald to Julius A. Stratton, Aug. 19, 1960, Box 23, f. 8, AC 134.

[22] Norman Dahl, "The Collaborative Program at the Indian Institute of Technology/Kanpur," Kanpur Indo-American Program, Final Report, 1962–1972, Institute Archives and Special Collections, Kanpur Indo-American Program, Box 1 (hereafter cited as MIT AC 334).

[23] W. W. Buechner, Norman Dahl, and Louis Smullin to Julius Stratton, Feb. 13, 1961, MIT AC 134, 8/4.

[24] Dahl, "The Collaborative Program" (cit. n. 22), 8.

[25] Norman Dahl, "The Kanpur Indo-American Program," *Technology Review* 67 (June 1965): 22.

> engineering. The capable, modern, imaginative engineer with initiative is a misfit, a man a little ahead of his time who must have courage, perseverance and patience in the face of endless frustration.[26]

That assessment perhaps said more about American prejudice than Indian experience. For the Americans, Kanpur seemed "the poorest, most backward, most unattractive part of India . . . With the exception of the few on the faculty who 'belong to' Kanpur, as the phrase goes, there is probably no one from the Director on down who would not prefer to live somewhere else—and who could not get as good or better a job somewhere else—in India."[27] For many Indians, however, IIT Kanpur was a place where they thought they could make a difference. The first round of faculty postings brought in a thousand applications, a fifth of them from the United States and Western Europe.[28] Two-thirds of the Indians chosen as faculty had earned their degrees in U.S. universities. Those without foreign degrees or experience would often be sent to one of the consortium universities for advanced training, and then paired with American counterparts on individual research projects once they returned. Turnover would prove to be far less than that at the other IITs.

Despite initial skepticism, MIT aeronautical engineer Robert Halfman, KIAP's second program leader, had to admit "that the faculty already gathered here is a really first-rate group without equal in India . . . [T]he word is really now going around among overseas Indians as well as within India that IIT/Kanpur is the place to go because that is where things are really happening."[29] From the start, undergraduate admission was dauntingly rigorous. The first hundred students came from a pool of 7,735![30] All told, IIT Kanpur would receive $14.5 million in U.S. aid for American "experts," fellowships for Indian faculty, and equipment.[31] The Indian government invested even more. With 1,000 undergraduates, 400 graduate students, and 150 faculty, IIT Kanpur was on the move.

Even though India had intended IIT Kanpur to draw on the U.S. model, "American style" had its drawbacks, especially during tense political relations between the United States and India, notably the second war between India and Pakistan over Kashmir in 1965 and U.S. arms sales and military assistance to then East Pakistan in 1971.[32] IIT Kanpur endured bitter debates over English (the language of instruction at IIT Kanpur); (unproven) accusations of CIA infiltration; late, lost, or damaged laboratory equipment; student strikes; and some Indian officials unaccountably (at least to the American faculty) enthusiastic about the Soviet models of technical education themselves being tried at Bombay.[33] For the most part, those disputes reflected limited American awareness of Indian history, politics, and academic culture—the

[26] Bob Halfman to Norman Dahl, Dec. 1, 1963, MIT AC 134, Box 23, folder 8.

[27] "Eighth Semi-Annual Progress Report, 1965," MIT AC 334, Box 1.

[28] Norman Dahl to KIAP Steering Committee, Sept. 12, 1962, MIT AC 134, Box 23, folder 8.

[29] Robert Halfman to KIAP Steering Committee, Sept. 28, 1966, MIT AC 134, Box 1, folder "Steering Committee Minutes 1966."

[30] "Report from the Fourth Five Year Plan," Nov. 1966, MIT AC 334, Box 2, folder "Fourth Five Year Plan."

[31] Rajguru and Pant, *IIT* (cit. n. 18), 30.

[32] Robert McMahon, *Cold War on the Periphery: The United States, India, and Pakistan* (New York, 1994); and Dennis Kux, *The United States and Pakistan, 1947–2000: Disenchanted Allies* (Washington, D.C., 2001).

[33] Shepperd Brooks to KIAP Steering Committee, Jan. 12, 1968, Kanpur Indo-American Program, Records, 1961–1972, MIT AC 334, Box 1, folder "Steering Committee."

British colonial legacy, a sometimes strident political neutrality, an overly bureaucratic and occasionally corrupt national educational system. Should it be all that surprising that Indian students, much like their counterparts in the developed world, would become increasingly willing to challenge conventional academic authority? (MIT itself would face far more serious campus demonstrations at home over CIA funding, classified research, and defense contracts.)[34]

Far more challenging to the success of the institute than petty resistance to American methods was a sense that IIT Kanpur might be pushing itself to the front rank of Indian engineering education on terms set by its American advisers, not by Indian engineering educators. If anything, perhaps the Indians had not been forceful enough in questioning American assumptions. After reading Halfman's "End-of-Tour-Report" (essentially a five-year evaluation), the USAID bureau chief for South Asia asked the $14.5 million question:

> How does A.I.D. manage to steer institutions in the direction of the West and orient personnel to the West, without educating the personnel away from their own environment?. . . Could we not hypothesize that the bringing of scholars regularly to this country from Kanpur might operate to alienate them from their own environment and contribute to the very thing that Dr. Halfman says India cannot afford, namely "research designed primarily to raise individual investigators to international reputations."[35]

Perhaps the biggest disappointment for the Americans was Indian industry's apparent indifference to IIT Kanpur. India's top educational adviser had predicted as much at an early planning meeting at MIT. "Industry in India," he said, "has not yet reached a stage of development or enlightenment that is sufficient to generate ideas within the technological institutes."[36] Would-be faculty consultants discovered that local companies "manufacture the way they have always manufactured. Or if they adopt a new process or a new machine, they usually bring process, machine, and even know-how in from the outside."[37] An "electronics park" to take advantage of IIT Kanpur's growing strength in electrical engineering—"With encouragement there might be repeated at Kanpur the type of industrial development that has occurred around M.I.T. in Boston and around Stanford in Palo Alto"[38]—went nowhere. So did a proposal to create a center of excellence in nuclear engineering. Pioneering programs in aeronautical engineering, computer science, and materials science, so effective at MIT, turned out Indian students overqualified for jobs at home and best prepared for graduate training and eventual employment abroad.

Yet by the end of the ten-year KIAP contract in 1972, the Americans and their Indian partners had accomplished more than anyone thought possible. Virtually from scratch, they had created one of "India's intellectual treasures." IIT Kanpur had an undergraduate enrollment of 1,600, a graduate enrollment of 400, and a faculty of 260, 132 of them Indian scholars recruited from abroad. Altogether, 122 American faculty spent time at IIT Kanpur, while 47 IIT Kanpur faculty and staff trained at KIAP institutions.[39]

[34] Dorothy Nelkin, *The University and Military Research: Moral Politics at M.I.T.* (Ithaca, 1972).
[35] Burton Newbry to Shepperd Brooks, Nov. 26, 1968, MIT AC 334, Box 1.
[36] "KIAP Summary Report, Sept. 5–6, 1961," MIT AC 134, Box 23, folder 8.
[37] "KIAP Seventh Progress Report," MIT AC 334, Box 1.
[38] "KIAP Fourth Progress Report," MIT AC 334, Box 1.
[39] "KIAP Final Report, 1962–1972," MIT AC 334, Box 1.

IIT Kanpur's computer science program had become the envy of India, thanks to its IBM 1620 (India's first), installed in 1963, and an IBM 7044, added three years later. IIT Kanpur's short courses, workshops, and conferences made it an internationally recognized center in computer science and trained the first generation of Indian programmers.[40]

Perhaps IIT Kanpur modeled itself too closely on MIT. Dahl moved on to the Ford Foundation and, from that broader perspective, had to acknowledge that despite its founding mission, IIT Kanpur had so far "been an irrelevant factor in the industrial and social progress of India . . . a kind of isolated island of academic excellence but not part of the mainstream of India's development."[41] In the short run, at least, IIT Kanpur accelerated rather than reversed India's "brain drain." Of the 840 undergraduates who had earned degrees by 1971, a quarter had gone abroad to complete their educations, while a fifth of the 576 master's students had done so, including the cream of the crop. None of the 111 Ph.D. graduates had taken a position abroad because the best prospective candidates had already left for U.S. universities.[42]

KIAP's founders had intended to create an Indian MIT, not merely an MIT in India.

> [It is] critically important for the faculty and staff to develop a pride in the Institute as an *Indian* institute of technology not as an imitation of some foreign technological institute. This entails an orientation toward problems confronting India and a realization that the development of an Indian technology for dealing with Indian problems can be both interesting and exciting. . . . It does no good to plan an ambitious program and then watch the best B. Tech., M. Tech., and M. Sc. graduates go off to foreign countries to complete their studies. . . . Technological institutions in the West have been successful primarily because they applied themselves to problems of local or national importance. The same model must apply to IIT Kanpur. Its constituency is India and the Indian people.[43]

In practice, though, IIT Kanpur had not yet established its independent identity as an *Indian* Institute of Technology attuned to local or national challenges in 1972, when KIAP wrote its final report. Nor has it done so since, sending up to four-fifths of its computer science graduates on to the United States. More than three decades after the founding of IIT Kanpur, its graduates remained "the only high-tech product in which India is internationally competitive."[44] As a common witticism in India holds, "When a student enrolls at an IIT, his spirit is said to ascend to America. After graduation, his body follows."

BIRLA INSTITUTE OF TECHNOLOGY AND SCIENCE

Rather than emulating contemporary MIT's concentration on engineering science, the supporters of the Birla Institute of Technology and Science (BITS) looked to MIT's past as the right model for India. They emphasized cooperative education and collaboration with local industry.

Industrialist G. D. Birla decided that his companies, and his country, needed a pri-

[40] "Report Prepared for the Fourth Five-Year Plan," Nov. 1966, MIT 84-59/2. Norman Dahl, "Revolution on the Ganges," *Technology Review* 69 (April 1967): 17.

[41] Dahl, "The Collaborative Program" (cit. n. 22), 29.

[42] Ibid., 114–5.

[43] J. J. Huntzicker, "Terminal Report," Jan. 1972, MIT AC 334, Box 1.

[44] P. V. Indiresan and N. C. Nigam, "The Indian Institutes of Technology: Excellence in Peril," in *Higher Education Reform in India,* ed. Suma Chitnis and Philip G. Albach (New Delhi, 1993), 334–63.

vate IIT and that MIT alone should provide the blueprint for the institute and train its faculty. A self-made "mogul" in the Carnegie and Rockefeller tradition, Birla parlayed his original jute mill near Calcutta into a powerful conglomerate, with holdings in textile and paper mills, aluminum and copper foundries, and light and heavy manufacturing. A political insider and long-time confidant of Gandhi (who would be assassinated in the garden at Birla House in New Delhi), Birla sought a middle way between Gandhi's self-sufficient villages and Nehru's state socialism and saw India's entrepreneurial spirit as the key to its industrial progress and eventual self-reliance.

To cultivate that spirit, and to train future engineers and managers for his own companies, Birla invested heavily in vocationally oriented education at all levels, from kindergartens to the Technological Institute of Textiles in the Punjab, with its own 600-loom mill.[45] As a final legacy, he proposed endowing an all-Indian institute of technology modeled on MIT. In so doing, comments his biographer, he "showed himself to be an enthusiastic participant in Nehru's project of nation-building with its emphasis on science, technology and modernization."[46] With no patience for middlemen, Birla wrote about his proposal directly and repeatedly to James Killian, chairman of MIT's board of trustees, until he got an answer. Killian finally provided a list of prospective consultants, headed by Thomas Drew, an MIT graduate in chemical engineering who had spent his professional career at Columbia University. Drew, nearing retirement, found the idea of advising, or perhaps heading, Birla's institute "to say the least, intriguing and I am in fact not so firmly wedded to Columbia that I could not be persuaded by a good cause."[47] Birla could be very persuasive. He hosted Drew that summer in India, where they discovered a shared conviction that what the country needed most were neither narrowly trained "technicians" nor "highly sophisticated research engineers" but "field and plant and applications engineers (as distinguished from 'desk engineers') able to take the responsibility of figuring out what needs to be done in the circumstances, [and] how to do it in the Indian scene with Indian materials and workmen."[48]

Birla next shopped his idea to the Ford Foundation's India representative, Douglas Ensminger. The Ford Foundation had recently begun funding European physics, as much to promote American values and cultural reintegration as to advance science.[49] Its only technology program in India, outside agriculture, had been on-the-job training for 500 young Indian production engineers in U.S. steel plants.[50] Ensminger immediately recognized in Birla's ideas an important opportunity for the Ford Foundation to broaden its programs to include industrial, as well as rural, development and thought the right expert "could—in a short time—help Mr. Birla sharpen and define his objectives. . . . in short, temper a wealthy industrialist's hopes and aspirations with

[45] Subhash Rele, "The Success Story of the Birlas," *Industrial Times of India,* Nov. 13, 1972, 4–9.

[46] Medha M. Kudaisya, *The Life and Times of G. D. Birla* (New Delhi, 2003), 393.

[47] James Killian to G. D. Birla, Jan. 19, 1962, MIT AC 134, 6/12 2; Thomas Drew to James Killian, Feb. 1, 1962, James Killian to Thomas Drew, Feb. 9, 1962, MIT Corporation, Office of the Chairman, Institute Archives and Special Collections, AC 125, Box 5, folder 35.

[48] Thomas Drew to G.D. Birla, Aug. 14, 1962, MIT 85-27, 5/35.

[49] John Krige, "The Ford Foundation, European Physics, and the Cold War," *Historical Studies in the Physical and Biological Sciences* 26 (1999): 333–61, provides a thoughtful analysis of Ford's strategy in Europe.

[50] Douglas Ensminger, "Why Did the Ford Foundation Get Involved in Training Five Hundred of India's Young Engineers in Steel Making?" April 24, 1972, Oral History, Ford Foundation Archives, New York City (hereafter cited as FF), section B9, 1–6.

the wisdom of the respected educationalist."[51] Ensminger's New York superiors dismissed the idea—"please, not technical education!"[52]—but Birla, as usual, had the right connections, in this case Killian and Julius Stratton, current MIT president and Ford Foundation trustee. Birla paid them a call in Cambridge when he dropped off his grandson for his freshman year at MIT in fall 1962. Stratton, in return, accepted Birla's invitation to visit India the following January. The Ford Foundation sent Drew and an MIT colleague back to India in the spring of 1963 to draw up detailed plans for transforming a lackluster complex of colleges supported by the Birla Education Trust, including the Birla Engineering College, into a worthy competitor of the IITs.

Drew faced a far more daunting challenge than had the IIT Kanpur team. IIT Kanpur started with a clean slate, a young dynamic director, freshly recruited faculty, the latest equipment, and lavish funding from USAID and the government of India. Birla had perhaps $3 million to invest, at least initially, with Ford willing to put in about the same, plus an entrenched faculty more concerned with job security than state-of-the-art research and teaching. Yet what BITS had that IIT Kanpur did not was a patron who truly understood Indian industry and its needs. Birla's vision of an Indian MIT, inspired by his American consultants, reached back to an earlier tradition. One trusted U.S. adviser told him what India needed was engineering, not engineering science.

> The five government engineering institutes, even with all their money and foreign technical assistance, are likely to fall short of the quality of engineering education that India needs. By "quality" I don't necessarily mean the ultra-modern, high-sophisticated space-oriented engineering that is now prevalent in United States engineering schools. India needs high-quality engineering education of the type that was prevalent in the better U.S. engineering schools in the period 1935–1950.[53]

The Ford Foundation, at the direct urging of Birla himself, asked MIT to serve as the formal American sponsor for BITS, to provide an advisory board, develop a curriculum, select equipment, upgrade the library, and recruit and train Indian faculty, essentially everything that KIAP had agreed to provide for IIT Kanpur.[54] To simplify the program administration, MIT gave Drew a courtesy appointment through its chemical engineering department. Dean Gordon Brown immediately grasped the implications. "The problem seems to boil down to this: There are two institutions in India that have now declared their desires to be developed along the lines of M.I.T. But there is only one M.I.T."[55] Having incurred one substantial obligation, could MIT do justice to a second? The original IIT Kanpur team did not think so. They considered a contract with BITS a tacit breach of contract with KIAP and an unacceptable drain on MIT resources since BITS seemed to have such little promise of becoming an "institution of excellence comparable to the goals we have set for Kanpur." They strongly urged "that MIT have no official connection with the BITS project."[56] Brown, though "troubled" by the possible conflict of commitment, took the longer view:

[51] Douglas Ensminger to George Gant, March 29, 1962, FF 0926/64-482/4.

[52] George Grant to Douglas Ensminger, April 5, 1962, FF 0926/64-482/4.

[53] Raymond Ewell to G. D. Birla, May 14, 1963, FF reels RR 0926/grant number 64-482 4.

[54] Thomas Drew, "Project for Assistance to the Birla Institute of Technology and Science at Pilani, India," July 11, 1964, MIT MC 24, 13.

[55] Gordon Brown to Edwin Gilliland, Aug. 5, 1964, MIT AC 134, 6/19 2.

[56] Norman Dahl, Robert Halfman, and Louis Smullin to Gordon Brown, Aug. 21, 1964, MIT MC 24, 13/521.

> India needs a good engineering school. Birla and the Ford Foundation in good faith are committed to a program that is well conceived, will make things better, and could surprise us. It seems to me that the price of being M.I.T., or being at M.I.T., or having the freedom ourselves to use M.I.T.'s name, imposes on us some moral responsibility to act in a statesmanlike and wise manner.[57]

Noblesse oblige, perhaps, but Brown's position carried the day. In August 1965, the Ford Foundation approved a two-year, $1.45 million grant to MIT for developing BITS, with the expectation of a renewal down the road.

MIT faculty, especially Kanpur veterans, considered BITS a bad bet. Asked to size up physics, a visitor commented, "the department can not be called a department even of bad physicists."[58] Louis Smullin, who had been a member of the original IIT Kanpur advisory team, thought that MIT could not hope to accomplish much with such relatively small resources. "Is it really clear that a company owned school isolated from the world within a company village can develop the freedom and spirit to lead Indian education?" he asked Gordon Brown. "Any lesser goal for BITS would be unworthy of MIT, as you instructed us when we went off to look over Kanpur in 1961."[59] Drew, however, appreciated Birla's more limited objectives and the predictable response of Indian faculty and students to perceived American condescension:

> I do not believe [Birla] supposes or wants an American MIT set down in India. In my judgment to attempt to develop such an *American* institution in India would be like trying to graft apples on a pine tree. We have not been asked to make such an attempt. We were asked to help devise in India an Indian technological school to produce graduates with the know-how to produce knowledge pertinent for India. . . . In many respects they consider us immature, rude, hypocritical barbarians who in certain respects happened to hit it lucky. To be viable in India an institution much be framed with Indian values in mind.[60]

If Kanpur looked unpromising to American eyes, BITS's location looked far worse. Birla had insisted on locating the institute in his ancestral village of Pilani, a tiny oasis in the vast desert 125 miles west of Delhi. The Americans wondered how such a place—"It reminds one of nothing so much as an old movie about North Africa, complete with camel caravans and hooded tribesmen"[61]—could possibly attract top faculty and students. Perhaps MIT could train future BITS faculty back in Cambridge or provide assistance through IIT Kanpur, but imagining BITS as an influential engineering school in its own right seemed preposterous. The Ford Foundation, however, would accept nothing less from MIT than the kind of energy and resources it was putting into IIT Kanpur.[62]

BITS clearly needed leadership, an MIT adviser willing to make BITS a top priority, as Dahl and Halfman had done at IIT Kanpur, and an Indian director with the vision and vigor of P. K. Kelkar. Electrical engineer David White, who had made five

[57] Gordon Brown to Norman Dahl, Robert Halfman, and Louis Smullin, Aug. 26, 1964, MIT AC 134, 6/19 2.

[58] Eugene Saletan to Gordon Brown, Nov. 13, 1967, FF Report # ED 67-23.

[59] Louis Smullin to Gordon Brown, March 28, 1966, MIT AC 134, 6/19.

[60] Thomas Drew to Members of the BITS Advisory Committee, Dec. 6, 1965, Arthur T. Ippen Papers, Institute Archives and Special Collections, MIT MC 11 IPPEN, Box 5, folder Gen. Correspondence.

[61] William Schrieber to Julius Stratton, May 7, 1966, MIT MC 24, 13/521.

[62] Howard Dressner to McGeorge Bundy, May 29, 1968, FF 4265 680/1.

shorter trips to BITS starting in 1964, accepted a two-year stint at BITS in 1968 as resident head of the MIT advisory group, replacing Drew, who had reached mandatory retirement age. Impatient with the pace of change, the Ford Foundation and its MIT advisers convinced Birla to reassign the popular, long-time director to another part of his industrial empire and hired in his place C. R. Mitra, former head of a private technical school in Kanpur.[63]

BITS's signature programs, in chemical and electrical engineering, closely followed the "practice school" model originally proposed by Drew and supported by White. Mitra pushed for a practice school program far more ambitious than anything MIT had done, as a requirement for all faculty and students. With its five-year undergraduate program, BITS had sufficient time in the curriculum for more than the usual industrial internship. Students, as part of small, interdisciplinary teams intended to model real-world experience, spent two months of "industrial training" during the summer after their third year, six months of "practice school" during the summer and first semester of their fifth year, and two months "design practice" after completing their formal coursework. Each BITS Practice School Station at one of the participating companies was a sort of miniature BITS, complete with professors (themselves learning current industrial practice), laboratories, libraries, and classrooms.[64] Starting with his Birla Industries connections, Mitra expanded the program to include the Central Electronics Engineering Research Institute (a Birla-supported national laboratory adjoining the campus), the National Physical Laboratory, and finally the National Institute of Oceanography.[65] Within a few years, the practice school option had essentially become a requirement, at least for the engineering students, with 95 percent enrollment.

By the numbers, BITS could hold its own with IIT Kanpur. In a decade of MIT–Ford Foundation support, it trained more than 3,000 undergraduates and more than 1,000 graduate students, while dramatically increasing and deepening its applicant pool. If not quite an "educational paradise in the desert,"[66] BITS nonetheless had an enviable placement record, with "BIT[S]ians" more likely to take jobs with Indian firms than the "IITans." Some 60 or 70 students in each class had job offers before graduating. Keeping faculty did prove challenging in the early years, and in any given year BITS would face a deficit of ten to fifteen positions. That turned around dramatically the year after Ford support ended, with 46 hires and only 23 departures. Like IIT Kanpur, BITS sent its best faculty for advanced training in the United States, all but one to MIT. Of the first twenty participants, sixteen returned and stayed at BITS, another enviable record.[67] Ford Foundation evaluators discovered an encouraging "esprit de corps" coupled with a "particularly practical direction that may be more difficult to accomplish in the IIT's." They proudly noted that the Indian government, despite having given no direct financial support, "was looking to BITS to provide a model for future development in education in engineering and science in India."[68]

BITS offered an opportunity, as IIT Kanpur did not, to build "a leading technolog-

[63] Douglas Ensminger, Oral History (cit. n. 50), 92–4.

[64] C. R. Mitra, "A Note on Practice School Programme," April 6, 1974, FF 4264/680 5.

[65] "BITS–MIT–Ford Foundation Report, 1964–1974," FF 5175/680 3.

[66] N. N. Sachitanand, "Pilani: Educational Paradise in the Desert," *Hindu Weekly,* Jan. 21, 1972, MIT MC 24, 15/591.

[67] "BITS–MIT–Ford Foundation Report" (cit. n. 65).

[68] John Sommer to Harry Wilhelm, April 2, 1971, FF 4264/680.

ical university in India" responsive to the country's goals, "to produce practicing engineers who will be in a position to graduate and to build industries in India, under Indian conditions."[69] With its emphasis on the Practice School and ties to Indian industry, BITS helped to educate Indian industrialists along with Indian engineers and so avoided the pitfall of (re)creating an American university in a foreign country while neglecting more pressing and appropriate local challenges.

THE ARYAMEHR UNIVERSITY OF TECHNOLOGY

The Aryamehr University of Technology (AMUT) gave MIT the scope to envision the future, a technical education where interdisciplinary research centers transcended traditional disciplinary departments. Established by imperial decree in 1965, AMUT marked an American turn in an Iranian higher education long modeled on the French system. The shah had put higher education near the top of his reform agenda. He had sent record numbers of Iranian students to the United States and built new, specialized universities in partnership with Harvard, Georgetown, and Columbia.[70] Still, he considered MIT an essential model for a rapidly industrializing Iran. When he appointed Seyyed Hossein Nasr (the first Iranian undergraduate at MIT) as AMUT's chancellor in 1972, the shah explained that he wanted an Iranian MIT, not an Iranian Harvard or Princeton, because Iran needed "a problem-solving type of education."[71] By the time AMUT's Tehran campus graduated its first class, just 257 students, Iran had already contracted with the U.S. consulting firm Arthur D. Little for a master plan for a far more ambitious campus in Isfahan, Iran's second city and leading cultural center. There AMUT could provide the expertise and leadership for a major industrial initiative anchored by a new Soviet-designed steel mill. Like India's leaders, the shah respected Soviet engineering but distrusted the politics of its engineers.

As "special consultant" to Arthur D. Little, Gordon Brown would have an opportunity to put into play the ideas about research centers he had been promoting at MIT—without much success—for a decade. Brown was particularly impressed with what engineering dean George Bugliarello had done at the University of Illinois at Chicago Circle campus to encourage "a much-needed degree of flexibility to cope with changes that are certain during the next decade or so as interdisciplinary work becomes more and more necessary" and so avoid "the compartmentalization and rigidity that the customary organization into Electrical, Mechanical, Civil, Chemical Engineering, etc., imposes on an institution."[72] Brown, AMUT's vice-chancellor, Mehdi Zarghamee, and Arthur D. Little's project leader met with Bugliarello, and incorporated many of his ideas into their master plan for Isfahan. In his handwritten notes, Brown outlined a basic organizational scheme for AMUT that included six divisions—materials, energy, information, food, systems, and basic sciences—rather than departments.[73] The final master plan closely following Brown's outline:

[69] Charles McVicker, "BITS Grant Supplement Meeting," FF 0926/64-482 4.

[70] Barry Rubin, *Paved with Good Intentions: The American Experience and Iran* (New York, 1980), 149–51.

[71] Seyyed Hossein Nasr, former chancellor of the Aryamehr Univ. of Technology, interview by authors, Sept. 4, 1996.

[72] Gordon Brown to George Bugliarello, Feb. 4, 1972, MIT MC 24, Box 6, folder 232.

[73] Gordon Brown, "Goals," n.d., MIT MC 24, Box 6, folder 244; and Arthur D. Little Co., "Functional Master Plan for the Isfahan Campus," Oct. 1972, MIT MC 24, Box 6, folder 233, 38–44.

> The main idea in this organization of instruction is to organize the academic activities on the major technological problems of the country instead of the usual disciplines. The reality of the needs of Iranian society and the aspirations for Iran's accelerated development requires that their educational system should not be a copy of the obsolete aspects of western systems by a lag of twenty years; instead, it must be based on Iranian culture and societal characteristics.[74]

Chancellor Nasr, perhaps better than anyone, appreciated the challenge of integrating western technology with Persian culture. Though he had completed his undergraduate degree in physics at MIT, Nasr had found the history and philosophy of science more compelling than science itself and earned his doctoral degree in that field at Harvard. He then returned to Iran to teach and to immerse himself in the study of Islamic philosophy and history. As a member of a prominent and politically well-connected family (his father and grandfather had been physicians to the royal house), Nasr had close ties to the shah, who personally asked him to become chancellor of AMUT. Nasr agreed, with the stipulation that he could develop vigorous programs in Islamic history, philosophy, and culture to complement its engineering training.[75] "What I wanted to do as president of the university," Nasr explained, "was to create an indigenous technology in Iran, and not simply keep copying from Western technology."[76] He sought a culturally appropriate technology, with deep roots in the Persian traditions, a project the shah viewed with considerable skepticism. Where Nasr intended to embrace history, the shah preferred to bury it in the hope of insulating the shock troops of his "White Revolution" from radical politics.[77]

Nasr interpreted his charge at AMUT as proving to the shah that the university could train engineers who could compete on a world level without abandoning their cultural values. He had been a student at MIT during the years when strengthening the humanities and social sciences had first become a priority and drew a completely different lesson than had Brown and his colleagues. MIT administrators considered the humanities a matter of broadening the horizons of future engineering leaders and corporate managers. Nasr believed that in the Iranian context the humanities were a question of national identity and purpose, the bedrock of a technical education, not a cultural veneer.

Brown returned from his first visit to Iran in 1972 convinced that the study of "technological and social systems" at AMUT might actually blunt growing student unrest, "in a country that is somewhat rigid and under the direction of one man—the Shah—who does not tolerate student radicalism or anything that could be called subversion. They executed several students after the university strike last June."[78] Brown's first-hand encounter with Iran found expression in the master plan's conclusion that "student disturbances pointed to the necessity for higher education to become more closely integrated with the social and economic life of the country and responsive to the citizens that it serves" and the hope that "students will enjoy an exciting educational experience and in coming into grip with the societal problems, be it technical, social, or economic, face the reality of the country's problems and shake off the dis-

[74] Arthur D. Little Co., "Functional Master Plan" (cit. n. 73), 38.

[75] Seyyed Hossein Nasr Foundation, http://www.nasrfoundation.org/bios.html.

[76] Nasr interview (cit. n. 71).

[77] James Bill, *The Eagle and the Lion: The Tragedy of American-Iranian Relations* (New Haven, 1988), provides a good overview of the White Revolution and its consequences.

[78] Brown to Bugliarello (cit. n. 72).

torted views of what is happening in the country. It is hoped that this system will be successful in diminishing the student problem."[79]

On the shah's direct instructions, Nasr sought an active partnership with MIT. He contacted MIT president Jerome Wiesner (a classmate from undergraduate days) about faculty sabbaticals at MIT for AMUT professors, sending AMUT graduates to complete their graduate training at MIT, and joint research programs between the two schools. He scheduled a visit to MIT to discuss his proposals with top administrators and in turn invited Wiesner and his wife to Iran to visit the cultural sites and to meet the shah.[80] In briefing Wiesner for his discussions with Nasr, Brown betrayed a strangely parochial view for someone of such international experience. Perhaps tongue in cheek, he urged Wiesner to read *The Adventures of Hajji Baba of Isfahan* (a classic piece of nineteenth-century British "orientalism") and, more ominously, warned him: "The matter of getting paid by Iran can be a sticky problem as your business associates have learned to their dismay. Persians love to bargain and haggle. It is a way of life—a game—for them. We are amateurs."[81] With oil prices soaring in the wake of the 1973 OPEC oil embargo, and Iran exporting oil in record quantities, there potentially would be plenty to haggle about.

In June 1974, Wiesner spent a week in Iran discussing the proposed MIT-AMUT agreement. He toured the campus sites, met with top government ministers and deputies, and had an hour's audience with the shah. Wiesner returned upbeat:

> The general mood in Iran at the moment is one of optimism, expansionism and general ebullience, based of course on the vastly increased funds available to the government for social development. It is obvious that everyone expects the rather successful industrialization of Iran will now move considerably faster and that the accomplishment of many social dreams having to do with education, social development and the elimination of illiteracy and poverty can be vastly speeded up.[82]

Over the summer, Brown (as official MIT liaison) and Zargamee (as AMUT vice-chancellor) drafted a formal understanding of collaboration between the two schools. They expected AMUT to "educate a group of elite engineers who would become the key instruments of the future economic and social development of Iran" and in the process "to accelerate the transfer of science and technology into the societal fabric of Iran to ameliorate the pressing industrial, economic, social, and human problems of a fast-paced industrializing society."[83]

Oil wealth inspired ambitious thinking. Iran seemed a natural sponsor to help turn MIT's new Energy Laboratory, established in the wake of the first energy crisis and headed by BITS veteran David White, into a "super international energy study center."[84] Wiesner also asked his faculty to prepare short proposals on centers for geophysical research and oceanography for Iran's consideration.[85] He appointed the head

[79] Arthur D. Little Co., "Functional Master Plan" (cit. n. 73), iv–v, 38.

[80] Hossein Nasr to Jerome Wiesner, April 19, 1973, MIT 85-12, 27/Iran.

[81] Gordon Brown to Jerome Wiesner, March 31, 1973, MIT 85-12, 27/Iran.

[82] Jerome Wiesner, Memorandum to Academic Council, June 20, 1974, MIT MC 24 6/215.

[83] An Agreement for a Program of Collaboration between Massachusetts Institute of Technology and Aryamehr University of Technology, June 19, 1974, MIT MC 24, Box 6, folder 244.

[84] I. H. Henry to James Killian, William Pounds, Gordon Brown et al., Jan. 15, 1975, MIT 81-27, 21/Kuwait.

[85] Jerome Wiesner to Hossein Nasr, July 8, 1974, MIT 85-12, 27/Iran—Oceanography Project.

of the Sloan School of Management as coordinator for MIT's educational and research efforts in Iran and hired a former American ambassador as a consultant on Middle Eastern affairs. The ambassador suggested that an AMUT for Kuwait would be the perfect way to "open the door to other highly and mutually profitable MIT associations with the Klondyke on the Persian Gulf in the future."[86] At the request of the shah's sister, MIT even committed itself to planning the Shiraz Technical Institute as a "'lighthouse' institution for hands-on technical education in Iran," though MIT had so little experience with vocational training that it subcontracted virtually the entire venture to the Wentworth Institute of Technology. MIT did agree to advise on curriculum design and oversee the project, at $300,000 a year for five years.[87] With a draft proposal on the table for a $50 million "pioneering association in energy research" to be supported by Iran at MIT's Energy Lab, someone might well have asked who would be assisting whom?[88]

By far the most controversial collaboration involved training Iranian nuclear engineers. In July 1974, the Iranian counselor for cultural affairs contacted MIT's departments of Physics and Nuclear Engineering about arranging a special master's program for students selected by Iran's Atomic Energy Organization.[89] The department chairs thought they could accommodate the sixty new students (thirty a year) Iran wanted to send, as long as it was willing to pay a slight premium. MIT nuclear engineers encouraged AMUT to consider a minor in nuclear engineering within its energy center, with close ties to the nuclear research reactor being planned for Isfahan.[90]

They did not anticipate the political fallout from their colleagues. The $1.3 million contract with Iran enraged MIT faculty and students opposed to the shah and to nuclear proliferation. Angry editorials appeared in the campus newspaper, and students and faculty mounted a sit-in protest at the Department of Nuclear Engineering.[91] Computer science professor Joseph Weizenbaum wrote a long article condemning the collaboration, under the inflammatory title "Selling MIT: Bombs for the Shah."[92] Brown responded with a revealing personal letter to Weizenbaum setting out the administration's point of view. "Because I respect the integrity and value system of our faculty," Brown wrote, "I am relieved to learn that we will have a chance to instill our value system into the minds of the Iranian students . . . to give them the resolve to see to it that nuclear technology is only used for peaceful purposes." Brown maintained that if MIT did not supply the training, others would, and MIT would then "*not* be a part of the establishment in Iran that within the next decade, will bring nuclear fission power under adequate operational control. We can ensure that the Iranians can be educated to the highest standards of competence and integrity." Brown concluded, "By working within the system, some of us can be part of the action—a member of the

[86] Jerome Wiesner to Walter Netsch, Jan. 23, 1975, MIT 85-12, 26/Iran (2); A. H. Meyer to James Killian, Dec. 22, 1974, MIT 85-27, 21/Kuwait.

[87] M. C. Flemings, "Summary and Current Status, MIT Project for the Shiraz Technical Institute, Shirza, Iran," Jan. 4, 1979, MIT 85-12, 26/Iran.

[88] Jerome Wiesner to Hossein Nasr, March 7, 1975, Jerome Wiesner Papers, MIT Archives and Special Collections, MIT 85-12, Box 27, folder Iran—Energy (hereafter cited as MIT 85-12).

[89] K. F. Hansen to Alfred Keil, Jan. 23, 1975, MIT MC 24, Box 6, folder 243.

[90] Manson Benedict to Jerome Wiesner, Feb. 5, 1975, MIT 85-12, Box 26, folder Iran (3).

[91] "Debate Over Iran in Nuclear Engineering Department," MIT MC 24, Box 6, folder 243.

[92] Joseph Weizenbaum, "Selling MIT: Bombs for the Shah," *The Tech,* March 7, 1975, 1; and *The Tech,* April 15, 1975, 5.

club so to speak. But we will not be admitted if we shut the door in their face."[93] Unconvinced, Weizenbaum wondered if "'insiders' have the greatest chance to affect changes and influence events." After all, he pointed out, "often the initiation fee of the clubs one must join in order to become an insider is precisely that one must adopt the very rules, standards, and modalities of action that at the outset one wished to change."[94] Even faculty who did not share Weizenbaum's opinion that "identification with Iran identifies us with torture" had strong misgivings about accepting "special students" likely screened for "political reliability."[95] What would they have thought had they known that MIT's Draper Laboratory (the world's leading center for missile guidance and control technology) had been negotiating a separate contract to provide a comparable facility for AMUT?[96]

No one at MIT imagined that the programs it was designing for the shah would soon fall into the hands of Islamic revolutionaries. No one would have believed how many of the Iranian students and faculty it was training would support that revolution. For historian of science Nathan Sivin, one of Nasr's campus talks had raised serious questions about whether MIT fully understood what it might be getting into. He told Wiesner that he and Brown

> have had a couple of conversations on . . . the institutional relations Hossein Nasr has been mediating . . . I have a very high regard for Gordon's judgment with regard to American society and the role of science and engineering in it. I have felt the need to convince him of the complexity of what might be called the social relations of science and engineering in societies that are still largely traditional. In particular, it seems to me extremely important to gather the widest possible cross-section of Iranian points of view before committing the good name of MIT in what I would assess as extremely unstable circumstances.[97]

Wiesner did not disagree, though he fell back on the Brown defense, that whatever his personal distaste of the shah's rule, MIT and the United States had more to gain by taking advantage of the "opportunity to play a constructive or supporting role in Iran" than by ignoring or undermining it, in the same way, and for the same reasons, that the United States maintained relations with the repressive Soviet Union.[98] Wiesner got a similarly astute assessment from a member of MIT's board, who understood the shah's deep distrust of higher education: "He knows he can't accomplish his mission without highly trained and sophisticated intellectual capital . . . On the other hand, his personal experience has alienated him from understanding—or even tolerating—the independence of those who think for themselves."[99]

The Islamic revolution that toppled the shah came as a shock to MIT, especially since AMUT became a leading center for revolutionary student activity. Nasr, who had resigned the chancellorship in 1975 after three stressful years, had seen it coming:

[93] Gordon Brown to Joseph Weizenbaum, March 28, 1975, MIT MC 24, 6/243.
[94] Joseph Weizenbaum to Gordon Brown, April 1, 1975, MIT MC 24, 6/243.
[95] George Rathjens to Walter Rosenblith, March 21, 1975, MIT 85-12, Box 27, folder Iran—Nuclear Engineering.
[96] Pat Bevans to Ed Porter, "Draper Laboratory Educational Program for Iran," June 18, 1974, MIT 85-12, Box 26, folder Iran (4).
[97] Nathan Sivin to Jerome Wiesner, Feb. 26, 1974, MIT 85-12, Box 27 (hereafter cited as MIT AC 12).
[98] Jerome Wiesner to Nathan Sivin, March 5, 1974, MIT 85-12, Box 27.
[99] Osgood Nichols to Jerome Wiesner, June 2, 1975, MIT 85-12, Box 27, folder Iran.

> Technology is not value free. It brings with it a kind of culture of its own. And so once you get into it on a high level you can become very easily alienated from your own culture and that creates a breeding ground for the worst kind of political activity. And that was also one of the reasons why the Shah paid so much attention to the new university. He said we must do everything possible to have our own scientists and engineering, to create our own technology, without this social and political explosion.[100]

AMUT had delivered what Nasr had promised—top-notch engineers grounded in Iranian culture, but engineers who, contrary to his intentions, interpreted revolutionary politics not as a variation of modernization but a repudiation of it. The faculty, traumatized by the revolution and tainted by association with the shah, left; 213 out of 230 went elsewhere, 102 of them to the United States.[101] The revolutionary government subsequently split AMUT into two separate universities: Sharif University of Technology in Tehran, renamed for a "martyred" electrical engineering student, and Isfahan University of Technology. Both suffered through the early years of Iran's "cultural revolution," which temporarily closed the universities and stressed ideological purity and egalitarianism over academic excellence.[102] Some exiles did return. Zarghamee, briefly jailed as a supporter of the shah, recalled, "At the time of the Revolution there was suddenly a very significant surge of interest in returning to Iran. Everybody became a revolutionary and they went back and wanted to get something done." One of the students sent to MIT became minister of science, many others entered government service at all levels, some took their professors' places. "So what was the impact of MIT?" Zarghamee reflected, "Well, it strengthened the Revolution."[103]

AMUT turned out to be a better student than MIT had imagined. Sharif University of Technology has grown into a major research university on the MIT model, with 8,000 students selected by competitive examination, and with many of the research centers (energy, communications, materials, ocean engineering, structural, and earthquake) its MIT advisers had originally envisioned.[104] Isfahan University of Technology, with 7,000 undergraduates and 2,000 graduate students, has followed a similar path, with research centers in information technology, steel, subsea exploration, and robotics.[105] Like its mentor, it has become the center of high-tech industry, notably in the defense sector. Under the Ayatollah Khomeini, Iran established a new center for nuclear research in Isfahan, which also became home to Iran's major missile, aircraft, munitions, and chemical weapons plants.[106]

CONCLUSION

MIT did not so much fail to export the MIT "idea" as fail to understand the full implications of exporting its brand of technical education to the developing world. Gordon Brown certainly gained an appreciation of the challenges. After reading a pro-

[100] Nasr interview (cit. n. 71).
[101] Sharif University of Technology Association, http://suta.org/.
[102] Farhad Khowrokhavar, with Shapour Etemad and Masoud Mehrabi, "Report on Science in Post-Revolutionary Iran—Part 1: Emergence of a Scientific Community?" *Critique: Critical Middle Eastern Studies* 13 (Summer 2004): 209–24, provides the best overview of science under the revolution.
[103] Mehdi Zarghamee, interview by authors, Sept. 12, 1996.
[104] Http://www.sharif.ac.ir/en/research/.
[105] Http://www.iut.ac.ir/.
[106] Http://www.fas.org/nuke/guide/iran/facility/esfahan.htm.

posal from Vanderbilt's dean of engineering about "Exporting Engineering Manpower," he mused:

> My experiences in India, Singapore, and last week in Tehran convince me that the problem is extremely complex, different in every country, and not one that will be solved by sending boys on a man's errand. In the past, I believe the U.S. has fragmented its attack on the problem, failed to plan for a five- to ten-year involvement, failed to understand the infrastructure or the "software" side of the society in which we were working, provided too little help for too short a time, and often of the wrong kind.[107]

Yet for Brown and his MIT colleagues, "software" was not essentially different from "hardware." Politics, cultural traditions, and social patterns remained obstacles to be overcome, problems to be defined and solved. Lacking perspective on the political and social changes swirling around them, the Americans tended to see only "resistance to [technical] change,"[108] rather than alternative paths to technological and national development.

MIT could successfully plan technical institutes closely patterned on itself, and it could train engineering educators to staff and administer them. It could not, however, escape the limitations of its own model. The very strengths that had given MIT its international stature could end up being weaknesses when put into practice elsewhere. An education designed to prepare undergraduates for the best American academic programs did just that. The IITs' original motto, "Dedicated to the Service of the Nation," led to the inevitable question, "Which nation?"[109] And no wonder, when four-fifths of the IITs' graduating computer science majors complete their educations and subsequently make their careers in the United States. The roster of IIT alumni reads like a Who's Who of top American engineers, entrepreneurs, and venture capitalists. Close to half of all IIT graduates, 125,000 strong and counting, live and work outside of India, 35,000 of them in the United States.[110] Silicon Valley alone employs an estimated 200,000 nonresident Indians, including the cream of the IITs.[111] Even more disappointing is that the IITs, for all their success in training future engineers and entrepreneurs, have contributed so little to their larger mission. Aptly enough, "The Role of IITs in Nation Building" was a key theme in the conference marking their golden anniversary. Despite a level of technical excellence no one could have imagined a half-century ago—getting into an IIT is now ten times tougher than getting into MIT, just 2,500 places for 200,000 hopefuls—the IITs have not provided much national leadership for India.[112] BITS has had more success than Kanpur in keeping its graduates in India, though perhaps at the cost of a lower international profile.

Sharif University of Technology and Isfahan University of Technology certainly

[107] Gordon Brown to Charles Goshen, Jan. 26, 1972, MIT MC 24, Box 5, folder 175.

[108] Gordon Brown to Roger Malek, Dec. 29, 1971, MIT MC 24, Box 6, folder 232.

[109] Rajguru and Pant, *IIT* (cit. n. 18), 6.

[110] Sandipan Dep, *The IITians: The Story of a Remarkable Indian Institution and How Its Alumni Are Reshaping the World* (New Delhi, 2004), 8, 57. Dep, an IIT-Kharagpur graduate, tracks the careers of many of the most successful IIT graduates and offers some shrewd insights into the strengths and limitations of the IIT experience.

[111] AnnaLee Saxenian, *Local and Global Networks of Immigrant Professionals in Silicon Valley* (San Francisco, 2002), based on an extensive Internet survey, is the best study of the Indian (and Taiwanese) high-tech community in Silicon Valley.

[112] Dep, *The IITians* (cit. n. 110), 310–7.

did their share of nation building, though not for the kind of nation AMUT's supporters had had in mind. Under an Islamic republic, these schools continued to send their faculty to American universities, including MIT. Yet as revolutionary ardor gave way to the harsh realities of unemployment and underemployment, Iran faced a brain drain as serious as India's. The numbers may be under dispute—one International Monetary Fund report ranked Iran first in lost scientific manpower, a figure Iran has contested[113]—but Iran loses a distressing amount of its top scientific and engineering talent to the developed world.

MIT's leaders saw their institution at the apex of an international system of expertise. Their assumption was that junior MITs would follow their example and so become nodes in an international network of scientific and engineering expertise. What they did not factor in was the asymmetry of the international community, which gave every incentive to graduates of these schools to pursue better opportunities in the developed world. MIT's engineers understood the world through the lens of modernization theory. The history of MIT in India and Iran suggests both the strengths and the limitations of that view.

[113] Http://www.parstimes.com/news/archive/2004/rfe/brain_drain.html.

SCIENCE AND U.S. FOREIGN POLICY

“An Effective Instrument of Peace”:
Scientific Cooperation as an Instrument of U.S. Foreign Policy, 1938–1950

By Clark A. Miller[*]

ABSTRACT

The profound transformation of postwar world affairs wrought by science encompassed both *competition*—driven by the atomic bomb and the growing centrality of science and technology to military as well as economic security during the cold war—and *cooperation,* driven by the desire to use science, in the words of the 1951 Berkner report, as “an effective instrument of peace.” Of the two, the former has garnered greater attention among scholars and among public audiences; the latter, however, has also significantly influenced the organization and conduct of world affairs. In the years since World War II, scientific and technological cooperation among governments has become a prominent fixture on the global stage, in programs of development, in the existence of powerful international expert institutions, and in the day-to-day business of international diplomacy. Indeed, scientific and technological cooperation has transformed the institutional apparatus of the state for foreign policy, supplementing, and on occasion displacing, diplomacy with programs of technical assistance, coordination, and harmonization. This paper explores the early phases of this transformation, globally and in the foreign policy organs of the state, in the mobilization and deployment of intergovernmental scientific cooperation as an instrument of U.S. foreign policy between 1938 and 1950.

INTRODUCTION

The years immediately following World War II brought a geopolitical and organizational transformation of world order. Geopolitically, victory in the war and monopoly over the atomic bomb left the United States temporarily dominant, militarily and industrially, until the Soviet Union exploded its own nuclear weapon in 1950. Rivalry between the two emerging superpowers consolidated the division of Europe and, as it spread to other parts of the world, dismantled Europe’s colonial empires, creating a rising tide of newly independent states. Organizationally, the defeat of Germany and

[*] Robert M. La Follette School of Public Affairs, University of Wisconsin–Madison, 1225 Observatory Dr., Madison, WI, 53706; miller@lafollette.wisc.edu.

The documentary record on which this paper draws is taken primarily from the records of the U.S. Department of State and the Interdepartmental Committee on Scientific and Cultural Cooperation. These documents can be found in the U.S. National Archives and Records Administration, at their College Park, Maryland, facility. Except where specifically noted otherwise, they are found in Record Group 353, Records of Interdepartmental and Intradepartmental Committees (Department of State), Section 3, Interdepartmental Advisory Council on Technical Cooperation and Its Predecessors (hereafter cited as NARA-RG353.3). The Interdepartmental Committee on Scientific and Cultural Cooperation is hereafter cited as SCC; the Interdepartmental Committee on Cooperation with the American Republics is hereafter cited as CAR.

Japan brought a new effort to institutionalize relations among nations. Transforming the United Nations from a wartime alliance into a multilateral organization for securing a peaceful and prosperous postwar era, the United States and its allies established the General Assembly and the Security Council as its central political institutions. At the same time, between 1943 and 1950, governments created a suite of expert institutions known collectively as the United Nations specialized agencies, including, among others, the Food and Agriculture Organization; the International Monetary Fund; the World Bank; the United Nations Educational, Scientific and Cultural Organization (UNESCO); the International Civil Aviation Organization; the World Health Organization; and the World Meteorological Organization. Many of these institutions remain central in the management of world affairs today.[1] The late 1940s also saw the United States establish extensive programs of economic and technical assistance, first in Europe, with the Marshall Plan, and later in what Harry Truman would label in his 1949 inaugural address the world's "underdeveloped countries." The end of the early postwar era, and the onset of the cold war, also brought the establishment of a new set of formal alliances radiating outward from the United States and the Soviet Union, spanning much of the globe in spheres of military and industrial competition.[2]

Science and technology contributed deeply to this transformation. Part of this influence arose from the use of science and technology as instruments of national security. The atomic bomb, and later ballistic missiles, fundamentally altered the calculus of power among nations. Less well appreciated, however, is the extent to which the postwar transformation of world order also derived from the contributions of science and technology to a fundamental shift in the practice and conduct of global diplomacy and in the organization of the state for world affairs. This latter transformation was driven by the rapidly expanding presence of scientific and technical experts in diplomatic affairs. On issues ranging from arms control and the stabilization of financial markets to public health and airline navigation, they lent their expertise to the day-to-day business of foreign policy during the immediate postwar era, supplementing, and sometimes even displacing, diplomats.[3] Especially in the UN specialized agencies and in

[1] The United States initiated or actively promoted the creation of these organizations in a series of conferences beginning with the UN Conference on Food and Agriculture in 1943 and the Bretton Woods Conference in 1944. By 1947, agreements had been signed constituting the majority of these organizations, although some would not come into force until the late 1940s or early 1950s.

[2] On the institutionalization of postwar international organization, see, especially, G. John Ikenberry, *After Victory: Institutions, Strategic Restraint, and the Rebuilding of Order after Major Wars* (Princeton, 2001); and John G. Ruggie, *Multilateralism Matters: The Theory and Praxis of an Institutional Form* (New York, 1993).

[3] The precise balance among diplomatic and expert delegates often varied, across fields and at specific meetings. In meteorology, for example, U.S. legal experts were present during the negotiation of the World Meteorological Convention, although the primary negotiations took place among the heads of national weather services. Subsequent to the creation of the World Meteorological Organization, U.S. representation to the organization was housed fully within the U.S. Weather Bureau, and the chief of the U.S. Weather Bureau served as the organization's first president. See Clark A. Miller, "Scientific Internationalism in American Foreign Policy: The Case of Meteorology (1947–1958)," in *Changing the Atmosphere: Expert Knowledge and Environmental Governance* (Cambridge, 2001), ed. Clark A. Miller and P. Edwards. In postwar arms control, by contrast, diplomats retained control over most activities, even as scientists began to join in meetings. Separate technical meetings became an important part of the arms control process, but even here diplomatic representatives kept a careful eye on the proceedings. See Kai-Henrik Barth, "Science and Politics in Early Nuclear Arms Control Negotiations," *Physics Today* 51 (1998): 34–9. See also Harold K. Jacobson and Eric Stein, *Diplomats, Scientists, and Politicians: The United States and the Nuclear Test Ban Negotiations* (Ann Arbor, Mich.,

programs of economic and technical assistance, scientists, engineers, economists, agronomists, and other experts became frontline participants in the negotiation, creation, and management of new global institutions and policy programs.

This shift toward growing involvement of experts in the conduct and organization of world affairs both reflected and drove state-level changes in the organization of diplomacy. Especially in the United States, which led much of the effort to institutionalize expertise in world affairs, agencies other than the Department of State often took the lead in managing U.S. policy with respect to the expert institutions of the United Nations and the burgeoning technical assistance programs. Building on trends during the Progressive and New Deal eras that had seen a wide range of federal agencies acquire new policy responsibilities through the mobilization of science and expertise, these agencies now asserted important new roles in conducting U.S. policy abroad, ending the State Department's traditional monopoly over U.S. foreign policy.[4]

The growing presence of experts in diplomatic affairs after 1945—and their roles in transforming science and technology into instruments of multilateral cooperation and governance—was a direct result of a novel commitment to *scientific internationalism* in the ideological underpinnings of postwar U.S. foreign policy. Scientific internationalism can be understood as the idea that international cooperation in science contributes in important ways to the furtherance of broader goals of international peace and prosperity. This idea, which dates to the nineteenth century, after World War I helped to underpin certain programs of the League of Nations, such as the International Committee for Intellectual Cooperation. In the aftermath of World War II, however, U.S. foreign policy officials adopted this idea and transformed it from a well-meaning, if essentially marginal, idea within the organization and activities of the League into a central tenet of postwar international organization. At the same time, they altered the idea itself, shifting international cooperation in science from the realm of cultural cooperation among nations (understood as peoples) to the realm of geopolitical relations among states. For U.S. foreign policy officials after the war, scientific cooperation was less important as a form of cultural and intellectual exchange and friendship than as a means for identifying, analyzing, and solving global policy problems and promoting technological and economic growth as the foundation for democracy and national security.

Elsewhere, I have described the broad outlines of scientific internationalism as it was implemented in U.S. foreign policy after World War II, its contributions to shaping the postwar order, and its influence on the character and organization of the United Nations specialized agencies between 1945 and the International Geophysical Year

1966). By the late 1990s, scientists formed a large portion of the representatives to international environmental organizations such as the Intergovernmental Panel on Climate Change (IPCC). Even at meetings of international climate negotiators, where the U.S. delegation was headed by a representative of the State Department, the other twenty-plus members of the delegation were technically trained experts from other agencies of the federal government. See Clark A. Miller, "Challenges to the Application of Science to Global Affairs: Contingency, Trust, and Moral Order," in Miller and Edwards, *Changing the Atmosphere* (cit. n. 3); and idem, "Climate Science and the Making of a Global Political Order," in *States of Knowledge: The Co-production of Science and Social Order*, ed. Sheila Jasanoff (London, 2004), for detailed discussion of the role of science in the IPCC and the climate negotiations.

[4] On the growing role of science, technology, and expertise in the transformation of early twentieth-century U.S. policy-making, see, e.g., Samuel P. Hays, *Conservation and the Gospel of Efficiency* (Oxford, 1959).

(1957–1958).[5] To date, however, the uptake and transformation of scientific internationalism among U.S. foreign policy officials remains somewhat of a mystery. When did U.S. foreign policy officials begin to adopt and adapt the idea that scientific cooperation among governments could serve as a "positive instrument for the development among nations and peoples of the soil and climate essential to the growth of peace"?[6] Who were its advocates among U.S. policy officials? What evidence did they offer on its behalf, in support of its broader utility to U.S. foreign policy objectives? How did their justifications evolve, over time, into a coherent and stable logic for supporting international scientific and technological institutions and programs? Finally, how did they coordinate the adoption and implementation of this new form of scientific internationalism across a vast array of policy domains, from health and weather to civil aviation, agriculture, and finance?

To answer these questions, this paper examines in detail the history of a relatively little known State Department organization known after the war as the Interdepartmental Committee on Scientific and Cultural Cooperation. Created by President Franklin D. Roosevelt to help secure Latin American cooperation with the war, this committee coordinated the first systematic U.S. technical assistance programs to Latin America from 1938 to 1945. As its name suggests, it drew at least part of its inspiration from the idea that intellectual exchange, like cultural exchange more broadly, could help to build mutual friendship among nations. During the war, however, the committee became a site of ideological debates between the State Department and other federal agencies, which ultimately won backing for a new idea, that scientific and technological cooperation was important less for its ability to generate international goodwill than for its ability to confront practical policy problems. Under the committee's programs, hundreds of U.S. government scientists and technical experts worked with Latin American counterparts to address such problems as fisheries management, the creation of statistical agencies, and the improvement of agricultural production.

By war's end, the program had emerged as a key exemplar of the possibilities of scientific and technological cooperation as a foundation for broader international policy coordination and problem solving. Moving quickly to secure its position in postwar foreign policy, the committee took on new responsibilities after 1945, including the coordination of U.S. policy vis-à-vis both the new United Nations specialized agencies and technical assistance programs in Europe, the Middle East, and Asia. Although the committee would ultimately be dissolved in the government reorganizations of foreign policy that took place in the late 1940s and early 1950s, its decade of work would see the emergence of key ideas underpinning the creation of the United

[5] Clark A. Miller, "The Globalization of Human Affairs: A Reconsideration of Science, Political Economy, and World Order," in *Rethinking Global Political Economy: Emerging Issues, Unfolding Odysseys,* ed. Mary Ann Tetreault, Robert Denemark, Kurt Burch, and Kenneth Thomas (New York, 2003); and Miller, "Scientific Internationalism" (cit. n. 3). See also Anne-Marie Burley, "Regulating the World: Multilateralism, International Law, and the Projection of the New Deal Regulatory State," in Ruggie, *Multilateralism Matters* (cit. n. 2), 125–56; and G. John Ikenberry, "A World Economy Restored: Expert Consensus and the Anglo-American Postwar Settlement," *International Organization* 46 (1992): 289–322.

[6] Olcott H. Deming, "In the Minds of Men: Speech Given before the Alumni and Faculty of Rollins College," Feb. 24, 1946, Folders—Speeches; Box #25; Reports; Project Reports, Quarterly; Library of Congress—Statistical Data; Entry #14; Subject Files, 1938–53; NARA-RG353.3 (hereafter cited as SCC Speeches).

Nations specialized agencies, as well as the Marshall Plan, Truman's Point Four Program, and, in subsequent years, the U.S. Agency for International Development.

Indeed, the committee's profound influence on the shape of postwar world affairs remains visible today, over a half-century later. Although frequently eclipsed in the 1950s and 1960s by the events of the cold war, the United Nations specialized agencies had reemerged by the 1990s as central sites in the management and regulation of economic and technological globalization. So powerful have many of these expert agencies become that antiglobalization protesters have labeled them a de facto world government. Experts also continue to pervade diplomatic relations. Attend a meeting of the United Nations Framework Convention on Climate Change or other treaty organization and one is likely to find the few diplomats in the room outnumbered by their technical colleagues from ministries of agriculture, energy, transportation, commerce, and the environment. Perhaps more ambivalently, science and technology remain today a key metric against which the progress of civilization is measured, despite fifty years in which the project of modernization, as described by James Scott and others, has far too frequently operated within a culture of scientific and technological hubris and given rise to tragic human consequences.[7]

ENROLLING OUR LATIN NEIGHBORS, 1938–1945

Governments had long been embarked upon limited programs of international scientific cooperation, including in colonial science policies as well as technical exchanges and missions between countries, before World War II.[8] After the war, however, United States foreign policy invested these activities with a broader meaning and significance for world order. In so doing, U.S. foreign policy officials were building on efforts to promote intellectual cooperation and exchange among the nations of the world in the late nineteenth and early twentieth centuries, many of which had led to the creation of international organizations and had been incorporated into the League of Nations. There are important differences, however, between international scientific cooperation during these two periods. First, many prewar international organizations operated as private entities, in which individuals, not governments, cooperated with one another, for professional, not geopolitical, reasons.[9] Second, much of the prewar activity was justified by a logic that connected scientific exchange to broader forms of intellectual and cultural exchange. By contrast, postwar justifications connected scientific cooperation to technological development, economic growth, and national security. Postwar programs of scientific and technological cooperation were not only professional but also geopolitical initiatives that involved direct government-to-government cooperation in an effort to secure a peaceful and prosperous world order.

[7] James Scott, *Seeing Like a State: How Certain Schemes to Improve the Human Condition Have Failed* (New Haven, 1998); Stephen Lansing, *Priests and Programmers: Technologies of Power in the Engineered Landscape of Bali* (Princeton, 1991); Vandana Shiva, *The Violence of the Green Revolution: Third World Agriculture, Ecology, and Politics* (London, 1991); James Ferguson, *The Anti-Politics Machine: "Development," Depoliticization, and Bureaucratic Power in Lesotho* (Cambridge, 1990).

[8] Merle Curti and Kendall Birr, *Prelude to Point Four: American Technical Missions Overseas, 1838–1938* (Madison, Wis., 1954); William K. Storey, *Science and Power in Colonial Mauritius* (Rochester, N.Y., 1997); idem, "Plants, Power, and Development: Founding the Imperial Department of Agriculture for the West Indies, 1880–1914," in Jasanoff, *States of Knowledge* (cit. n. 3).

[9] See, e.g., the discussion of the differences between the prewar International Meteorological Organization and the postwar World Meteorological Organization, in Miller, "Scientific Internationalism" (cit. n. 3).

These two shifts in logic were forged in the development of U.S. Latin American policy from 1938 to 1945. When Franklin D. Roosevelt came to office in 1933, his immediate concerns were on the depression that haunted the American economy. During his first hundred days, however, Roosevelt took time to put in place new foreign policy objectives for Latin America. He announced, during his inaugural address in January, that his foreign policy would make the United States into a "good neighbor" to the countries of the world, and in April, he articulated his vision for better neighborly relations with Latin America. The following summer, the president put this vision into action, refusing to use American treaty rights to intervene militarily during a series of changes in Cuban government, thereby winning popular and official admiration in many of the Latin American republics.[10]

Increasingly referred to as Roosevelt's Good Neighbor Policy, inter-American relations acquired further importance in the president's eyes with the rise of tensions in Europe. Roosevelt now upped the ante, moving to enroll Latin America in a security agreement that would prevent war from spreading to the Western Hemisphere. U.S. and Latin American leaders met in 1936 in Buenos Aires and again in 1938 in Lima to lay down ground rules for inter-American cooperation should Europe plunge into active war. Following the Lima meeting, Roosevelt indicated his desire to build upon these dramatic successes through "a concrete program designed to render closer and more effective the relationship between the Government and people of the United States and our neighbors in the twenty republics to the south." In response, Roosevelt's advisers set up a new committee, appropriately, if unimaginatively, named the Interdepartmental Committee on Cooperation with the American Republics. The committee set to work in May 1938 to identify projects and activities "in which the Government of the United States is in a position to cooperate with the other American republics for their mutual advantage." Composed of key government leaders, the committee was chaired by Sumner Welles, undersecretary of state. Its November 1938 report, detailing a proposed slate of projects, was submitted to Roosevelt under the signatures of his key cabinet officials.[11]

The program laid out in the committee's first report proposed approximately $1 million in projects to be carried out by each of the thirteen agencies involved, with the largest contributions to the Department of Agriculture ($350,000), the National Emergency Council ($176,000), and the Public Health Service ($100,080). Proposed projects ranged broadly, including soil and forest surveys, training in meteorological techniques, construction of the inter-American highway, loans of experts in fisheries, improvement of statistical services, films for distribution to both U.S. and Latin American audiences, fellowships for librarians, collections of folk music, ethnologi-

[10] See, e.g., the account of Roosevelt's prewar policies laid out by Assistant Secretary of State Sumner Welles. Welles, *The Time for Decision* (New York, 1944).

[11] Sumner Welles, Henry Morgenthau, Frank Murphy et al., "Report of the Inter-Departmental Committee on Cooperation with the American Republics Together with the Program of Cooperation Endorsed by the Committee," Nov. 10, 1938, Folder— History and Development of the Committee, Box 1, Entry #13, Records Relating to the History and Development of the Council and its Predecessors, 1938–1953, NARA-RG353.3 (hereafter cited as Council History Documents). The report was signed by the secretaries of interior, agriculture, commerce, and labor, the assistant secretary of the treasury, the librarian of Congress, the secretary of the Smithsonian Institution, the chairman of the Federal Communications Commission, the vice-chairman of the U.S. Maritime Commission, the president of the Export-Import Bank, the executive director of the National Emergency Council, and the chairman of the Civil Aeronautics Authority.

cal studies of "aborigines of the New World," establishment of a central translating office in the State Department, research on tropical diseases, and studies of means to simplify passport and landing requirements and liberalize tourist regulations.[12]

Justifying these projects, the report noted that they included "studies, investigations and enterprises to be carried out in such of the American republics as are desirous of engaging in them," as well as "projects for extending the educational, scientific and technical facilities of the several agencies." Attention was given to "intergovernmental cooperation," "expert assistance," "offer of service training for accredited foreign officials," "exchanges of students and professors," and translations of U.S. government publications, "especially those relating to public health, educational, scientific and technical matters, commerce, conservation, et cetera." The projects were assembled with an eye toward "the increasing importance of cultural relationships" and "the premise that the republics of the New World have the same aspirations, that the welfare of the community of American nations demands their increasingly close and friendly association, and that through a program of practical, reciprocal cooperation the fulfillment of our common American ideals can be brought appreciably closer to achievement."[13]

The committee took immediate advantage of Public Law 535 (1938), later amended by Public Law 63 (1939), which authorized the president to detail U.S. government employees for temporary service for periods of up to one year to the government of an American republic "if such government is desirous of obtaining the services of a person having special scientific or other technical or professional qualifications." Congressional appropriations for other kinds of projects were delayed for a year, however, until the passage of Public Law 355 (1939). This new law allowed the United States "to utilize the services of the departments, agencies, and independent establishments of the Government in carrying out . . . reciprocal undertakings and cooperative purposes" set forth in the 1936 and 1938 inter-American treaties.

Based on the committee records alone, the impact of these projects in Latin America and their reception by Latin American leaders and publics is unclear. In Washington, however, the committee's work received positive, if not stellar, reviews. Between 1940 and 1944, the committee budget grew by an order of magnitude to $4.5 million.[14] By 1944, the committee had coordinated seventy-two details of U.S. government experts to other countries, averaging six months per detail, at an additional cost of $270,000. Clearly, the committee's work satisfied its members, its supporters in the Roosevelt administration, and its congressional funders that it served a useful purpose. The war had not crossed the Atlantic into Latin America, and that alone might offer sufficient reason for continuing to support anything that looked like it had any value.

[12] CAR, "Description of the Program Endorsed by the Committee Together with Estimates of the Appropriations Required to Put the Program into Effect," Nov. 10, 1938, Council History Documents.

[13] Welles et al., "Report" (cit. n. 11). Committee records unfortunately provide no direct evidence of how Latin American governments and citizens responded to the committee's activities. However, committee records contain no materials that would indicate that either governments or individuals in Latin America took exception to the committee. Moreover, there is plenty of evidence in the committee records that Latin American governments availed themselves extensively of the committee's programs. U.S. law required that all committee projects be initiated by requests from Latin American governments.

[14] Appropriations for sixteen committee projects totaling $370,000 were legislated beginning in fiscal year 1940/41. In 1942, Congress appropriated $600,000 to the committee, and in 1943, committee appropriations grew again to $1,685,000. By 1944, congressional appropriation for the committee had risen to $4.5 million (more than $50 million in 2003 dollars).

More important for our purposes here is the evolving rationale offered by the committee in justifying its continued work. During the committee's first few years of operation, its documents highlighted the importance of "practical, reciprocal cooperation" in identifying projects. The two largest categories of details were agricultural and fisheries experts (thirteen and twelve, respectively).[15] Committee reports indicated that projects were chosen out of a common belief "in the efficiency of practical day-to-day collaboration" and that the goals of "friendship in a peaceful American world" required that they be "cooperative" and "of practical value" and convey "reciprocally mutual benefits." In this fashion, the committee sought to undermine two potential criticisms: that the projects were being undertaken for propaganda purposes, and that they were simply giveaways of U.S. money and resources to Latin America. Thus, committee selection rules insisted that both the United States and its partner benefit from potential projects. U.S. benefit was determined and ensured by the sponsoring agency, but benefit to the host country was more difficult to determine and ensure. To address this concern, Public Law 63 (1939) required that any detail of U.S. experts to a foreign government be preceded by a formal request for the project by the host country. Likewise, host governments were encouraged, and in some cases required, to share in the costs of many committee projects. In short, at least in its early years, the committee went out of its way to choose projects it could defend as having clear, tangible benefits for all countries involved and that it could argue would bring U.S. and Latin American partners together in a spirit of goodwill, with the intention of building "closer relationships with the other American republics."[16]

In late 1941 and early 1942, however, two events combined to change the committee's outlook. The bombing of Pearl Harbor in December 1941 radically altered the stakes of inter-American cooperation and quickly brought nearly all American republics into a state of war with Germany and Japan. Soon thereafter, in spring 1942, the committee's formal housing in the State Department was moved from the Division of American Republics to the Division of Cultural Relations as part of a larger departmental reorganization. Although the committee always had officially been an interdepartmental one, with twenty-seven member agencies by 1942, its chair and staff had resided within the State Department.

Given the evolving logic of the committee's work, geared as it was toward improving relations between the peoples and governments of the Americas, this move seemed, at first blush, obvious. The State Department's Division of Cultural Relations (DCR) had been created in 1938 "for the purpose of encouraging and strengthening cultural relations and intellectual cooperation between the United States and other countries." However, DCR had operated at a fairly small scale and received little notice until the outbreak of war. By 1942, the budget of the Interdepartmental Committee on Cooperation with the American Republics far outstripped that of DCR. By moving the committee to DCR, the State Department would combine under one umbrella all its

[15] CAR, "An Account of Its Organization and Present Activities," 1943, Council History Documents; "Act of May 25, 1938 as Amended by the Act of May 3, 1939 (Public Law No. 63, 76th Congress), Recapitulation of Statistical Data," Oct. 14, 1944, Council History Documents; SCC, "Program for 'Cooperation with the American Republics,'" Aug. 15, 1947, Council History Documents.

[16] SCC, "Objectives of the Program of the Interdepartmental Committee on Scientific and Cultural Cooperation (SCC) and Criteria Which Have Been Used in Selection of Projects," Nov. 30, 1948, Folder—Memos #2–45. Box #30, Records of the Full Committee, Entry #26, Interdepartmental Committee on Scientific and Cultural Cooperation, NARA-RG353.3 (hereafter cited as SCC Memos).

efforts to strengthen cultural relations between the United States and other countries of the world and also dramatically increase the DCR budget.[17]

Within the State Department, the move caused few complaints. Rhetoric in defense of the committee underwent a subtle shift, reflecting the emphasis at DCR on cultural cooperation and understanding. Sumner Welles and Assistant Secretary of State G. Howland Shaw spoke of the committee in terms that emphasized its "scientific, economic, cultural, and social endeavor" and pointed to its value in generating "better understanding of our ways, means, and methods of life" and improving "inter-nation and inter-people understanding." Consistent with DCR's mission, the new rhetoric of justification stressed the construction of a positive image of the United States and the export of that image abroad. As Welles framed the committee's mission at a meeting in August 1942, "Effective international cooperation cannot exist unless there is an appreciation and understanding in each country of those problems of other countries which arise from national customs, traditions, achievements, and philosophies of life. . . . We have the task of learning to appreciate and understand the viewpoints, traditions, and customs of our neighbors in the other American Republics and of making it possible for them to see our problems and ways of life—not by propaganda or proselyting, but rather by the joint execution of useful undertakings and through the personal associations incident thereto."[18]

For many of the other members of the committee, however, relations with their new hosts in DCR quickly soured. While the rhetoric's subtle shift toward cultural friendship was not a problem, per se, DCR leaders had a tendency to overweight cultural projects in budgetary terms, too. Committee members from other government agencies complained that too much money was being spent on films and libraries and not enough on science, and they expressed dismay that the program was straying from what it did best. In their views, the committee had derived significant early success from its emphasis on projects of practical value, especially in areas of scientific and technical expertise and competence, and that success was now in danger of evaporating. The shift to DCR also gave rise to a second set of questions about the administrative positioning of the committee. Initially, the committee had operated as an independent program in the office of the undersecretary of state. The shift to DCR lowered the committee two tiers within the State Department organization, well below not only its previous position in the department but also its organizational status with respect to most other federal agencies as well.[19]

By the end of 1943, dissatisfaction with the new arrangements led to organizational struggles within the State Department. In January 1944, the State Department returned the committee to independent status within the Office of American Republic Affairs and assigned it a permanent chairman and an executive secretary, reestablishing its former status and significantly boosting its institutional infrastructure. Two years of deliberations under DCR had honed what most members outside the State Department saw as the committee's core mission—scientific and technical cooperation—and after the 1944 reorganization, that mission was given new prominence,

[17] "Administrative Relations of the Secretariat of the Interdepartmental Committee on Cooperation with the American Republics within the Department [of State]," Council History Documents.

[18] Undersecretary of State Sumner Welles, "Meeting of the Interdepartmental Committee on Scientific and Cultural Cooperation," Aug. 12, 1942, quoted in SCC, "Objectives" (cit. n. 16).

[19] "The Interdepartmental Committee on Cooperation with the American Republics: An Analysis of Its Administration," 1944, Council History Documents.

independent of the goal of improving cultural relations.[20] In his February 1944 "Report to the President on Closer Relationship between the American Republics," Acting Secretary of State Edward Stettinius (who had replaced Cordell Hull) outlined the new logic of the committee's operations:

> Programs of this character are an effective means of achieving international, hence national, security. Measures which spread an understanding of the democratic way of life and diffuse scientific knowledge useful in organizing it may be made the support of political and economic peace measures. In this connection it should be emphasized that the amelioration of the lives of common men is actually achieved only as they learn new ways of doing things. Thus the cooperative program may provide means of creating necessary conditions for orderly and peaceful development. In providing the world's peoples with the means of doing better for themselves, the American people will be creating conditions favorable to the development of their own way of life; and in this prospect alone is true national security.[21]

For the first time in defending the committee, Stettinius articulated in this statement the connections between scientific cooperation, development, and peace and security that would be highlighted in postwar programs of technical assistance. What was important, in Stettinius's logic, was the idea that practical, technical advice would provide people with the means of securing a better living—and therefore with the foundations for peace. For the first time, too, as foreign policy officials began to look toward war's end, the glimmerings of the idea appeared that what was necessary for American security was a significant reordering of world affairs—as well as the role that scientific and technological cooperation might play in bringing that transformation about through stable, peaceful means.

Struggles over the committee's organization continued throughout 1944. Another State Department reorganization, carried out on December 20, 1944, reformulated DCR as the Division of Cultural Cooperation and transferred the committee to this new organization, but this time at the highest levels. To protect the committee's strong focus on science, its name was changed. It would be known for the last few months of the war as the Interdepartmental Committee for Cultural and Scientific Cooperation, for the first time adding an explicit reference to science in the committee's mission.[22] The State Department also appointed Raymund L. Zwemer as its new executive director and, at the same time, as associate chief of the Division of Cultural Cooperation, giving the committee new clout within the division. Son of a missionary preacher, Zwemer had a Ph.D. from Yale and was a professor of anatomy at Columbia University. He had spent the previous two years giving lectures on medical topics in Latin America through the committee's programs and so was uniquely suited to lead the committee under its new mandate, especially, as we will see, in the crucial years of transition that followed.[23]

[20] SCC, "Program" (cit. n. 15).

[21] Acting Secretary of State Edward R. Stettinius, "Report to the President on Closer Relationship between the American Republics," Feb. 21, 1944, House Document 474, 78th Congress, quoted in SCC, "Objectives" (cit. n. 16).

[22] *United States Government Manual* (Washington, D.C., 1945). See also SCC, *Activities of the Interdepartmental Committee on Scientific and Cultural Cooperation,* Dept. of State, Inter-American Series, no. 31 (Washington, D.C., 1946).

[23] Joseph Sullivan, with Paul Colton, *Raymund L. Zwemer: A Register of His Papers in the Library of Congress* (Manuscript Division, Library of Congress: Washington, D.C.), 1996. Zwemer's position

Zwemer's appointment as executive director began an important period of transition in the committee's work. By late 1944, the committee had achieved a great deal. Its leadership now rested in the hands of a well-respected scientist, and its position within the State Department was secured at the highest levels of the Division of Cultural Cooperation. The committee had a proven record of successful organization (since 1938) and implementation of projects (since 1940). It had the backing of Secretary of State Stettinius, who as described above had defended the program to Congress as "an effective means of achieving international, hence national, security." Perhaps most importantly, the committee had developed an institutional and intellectual framework for pursuing scientific and technological cooperation in the service of U.S. foreign policy. Key to the committee's institutional success was an effective interagency coordination mechanism that allowed the committee to both call upon the skills, expertise, and resources of a wide range of federal government agencies and coordinate their deployment to various Latin American countries.[24] Just as importantly, for purposes of my argument here, its members had articulated a new logic that distinguished scientific and technical cooperation from other forms of cultural and intellectual exchange and that highlighted the unique capacity of scientific and technical cooperation to achieve certain kinds of outcomes in international cooperation not possible through other kinds of activities. All of these successes in Latin America would turn out to be important for the committee's subsequent emergence as a source of policy leadership in U.S. foreign policy for science after the war's end.

For a few months in early 1945, the impending end of the war and the upcoming United Nations conference in early April occupied much of the attention of Roosevelt and his foreign policy leadership, including many of the committee's members. Roosevelt's death, in early April, also affected the committee's work, although the number of projects and the level of financial support for that work continued to grow. Committee meetings became further and further apart. As with many wartime programs, questions arose as to whether the end of the war would bring the termination of its activities. All of that would change, however, with Truman's announcement in August 1945 that the United States had dropped an atomic bomb on Japan.

INTERLUDE: INTERNATIONAL CONTROL OF ATOMIC ENERGY, 1945–1946

As Paul Boyer has described at length in *By the Bomb's Early Light,* public reaction to the news that U.S. scientists had developed an extremely powerful new weapon was immediate and fearful.[25] The role of science in world affairs became a central subject for conversation at the highest levels of U.S. policy and raised questions among the wider American public—who shared a wary fascination with nuclear science, its use,

---

as executive director was an important starting point for a long career in public service. After leaving the committee in 1947, Zwemer would go on to become executive secretary of the National Academy of Sciences and the National Research Council, chief of the Science and Technology Division of the Library of Congress, and chief of the Division of International Cooperation in Scientific Research for UNESCO.

[24] The historical importance of this mechanism can hardly be overstated. During the postwar era, interagency mechanisms would be used to coordinate U.S. policy vis-à-vis a wide range of international issues. Today, such institutions are legion and a key element of governance. At the time, however, it was one of the earliest such committees created and one of only a handful in existence.

[25] Paul Boyer, *By the Bomb's Early Light: American Thought and Culture at the Dawn of the Atomic Age* (Princeton, 1985).

and its implications for world order that continues to the present. What were the implications of the enormous destructive potential of the atomic bomb for relationships among states? Had science ended war forever? Or had it unleashed a new race for power that would plunge the United States back into war and lead to the destruction of humankind? How soon would other countries, especially the Soviet Union, develop atomic weaponry of their own?

These questions not only placed science at the forefront of debates over the future of world affairs but also gave new weight to the presence of scientists in discussions of U.S. foreign policy. As numerous historians have described, postwar debates over atomic energy gave rise to increasingly close ties between scientists and the security state. Physics, perhaps, epitomized this relationship most clearly. No other discipline was as deeply implicated in the postwar transformation of the sciences into instruments of national security. In the wake of their wartime successes in radar, proximity fuses, and the atomic bomb, physics laboratories became "our first line of defense," and a great deal of physics research retreated behind the walls of secrecy. As Robert Patterson, U.S. secretary of war, put it in 1945, "A nation that lags in the laboratory will not only have no chance of victory in a future war . . . it will not survive."[26] The military invested heavily in science, building its own "national laboratories," transforming the modern research university, and creating what Eisenhower would later decry as the military-industrial-academic complex. Physicists, in turn, became high-profile political advisers to the military, the State Department, and the presidency. Other countries followed suit, forging tightly knit communities linking physicists to the national defense establishment.[27]

The growing ties between science and the state were equally evident in fields other than military research. As I have described above, the wartime work of the Interdepartmental Committee on Scientific and Cultural Cooperation and its member agencies had already begun to involve scientists and experts from numerous fields in U.S. diplomatic affairs. This work expanded from Latin America to the globe following World War II, growing dramatically in scale and involving experts from agronomy and economics to meteorology and health. Before I turn to the committee's postwar work, however, I want to take a brief interlude to examine the broader context of its work. While the committee served as a major site of policy coordination for scientific and technical cooperation after the war, the dropping of the atomic bomb also brought calls for international scientific cooperation in atomic physics. The fate of calls for international cooperation in atomic energy offers a useful comparison for the committee's work, helping to highlight what was unique about the committee's justificatory logic for supporting scientific and technical cooperation as a key element in U.S. foreign policy. Indeed, support for international control of atomic energy was pitched in a framework almost exactly the reverse of the committee's logic. Instead of scientific and technical cooperation being seen as a strategy for laying the groundwork for

[26] Cited in Michael A. Dennis, "'Our First Line of Defense': Two University Laboratories in the Postwar American State," *Isis* 85 (1994): 427–55.

[27] Daniel J. Kevles, *The Physicists: The History of a Scientific Community in Modern America* (Cambridge, Mass., 1987); Dennis, "'Our First Line of Defense'" (cit. n. 26); Stuart W. Leslie, *The Cold War and American Science: The Military-Industrial-Academic Complex at MIT and Stanford* (New York, 1993); James Killian, *Sputnik, Scientists, and Eisenhower: A Memoir of the First Special Assistant to the President for Science and Technology* (Cambridge, Mass., 1977); Etel Solingen, ed., *Scientists and the State: Domestic Structures and the International Context* (Ann Arbor, Mich., 1994).

broader forms of political cooperation, political cooperation came to be seen as a precondition for technical cooperation in the field of atomic energy. The comparison also helps to explain why the committee's work was so important. Despite the active, high-level support of powerful figures such as Vannevar Bush from 1944 to 1947, U.S. and international foreign policy leaders ultimately rejected international scientific cooperation in atomic physics.[28] That they not only accepted the idea of scientific and technological cooperation in other fields but also embraced it as a central pillar of postwar international organization is a result of the committee's significant influence and accomplishments.

One of the most hotly contested debates about science and world affairs in 1945 and 1946 centered on the desirability of a continued U.S. monopoly over the atomic bomb. Should the United States seek to maintain its monopoly for as long as possible? Or should atomic energy, including the bomb, be handed over to the control of an international agency to ensure that it would be used only for peaceful purposes? Among U.S. scientific leaders, these questions were part of a broader set of concerns about the new position of science as an essential component of national security and a source of potentially significant conflict as nations sought primacy over their rivals in the laboratory. One problem was the long-term potential for the United States to lag in defense-related science. As the 1951 *Science and Foreign Relations* report (referred to as the Berkner report, for committee head Lloyd Berkner) put it, "American preeminence as demonstrated thus far is in the application of scientific discovery."[29] U.S. scientists and officials feared that the United States would not be able to compete successfully in science unless they were assured access to foreign science, since that was where the best basic science was being done.

Another problem was the possibility of a dangerous arms race prompted by the U.S. lead in the ability to construct atomic bombs. As early as September 1944, Vannevar Bush argued to Stimson and Roosevelt that the United States could stave off Russian efforts to secretly develop the bomb, and thus preempt a military arms race following the war, only by promoting widespread international exchange of scientific material. Disturbed by the possibility that Roosevelt would pursue a secret Anglo-American agreement to retain bilateral control over the new technology, Bush suggested consideration of an international organization to share control over atomic energy among all nations.[30] In October 1944, as he sought to dissuade Roosevelt from concluding a secret deal with the British that would lock out the Russians, Bush wrote to his colleague James Conant of the potential to use international exchange among biological warfare experts as a first step toward exchange among atomic scientists. Foreshadowing events to follow, however, Bush could not persuade the president of the value of scientific cooperation to the problem of atomic energy, and Roosevelt went ahead with his secret agreement with the British.

Bush did not give up on his idea that international scientific exchange and cooperation might provide an important impetus toward an international political solution to the problem of the bomb, and the end of the war brought new efforts in this direction. In November 1945, Bush reiterated his proposal for a United Nations scientific

[28] Not until the mid-1950s would Eisenhower resuscitate the idea of international cooperation among atomic energy experts in his proposal for the International Atomic Energy Agency.

[29] Lloyd Berkner, *Science and Foreign Relations* (Washington, D.C., 1951), 3.

[30] Richard Hewlett and Oscar Anderson, *The New World: A History of the United States Atomic Energy Commission,* vol. 1, *1939–1946* (University Park, Pa., 1962), 328.

agency to promote full dissemination of scientific information in all fields. He was not alone in calling for scientific cooperation in the field of atomic energy. Key scientific leaders advocated international scientific cooperation as a central element of a plan for international control of atomic energy. In the same month, the Federation of Atomic Physicists formed to advocate for a restoration of the free flow of scientific information among countries. In December, the federation released a memorandum detailing the feasibility of international control over atomic energy monitored by inspections carried out by "an international laboratory" made up of scientists from many countries.[31]

Bush and the atomic scientists ultimately proved unsuccessful in their efforts. The postwar record makes clear that U.S. control over "atomic secrets" remained the primary objective of postwar U.S. security policy. For the most part, American foreign policy leaders were not persuaded by the need for international control and opposed the idea of experimenting with an exchange of scientific ideas as a precursor to broader political cooperation. For military planners and their allies, science remained a key source of competition and potential conflict in world affairs, and those who favored retaining America's monopoly over the atomic bomb as long as possible consistently won out over those proposing scientific exchange and cooperation.

As a result, U.S. postwar policies insisted that international scientific exchanges in the field of atomic energy could only follow the achievement of a successful settlement of the broader question of international *political* control of atomic energy. The directionality of the logic here is important. Scientific cooperation would be allowed only following the settlement of the political problems of atomic energy. The idea that scientific cooperation might serve as a basis on which to build a subsequent era of political cooperation was rejected out of hand. The Truman-Attlee-King declaration of November 1945 seemed to suggest at first that the United Nations Atomic Energy Commission, which would begin work in January 1946, might start with consideration of an international scientific exchange and move on to consideration of safeguards (against other countries secretly developing atomic bombs) and the elimination of atomic weapons. However, later clarifications of the policy made clear that the agreement ruled out an exchange of key information on the military and civilian application of atomic energy until safeguards were in place. Even this went too far for key members of the administration and Congress, who opposed any release of atomic secrets without prior agreement on safeguards. By the time Secretary of State James F. Byrnes was ready to make a proposal to the Soviets in late December, Truman had been persuaded to change his position and now insisted on the primacy of safeguards.

Even the atomic scientists seemed to agree, by 1946, that this was the correct ordering of the logic. Working parallel to Bush outside the administration, the atomic scientists movement had become the most important group advocating publicly for international control of atomic energy. Through Robert Oppenheimer, they had access to high-level policy circles. Oppenheimer worked to infuse their ideas into the Acheson-Lilienthal report, drafted in early 1946.[32] The report called clearly for the release of U.S. scientific information only subsequent to the establishment of an effective

[31] Alice Kimball Smith, *A Peril and a Hope: The Scientists' Movement in America, 1945–47* (Chicago, 1965), 257.

[32] Smith, *A Peril and a Hope* (cit. n. 31), 455–60; Hewlett and Anderson, *The New World* (cit. n. 30), 531–54.

Atomic Development Authority with complete control over nuclear materials, unfettered inspections, and safeguards against national or private misuse of atomic energy. Notably, the report does not suggest that the proposed authority function as an expert agency. In this the Acheson-Lilienthal report differed markedly from other plans for international agencies being pursued by the United States in 1946 and 1947, which will be discussed below. The Acheson-Lilienthal report characterized the organization as an international corporation, requiring extensive experience in international business and government relations, with ownership over all nuclear materials and the sole right to license atomic technologies. Expertise was clearly important in carrying out the organization's functions, but the Acheson-Lilienthal report insisted that it would be a mistake to "overemphasize the advantages that may arise from the free association of the Authority's scientists and experts with those engaged in private or national undertakings." The report denied that scientists would naturally opt for international over natural or private allegiances. Cultivation of such feelings, it noted, would require "serious effort" on the authority's part.[33]

The primacy of safeguards and international control over scientific exchange continued to be U.S. policy through the opening of negotiations at the United Nations in 1946, and the subsequent development of the Baruch Plan only solidified this position.[34] The Baruch Plan drew heavily on the Acheson-Lilienthal report, especially with regard to the proper ordering of political and scientific cooperation. Up front, the proposal required a system of international monitoring and inspection that would enable verification that no countries were pursuing independent atomic energy programs. Bernard Baruch insisted that the Soviets would have to agree to a provision for sanction and punishment, up to and including war, should a nation violate the agreement, before the United States would be willing to share sensitive atomic secrets or give up its atomic weapons. He even insisted that the United Nations be restructured to eliminate its veto provisions to ensure that such punishments could be carried out. This last provision was consistently denounced by the Soviet Union as impossible and ultimately sank the negotiations. With ongoing Soviet rejection of the Baruch Plan throughout 1946 and 1947, American policy on atomic science became increasingly clear. The United States based its policy for nuclear weapons on sole possession of the scientific, technological, and material basis of atomic energy.[35]

THE POSTWAR PUSH FOR INTERNATIONAL SCIENTIFIC AND TECHNOLOGICAL COOPERATION, 1945–1947

The failure of Bush and the atomic scientists to achieve international cooperation in atomic energy stands in marked contrast to the success of programs of scientific and

[33] Chester I. Barnard, J. R. Oppenheimer, Charles A. Thomas et al., *A Report on the International Control of Atomic Energy* (Washington, D.C., 1946), 41. Bush agreed. Countering the idea that scientists somehow were naturally inclined toward internationalism, he argued that active steps would have to be taken to ensure that atomic scientists maintained a truly international perspective if international control were to work effectively over the long term. (See, e.g., Elizabeth Hodes, "Precedents for Social Responsibility among Scientists" [Ph.D. diss., Univ. of Calif., Santa Barbara, 1982]). Compare this attitude with that of U.S. assistant secretary of state Garrison Norton, at the opening of negotiations to establish the World Meteorological Organization, described below and in detail in Miller, "Scientific Internationalism" (cit. n. 3).

[34] Hewlett and Anderson, *The New World* (cit. n. 30), 455–77.

[35] Gregg Herken, *The Winning Weapon: The Atomic Bomb and the Cold War, 1945–1950* (Princeton, 1981), 171–191; Hewlett and Anderson, *The New World* (cit. n. 30), 531–619.

technological cooperation in other fields after the war. Between 1945 and 1950, scientific and technological cooperation became one of the most important instruments of U.S. foreign policy in fields as far apart as finance, public health, agriculture, aviation, meteorology, and development. Unlike their counterparts in debates about atomic energy, U.S. foreign policy officials concerned about other security challenges to the postwar order were easily persuaded of the value of international scientific, technical, and technological cooperation to their goals and objectives. On frontline issues—including starvation and disease, the need for financial stability in international currency markets, the desire for a peaceful dismantling of European colonial empires, and efforts to combat Communism in Europe and developing countries—cooperation among experts became commonplace. The United States dramatically increased investment in international scientific and technical cooperation, both diplomatically and financially, giving rise to rapid growth in the scale and scope of programs. During this period, U.S. scientists and foreign policy officials orchestrated the establishment of major new international scientific and expert bureaucracies in numerous fields that collectively became known as the United Nations specialized agencies. The United States also established new programs of scientific and technical assistance that laid the groundwork for rapid growth in the field of international development in the 1950s and 1960s. This dramatic growth was all based on the proposition that international cooperation among experts would sow the seeds for subsequent economic and political cooperation and, hence, peace.

Much of the success of U.S. foreign policy in promoting scientific and technological cooperation can be attributed to the ideas and the work of the Interdepartmental Committee on Scientific and Cultural Cooperation. As we have already seen, during the war the committee had rejected the idea of scientific cooperation as a form of cultural and intellectual exchange in favor of the value of scientific and technical assistance on practical projects that contributed to social and economic development, political stability, and international security. After the war, this logic became the foundational justification for expanding international scientific and technological cooperation in the postwar era and, indeed, for the next half century. Moreover, the committee had established its own capacity to run such programs effectively and to inject its ideas into high-level policy debates. In the three years following the war, the size of the committee's budget would triple, and the committee's programs would help shape the Marshall Plan and Truman's Point Four Program, which would ultimately raise U.S. funding for scientific and technical assistance abroad by a factor of a hundred by the early 1950s. In 1946 and 1947, the committee would also acquire new responsibilities for coordinating U.S. policy toward the United Nations specialized agencies. The committee's role in this area would ultimately help not only to establish powerful new international expert agencies but also to transform U.S. foreign policy institutions, decentralizing many aspects of U.S. diplomatic relations from the State Department to other federal agencies.

The value of the committee's ideas and work in the new world of atomic diplomacy became apparent as soon as the dust settled from Hiroshima and Nagasaki. As the war in the Pacific drew to a close and gave way to concern about the impact of the bomb on world affairs, the focus of U.S. foreign policy turned to reassuring both the world community and domestic audiences that the United States had the ideas and the capacity to secure the postwar order. In this quest for security, the committee, whose self-defined mission was, by this point, to promote peace and security through international scientific cooperation, seemed to offer a persuasive answer. On August 31, 1945, three weeks after the two atomic bombs were dropped on Japan, the committee was renamed the Interdepartmental Committee on Scientific and Cultural Cooperation and transferred to the Office of the Director of the Office of International Information and Cultural Affairs. Its chairman once again returned to the level of assistant secretary of state, this time for public affairs. Far more importantly, the State Department began publicizing the committee's ideas, work, and accomplishments in a series of major public speeches and reports. The goal was not only to reassure the U.S. public that the Truman administration was aware of the need to secure a stable world and planning accordingly but also to restore the public's confidence that science could serve a positive role in world affairs—and so counteract fears generated by the bomb that indelibly connected science to the pursuit of war.[36]

On September 8, less than one week after the Japanese surrender, the State Department sent the committee's executive director, Raymund Zwemer, to speak to a nationwide audience on the Columbia Broadcasting System's (CBS) *Adventures in Science* program. In speaking to "the role of science in foreign relations," Zwemer opened by quoting Truman from the April meeting of the United Nations in San Francisco: "The world has learned again that nations, like individuals, must know the truth if they would be free—must read and hear the truth, learn and teach the truth. We must set up an effective agency for constant and thorough interchange of thought and ideas. For there lies the road to a better and more tolerant understanding among nations and peoples."[37]

Zwemer went on to explain that the United States had operated a program designed to promote "cooperative projects of a scientific nature between the United States and the other American republics" since 1938. Describing several specific projects carried out by the committee in the areas of geology, agriculture, meteorology, and anthropology, he explained that these projects increased knowledge, had practical benefits for both the United States and its neighbors, and built closer relations between the scientists and people of the Americas. Zwemer noted that the committee's work was expanding beyond Latin America to include China, the Near East, and Africa and that European scientific leaders were increasingly looking to the United States to replace Germany as a source of scientific knowledge, training, and instruments. Toward the end of the interview, the CBS anchor turned to the question of atomic science: "You have pointed out many of the values of scientific interchange. What about the dangers?" Zwemer responded, obliquely: "We are discussing only the free interchange of those scientific facts available to all scientists in their professional work. Science has always been international in character. Do you know of any scientific discipline, the study of which is restricted to one nation? . . . In the words attributed to James

[36] SCC, *Activities* (cit. n. 22). The opening paragraphs of this report state: "A fact demonstrated forcibly by the war just past is that international relations based on power politics offer no real security to any nation. Mankind is therefore faced with the choice of accepting a new basis for international cooperation or of risking annihilation. No nation can longer assure its safety by isolating itself or drawing apart from the rest of the world. As Franklin D. Roosevelt wrote on the eve of his death: 'Today science has brought all the different quarters of the globe so close together that it is impossible to isolate them one from another. Today we are faced with the preeminent fact that, if civilization is to profit, we must cultivate the science of human relationships—the ability of all peoples, of all kinds, to live together, to work together in the same world, at peace.'"

[37] Harry S. Truman, Speech to the United Nations, San Francisco, April 25, 1945. Quoted in Raymond Zwemer, CBS Radio Broadcast, *Adventures in Science,* Sept. 8, 1945, SCC Speeches.

Smithson, English founder of our Smithsonian Institution over 100 years ago: 'The man of science is of no country. The world is his country and all mankind his countrymen.' "[38] Zwemer's response acknowledged that atomic physics, by virtue of military secrecy, would likely be treated differently than other fields of science and technology. He also intimated that it was unrealistic to assume that the United States would be able to retain its monopoly on atomic physics over the long-term and that the internationalism of science might serve, if carefully nurtured, to defuse the dangers of atomic energy.

Olcott Deming, the committee's executive secretary, also took to the road to promote international scientific cooperation. In a speech to Rollins College in early 1946, Deming contrasted the United Nations Security Council—"a negative instrument of international cooperation" designed "with sweeping powers to stop aggression"—with the newly forming United Nations Educational, Scientific, and Cultural Organization—a "positive instrument for the development among nations and peoples of the soil and climate essential to the growth of peace." Deming quoted the newly written UNESCO Constitution: "It is in the minds of men that the defenses of peace must be constructed." He acknowledged, as UNESCO's title suggested, that science had a special role to play in constructing those defenses. Expounding on the committee's ideas on the subject, however, Deming characterized that role in terms that went well beyond the ideas that had given rise to UNESCO. Not only would scientific cooperation promote knowledge and mutual understanding, he indicated, but it would also "encourage those who have the techniques essential to an abundant economy to share these techniques with those anxious to help themselves in overcoming poverty, disease, and hunger—the parents of ignorance, despotism and war."[39]

Deming's qualification was important. The committee believed that the real importance of scientific cooperation stemmed from its ability to promote material progress as a foundation for social and political progress. Over the course of 1946, as Zwemer and Deming canvased on behalf of legislation to support UNESCO and to expand the committee's authority to build cooperative projects around the world, they developed this logic into a tightly conceived and argued package. Zwemer's speech to the American Chemical Society (ACS) in September 1946, in Chicago, epitomizes the evolving logic. Giving a keynote address to the collected assembly of the society on Friday evening, titled "The Role of the Government in Assisting International Cooperation between Scientific Groups," Zwemer opened: "It has been rightly said that scientists are probably the world's oldest internationalists." He went on to explain that, "until recently, most scientific collaboration between countries was carried out by scientists themselves independently of their governments. In this the scientific unions and large national Congresses performed services of widely recognized value." However, "during the war it became more and more evident that science and human affairs are inseparable. Industrial progress and high living standards depend in the final analysis upon the scientific development of the fullest use of natural resources." Without this, he argued, a faltering economy would give rise to ongoing war and conflict.[40]

As a consequence, Zwemer suggested, governments were increasingly beginning

[38] Ibid.

[39] Deming, "In the Minds of Men" (cit. n. 6), 1–3.

[40] Raymund L. Zwemer, "The Role of the Government in Assisting International Cooperation between Scientific Groups: Address at the Meeting of the American Chemical Society, Chicago, Illinois," Sept. 1946, SCC Speeches.

to cooperate in the pursuit of science, and scientists were increasingly "being drawn into political and international life." Two programs stood out, in particular: first, the burgeoning United Nations specialized agencies (of which he included the Atomic Energy Commission of the UN Security Council—although it did not officially have that status and was quite different from the others—the Food and Agriculture Organization, the Provisional International Civil Aviation Organization, the World Health Organization, and UNESCO), in which government experts were working together to address key problems of social and economic welfare; and, second, the programs of the Interdepartmental Committee on Scientific and Cultural Cooperation, in which the United States was actively promoting international scientific cooperation with other countries. These programs, Zwemer argued, had value "in promoting hemispheric solidarity," "improving health conditions," "improving standards of living in other countries," and "assisting in promoting a more stable world economy." He concluded with a brief reminiscence: "Our cooperative scientific and technical projects with the nations of this hemisphere have served in a way as a laboratory experiment. They have shown us that the kind of cooperation that can win a war can also be effective in building up a friendly neighborhood of nations. I trust that we can continue to build good neighborhoods throughout the world—a world which science has made too small for war."[41]

SCIENCE AT THE UNITED NATIONS, 1946–1947

Zwemer's inclusion in his remarks to the ACS of an extensive discussion of the UN specialized agencies and Deming's promotion of U.S. legislation and scientific activism on behalf of UNESCO reflected the changing mission of the committee in 1945 and 1946. The committee's high-level membership and public visibility on the issue of international scientific and technical cooperation made the committee an important source of leadership on new and emerging problems in U.S. foreign policy for science and technology. One such problem was coordinating U.S. policy vis-à-vis the UN specialized agencies. While the committee continued to operate an expanded array of technical assistance programs, to which I will return below, the committee now also pursued three additional important roles: as a source of ideas regarding the value of scientific and technical cooperation as an instrument of U.S. foreign policy, as a site of policy coordination and leadership across the United Nations expert agencies, and as a proponent of a strong role for federal agencies other than the State Department in postwar U.S. foreign policy.

The position of the committee at the highest levels of the U.S. administration gave it a unique ability to interject ideas in emerging policy debates. Frequently, these ideas focused on the value of scientific and technical cooperation to U.S. foreign policy objectives. Foreign policy officials began to weave ideas about scientific and technical cooperation into their arguments across a range of distinct policy domains and issues. In March and April 1946, for example, U.S. surgeon general Thomas Parran met in Paris with chief health officers from around the globe at the Technical Preparatory Meeting for the International Health Conference. To be convened later in the year, the International Health Conference was designed to establish the World Health Organization as the first specialized agency of the United Nations and as a prominent exemplar of

[41] Ibid.

the value of scientific and technical cooperation to the international community.[42] At the preparatory meeting, Parran submitted a draft proposed constitution for the organization. While much of the draft took the form of legalistic prose, two sections, the preamble and the proposed functions of the new organization, read as if their phrasing had been adapted directly from writings of the Interdepartmental Committee on Scientific and Cultural Cooperation.[43] The proposed preamble read: "international co-operation and joint action in the furtherance of all matters pertaining to health will raise the standards of living, will promote the freedom, the dignity and the happiness of all peoples, and will further the attainment of peace, security and understanding among the peoples of the world." Following this, Parran identified a number of proposed functions, many of which specified forms of scientific and technical cooperation: "establish and maintain an epidemiological and statistical service for the collection, analysis, interpretation and dissemination of information pertaining to health, medicine and related subjects"; "develop, establish and promote standards"; "promote research and develop the interchange of information"; "foster professional education . . . and training in health professions, through fellowships, study tours, and exchanges of visits."[44] Although modified and extended to varying degrees by the final International Health Conference, these core ideas were carried over into the final constitution of the World Health Organization, along with the requirement that "delegates should be chosen from among persons most qualified by their technical competence in the field of health."[45] The World Health Organization was to be, first and foremost, an agency to promote international cooperation among experts in the field of health.

Similarly, in 1947, U.S. assistant secretary of state Garrison Norton welcomed delegates to the negotiation of the World Meteorological Convention with words of high praise for science as an instrument of international cooperation. Invitations to the meeting had been sent specifically to the heads of national weather services, acting in their professional capacity as meteorologists. As I have described in greater detail elsewhere, Norton explained to the delegates that scientists were uniquely morally situated for international cooperation. They shared a "global outlook," an "appreciation of international cooperation," and an ability "in the search for scientific truths" to avoid "the more uncertain and selfish motives that complicate and hinder co-operation in some fields of international interest." Scientists, Norton suggested, shared "a universal language" such that "language differences and international boundaries present no barrier to your exchange of information." Just as importantly, their scientific advances would, in Norton's estimation, contribute to the ability to solve real problems in the world, from increasing the safety of international travel to promoting improved agricultural production, which would foster "an era of progress . . . a higher standard of living for all peoples . . . [and] the aims of permanent world peace and

[42] Although the Food and Agricultural Organization, Bretton Woods institutions, and the Provisional International Civil Aviation Organization had already been established, they would not formally join the United Nations as specialized agencies until a few years later.

[43] It is unclear whether Parran ever served personally on the Interdepartmental Committee on Scientific and Cultural Cooperation. The U.S. Public Health Service, however, which he headed, was one of the principal agencies that sent technical experts abroad under the committee's auspices, and Parran would have been highly familiar with the committee's work.

[44] United Nations World Health Organization Interim Commission, *Minutes of the Technical Preparatory Committee for the International Health Conference, Held in Paris from 18 March to 5 April, 1946* (Geneva, 1947).

[45] *The Constitution of the World Health Organization* (Geneva, 1946).

prosperity."[46] Like the World Health Organization, the World Meteorological Organization became a specialized agency of the United Nations, fostering high-level scientific and technical cooperation among nations, and remains today an important international expert institution.

The rapid growth of scientific and technical cooperation at the United Nations added to the responsibilities of the Interdepartmental Committee on Scientific and Cultural Cooperation. By 1946, the committee was increasingly being asked to coordinate policy development and implementation vis-à-vis the UN specialized agencies and other UN scientific and technical programs. The need for this coordination was evident. With the development of perhaps a dozen new international expert agencies, there was a significant need to ensure that U.S. policies with respect to these agencies were consistent from one to the next. Here, the committee's work was essential, and it adjudicated a number of important cross-agency issues.

Perhaps the most important such issue was the committee's role in ensuring that federal agencies other than the State Department maintained a strong presence in postwar diplomacy, especially with regards to the UN specialized agencies. For eight years, the committee had operated its own programs on the understanding that the best way to organize technical assistance abroad was not to create a vast new technical capacity within the State Department but to mobilize the expertise of the existing technical agencies of the federal government. Now the committee worked to ensure that the United States continued along the same path with respect to the new international expert agencies. Delegations to these new agencies were to be led by the relevant technical agencies. The Public Health Service would represent the United States at the World Health Organization. The Weather Bureau would represent the United States at the World Meteorological Organization. The Department of Agriculture would represent the United States at the Food and Agriculture Organization. And so forth.[47] Over the next decades, this insistence would radically transform the institutional architecture of U.S. foreign policy. While the State Department would retain nominal control over U.S. foreign policy, nearly all the federal agencies would establish international offices in their own right, and experts from these agencies would permeate the corridors of international diplomacy.

The decentralization of U.S. foreign policy for scientific cooperation may have helped cement the presence of federal technical agencies and their expert representatives in U.S. diplomacy, but it also added to the committee's policy coordination headaches. For each of the UN specialized agencies, at least two U.S. agencies were always involved in preparing U.S. positions—the State Department and the relevant technical agency—and under some circumstances the number was much higher. How

[46] Garrison Norton, "Address of Welcome by Mr. Garrison Norton, Assistant Secretary of State" in International Meteorological Organization, *Conference of Directors, Washington, September 22–October 11, 1947* (Geneva,1947), 372–6. By 1946, F. W. Reichelderfer—who headed the U.S. Weather Bureau, helped draft Norton's speech, and would become the first president of the World Meteorological Organization—was a member of the committee and was fully aware of its work. Reichelderfer's agency had sent several experts to Latin America under its auspices. See also Clark A. Miller, "Scientific Internationalism" (cit. n. 3). Note also that there is no hint in Norton's welcome of the concerns expressed by the Acheson-Lilienthal report that scientists would have to work hard to overcome the bonds of nationalism and service to their governments in favor of internationalist sentiments.

[47] For an explanation of how this worked in practice, see Ross B. Talbot, *The Four World Food Agencies in Rome* (Ames, Iowa, 1990). See also n. 3.

would U.S. government policy be developed, communicated, and enforced across the various agencies?[48] How would appropriate representation be obtained from relevant agencies? Here, again, the committee not only served as a central site for resolving such questions but also offered in its own structure a model for solving the problem in the future.

Writing in a proposal to establish a governmental mechanism for coordinating U.S. policy vis-à-vis the UN organizations, Robert Carr, executive secretary of the Executive Committee on Economic Foreign Policy, outlined the dilemma. He began his analysis by highlighting the complexity of interdepartmental coordination using the example of worldwide cotton surpluses. Any resolution of this issue would have consequences for domestic agriculture, which fell under the purview of the Department of Agriculture. Carr proceeded to note that the Department of Commerce would be interested in its consequences for international trade and the Department of Treasury in its consequences for international balance of payments. The Department of Justice might be concerned about the formation of a cartel in the textile industry, and the Department of Labor in the impacts of the agreement on the cost of living and employment. Carr concluded by noting the interest of the Department of State, "particularly from the point of view of promoting peaceful relations with other countries."[49]

The solution to these problems, Carr insisted, was to establish "interdepartmental machinery for the formulation of policy relating to many of the fields of activity covered by existing or proposed international organizations in the economic and social fields." He noted that a few such committees already existed.[50] In the future, Carr said, many more international organizations would likely emerge and need interagency coordination. To address this concern, he went on to recommend that "the [Committee on Scientific and Cultural Cooperation] be examined with a view to revising, if necessary, its membership and its terms of reference so that it can serve for policy coordination in the broad field of social, cultural, educational, scientific, and health matters."[51]

Carr's recommendations led to the reorganization of the committee in early 1946. A February 25, 1946, organizational chart detailed the reorganized committee structure, including the newly established Subcommittee on International Organizations. This new sub-committee was charged with coordinating government policy vis-à-vis the rapidly growing array of UN organizations operating in fields other than economics and finance. The membership of this subcommittee was selected from those departments with operational responsibility for coordinating U.S. policy with respect to the key international organizations in existence in 1946, with overall coordination

[48] Today, the U.S. government would establish an interagency task force or committee to prepare a U.S. position regarding the issue. Contemporary sources point to the SCC as one of the very first such committees. See Executive Committee on Economic Foreign Policy, "Proposal for Interdepartmental Participation in the Formulation of Policy for the Guidance of United States Representatives on International Economic and Social Organizations," Jan. 23, 1946, Box #32. Subcommittee Reports, 1943–49, Entry #29, Interdepartmental Committee on Scientific and Cultural Cooperation, NARA-RG 353.3.

[49] Ibid.

[50] He listed, specifically, the Executive Committee on Economic Foreign Policy, established April 5, 1944; the National Advisory Council on International Monetary and Financial Problems, established by the Bretton Woods Agreements Act of July 31, 1945; the Air Coordinating Committee, which coordinated aviation policies across the secretaries of war, navy, state, and commerce; and the Interdepartmental Committee on Scientific and Cultural Cooperation. Executive Committee on Economic Foreign Policy, "Proposal" (cit. n. 48).

[51] Ibid.

provided by the Department of State.[52] The subcommittee met regularly to consider a variety of topics. These included how U.S. policy for international organizations could best be coordinated; provision of policy guidance on U.S. relations with international organizations (e.g., whether or not to establish a national commission for the Food and Agriculture Organization); and coordination of U.S. policy toward UNESCO, including taking overall responsibility for the establishment of a national commission for UNESCO.[53]

SCIENCE AND TECHNOLOGY FOR THE DEVELOPING WORLD, 1946–1950

The emergence of the committee as a source of high-level policy coordination and its ability to inject ideas into important new U.S. foreign policy debates with respect to the United Nations specialized agencies was mirrored in other arenas of scientific and technical cooperation. Particularly with respect to U.S. technical assistance programs, the committee's responsibilities expanded rapidly after the war. As early as 1943, the committee and the State Department had sought approval from Congress to expand the committee's range of operations beyond Latin America. A small program had been approved for China and the Near East in the final year of the war. In 1946, the committee was assigned the task of formulating and managing the technical training programs for the Philippine Rehabilitation Act and the Greece-Turkey Aid Act, the former to help reinforce democracy in the wake of Japanese occupation, the latter to help prevent the spread of communism in Europe. By the end of 1946, however, despite three years of efforts to promote expanded legislation that would cover a broad range of other countries, only the U.S. House of Representatives had acted on the matter.[54] Prospects for a broader program of technical assistance began to change that winter, however, as the extent of economic debilitation in Europe became increasingly clear and many Europeans struggled to survive harsh weather and food shortages. European economic weakness also fueled fears of renewed Communist uprisings.

[52] Subcommittee membership included: the Department of Agriculture (responsible for the Food and Agriculture Organization), the Civil Aeronautics Administration (responsible for the Provisional International Civil Aviation Organization), the Foreign Security Agency (responsible for the World Health Organization), the Department of Labor (responsible for the International Labor Organization), and the Library of Congress (responsible for UNESCO). In addition to establishing this new subcommittee, membership lists for the committee were broadened to include specific subdivisions within member agencies that carried out committee projects. See SCC, "Activities" (cit. n. 22).

[53] SCC, "Committee Organization," Feb. 25, 1946, Council History Documents. The committee had already played an important role in U.S. policy toward UNESCO. The United Nations proposed to create an Educational and Cultural Organization to foster intellectual and cultural exchange in late 1945. This new organization fell within the mandate of the committee, which was tasked with coordinating the creation of a national commission that would oversee the organization's activities in the United States. Immediately, the committee insisted that U.S. negotiators in New York insert language into the draft text to make the organization a United Nations Educational, *Scientific,* and Cultural Organization, consistent with committee's own views on the relatively greater importance of scientific and technical cooperation.

[54] CAR, "Memorandum on the Origin of the Cultural Relations Program, Its Geographic Growth, Broad Accomplishments, Major Projects, and Personnel," Oct. 22, 1943, Council History Documents; Raymond L. Zwemer, "Interdepartmental Committee on Cooperation with the American Republics," *Department of State Bulletin* 9(274) (1944): 319, Council History Documents; CAR, "Replies Received from Members of the Interdepartmental Committee in Regard to Proposed Expansion of the Program of the Interdepartmental Committee on Cooperation with the American Republics under HR-5350," 1944; and CAR, "Meeting of Long-Range Planning Subcommittee," Folder—Subcommittees, Executive Subcommittee, Project Subcommittee, 1944, Box #27, Entry #17, Subcommittee Records, 1944–49, NARA-RG353.3.

Building on the ideas that had gone into the aid programs to the Philippines and Greece, U.S. secretary of state George Marshall announced plans in June 1947 for the largest peacetime aid program ever carried out by the United States. By September, Europeans had formulated a plan for putting American aid to use, and Americans began sending money, food, and expertise to Europe in an effort to stave off an even more calamitous winter.

The Marshall Plan proved a great boon to the committee, even as it also began to raise questions about whether the committee was up to managing the rapidly expanding magnitude of U.S. technical assistance programs and how power should be shared among the growing number of agencies operating in the foreign aid business. The committee and the European Cooperation Administration (which had been established to administer the Marshall Plan) tussled over European aid but ultimately settled amicably on a division of responsibilities in which the committee opened new assistance programs in many of the scientific and technical fields that it worked in Latin America. The dispute was a foretaste of future problems, however.

Besides adding significant new committee programs in Europe, the political push in Washington to fund the Marshall Plan also had positive spillover effects for the committee's plans in other areas of the globe. State Department officials and members of the committee began work early in 1947 to establish the basis for the Marshall Plan legislation. This work continued efforts to articulate the fit between the committee's activities, the growing emergency assistance programs sought by the United States, and the goals proclaimed by the charter of the United Nations. On May 22, 1947, the State Department released a policy statement on the committee's activities, noting: "The United States is today a storehouse of scientific and technical knowledge urgently needed by other countries to quicken their economic development." Quoting Secretary Marshall from a March 3, 1947, speech to the House Committee on Appropriations, the statement argued that through programs of "scientific, technical, educational, and cultural" exchange: "We are continuing to encourage those conditions which will lead to the development of a free and democratic way of life for the peoples of the world." The statement further highlighted the mandate of Article 55 of the UN Charter: "With a view to the creation of conditions of stability and well-being which are necessary for peaceful and friendly relations among nations based on respect for the principle of equal rights and self-determination of people, the United Nations shall promote: (a) higher standards of living, full employment, and conditions of economic and social progress and development; (b) solutions of international economic, social, health, and related problems; and international cultural and educational cooperation; and (c) universal respect for, and observance of, human rights and fundamental freedoms for all without distinction as to race, sex, language, or religion."[55]

Building on this logic, the committee leadership once again set out to sell a broad program of technical assistance as an aid to expanding democracy and containing Communist advances in poorer parts of the world. They argued bluntly that the old policies of American exceptionalism, in which the United States provided an example and guiding light for others desirous of liberty, were no longer enough. Speaking to the Mid-Atlantic Conference of the National Council of Jewish Women, Deming

[55] SCC, "Policy Statement for the Guidance of Agency Members in the Preparation of Their Fiscal 1949 Budgets for 'Cooperation with the American Republics,'" May 22, 1947, Council History Documents.

observed: "The United States is now engaged as never before in the task of 'selling democracy abroad.' This is accomplished both by setting a good example at home and letting the world know about it, and by supplying U.S. goods and U.S. know-how to other peoples so they may help themselves establish that economic and social stability necessary to the survival of democratic principles and world peace." Offering the example of the committee's Latin American programs, Deming continued: "The United States cannot indefinitely 'support' democracy throughout the world with money and materials, but it can, at relatively small cost, continue to support and transfer to the minds and hands of other people the special knowledge and skills needed by these peoples to establish healthy economies in a free society."[56]

The committee's efforts finally met with success on January 27, 1948, when the Eightieth Congress approved Public Law 402, The United States Information and Educational Exchange Act of 1948 (the Smith-Mundt Act). The act authorized the U.S. government to pursue cooperative scientific and cultural projects and educational exchanges with other countries of the world as "a permanent peacetime policy." Under the new law, "scientific and technical" projects included lending U.S. technical experts to other governments, training foreign experts, exchanging data and publications, and assisting in the collection of scientific and technical information. In the State Department's proposed plans for 1950, put together in December 1948, scientific and technical projects amounted to approximately one-half of the overall initiative (for a total of approximately $13.5 million), the other half being composed of cultural and educational exchanges. State Department policy described the objective of this scientific and technical program as: "to make available to other countries the scientific and technical knowledge and skill of the United States and its citizens, in order to assist them in their social and economic development."[57]

Even with the Smith-Mundt Act, however, many in the Truman administration remained dissatisfied with the level of commitment for programs of this sort, especially in the wake of growing Soviet intransigence on a broad range of issues. Increasingly, Truman himself felt that the United States was moving into a new phase of conflict with the Soviet Union that would take place around the globe, not just in Europe. Flush from a narrow, come-from-behind electoral victory in November, he and his advisers began planning for "a bold new program" to bring American aid to the world's poorest peoples. Truman modeled his ideas on the Marshall Plan, which had proved an enormous success in Europe, and on the committee's work in Latin America. Even the language used by the president to pitch the program in his inaugural address on January 20, 1949, was borrowed heavily from Deming's earlier speeches promoting the committee's work:

> Fourth, we must embark on a bold new program for making the benefits of our scientific advances and industrial progress available for the improvement and growth of underdeveloped areas. More than half the people of the world are living in conditions approaching misery. Their food is inadequate. They are victims of disease. Their economic life is primitive and stagnant. Their poverty is a handicap and a threat both to them and to more prosperous areas. For the first time in history, humanity possesses the knowledge and skill

[56] Olcott H. Deming, "Summary of Remarks before the Mid-Atlantic Conference of the National Council of Jewish Women," Oct. 27, 1947, SCC Speeches.

[57] SCC, "The Scientific and Technical Program under Public Law 402," Dec. 8, 1948, Folder—SCC Memos, Box #30, Records of the Full Committee, Entry #26, Interdepartmental Committee on Scientific and Cultural Cooperation, NARA-RG353.3.

> to relieve the suffering of these people. The United States is pre-eminent among nations in the development of industrial and scientific techniques. The material resources which we can afford to use for assistance of other peoples are limited. But our imponderable resources in technical knowledge are constantly growing and are inexhaustible. I believe that we should make available to peace-loving peoples the benefits of our store of technical knowledge in order to help them realize their aspirations for a better life.[58]

The Point Four Program, as it became known, signaled the ultimate success of the committee's ideas. The commitment of the United States to sharing its scientific and technical know-how in the pursuit of social and economic progress, political stability, and international security had now become a central component of the cold war diplomatic agenda. Not surprisingly, 1949 and 1950 were extremely busy years for the committee. Many committee members were assigned to the new organization set up to plan and oversee Point Four, the Advisory Committee on Technical Assistance (ACTA), which relied heavily on their experience in developing its initial programs.[59] In May 1949, ACTA's budget request for Point Four was submitted to the Bureau of the Budget for approval and presentation to Congress. This request authorized $68 million for the Point Four Program in 1950, ten times the budget the Committee on Scientific and Cultural Cooperation had been planning to request prior to Truman's announcement.[60] The budget request followed the same basic outlines as prior committee programs had long followed in Latin America. The only major differences were that public health ($17 million) had taken the lead from agriculture ($12 million) in terms of overall support and two new categories had emerged that would have enormous implications down the road: general economic development ($5.5 million) and reclamation, hydroelectric power, and flood control ($6.5 million).[61]

The creation of Point Four ultimately also signaled the committee's end. As the size and scope of technical assistance programs expanded rapidly, an interdepartmental committee with little staff no longer sufficed to manage them. What was needed was a new bureaucracy. After 1950, at the behest of Congress, the State Department undertook a series of major reorganizations of the U.S. government's technical assistance programs. These reorganizations folded the committee into the Technical Cooperation Administration and, in 1953, the Mutual Security Agency, which would, in turn, become the Agency for International Development in 1961.

CONCLUSION

The Interdepartmental Committee on Scientific and Cultural Cooperation left an impressive legacy of scientific and technological cooperation in world affairs. During its decade of operations, the committee oversaw the dramatic expansion of scientific and technical cooperation as an instrument of U.S. foreign policy. Its wartime activities in

[58] Harry S. Truman, "Inaugural Address," Jan. 20, 1949, *Public Papers of the Presidents of the United States* (Washington, D.C., 1964).

[59] SCC, "Implementation of Point Four Program of Technical Cooperation," Feb. 18, 1949, Folder—Memoranda and Documents Distributed to the Full Committee, Box #31—Records of the Full Committee, 1947–1950, Entry #27, Interdepartmental Committee on Scientific and Cultural Cooperation, NARA-RG353.3.

[60] The 1950 Act for International Development would ultimately authorize $35 million in funding through June 30, 1951.

[61] Interdepartmental Advisory Committee on Technical Assistance, "Explanatory Book for the Presentation of the Development Cooperation Program," May 18, 1949, SCC Memos.

Latin America served as a site in which members honed a new logic connecting international cooperation in science and technology to broad U.S. policy goals of promoting international peace and security through economic development and political stability. After the war, committee members continued to develop and deploy this logic in arguing for the expansion of technical assistance programs and in their development of U.S. policies with respect to the UN specialized agencies.

With the committee's end, however, a gaping hole was left in the State Department's ability to mobilize science in pursuit of U.S. foreign policy goals. Not surprisingly, scientists were the first to note the problem. Already, by 1950, scientists were complaining that their work was getting short shrift in the Point Four Program. Contributions for economic and financial assistance, they argued, were far larger than contributions for scientific and technical exchange. Prompted by these and other concerns, Lloyd Berkner was asked in 1950 to advise the State Department on how to reenergize its commitment to scientific and technical cooperation. The product of his work is the now famous report *Science and Foreign Relations* to which I alluded earlier.

By this point, however, the die was cast. The State Department was unwilling to relinquish significant influence in the new development agencies to other federal agencies, which had the necessary reserves of scientific and technical talent. At the same time, neither Congress nor the other federal agencies were willing to see significant duplication of that scientific and expert talent within the State Department. Congress saw little need to waste money in this regard, and the agencies were hardly enthusiastic about giving back to the State Department the hard-won positions they had achieved in foreign policy as a result of their monopoly over scientific and technical expertise. The agencies decided that their victories in achieving control over U.S. policy with respect to the UN specialized agencies were enough. Thus, while federal agencies often played central roles in the 1950s and 1960s in the technical assistance programs of the United Nations specialized agencies, their influence diminished over the growing development and foreign aid programs operated by the U.S. State Department.

Scientific cooperation would never again have the same cachet at the State Department as it did during the period from 1945 to 1950. Yet that brief period of flourishing of U.S. diplomatic support for international scientific and technical cooperation left an indelible mark on world affairs. These years laid the foundation of a new form of international organization, grounded in both a new suite of international expert agencies and a new organization of the state for foreign policy. The ideas and work of the Interdepartmental Committee on Scientific and Cultural Cooperation were instrumental to that transformation. The committee didn't make science the central focal point of postwar world affairs—the atomic bomb did that—but it did contribute significantly to making scientific and technical cooperation an important element in the management of global policy.

So it has been ever since. Extensive expert involvement remains the norm in world affairs in virtually all areas of world affairs. Even the atomic scientists finally succeeded in establishing an agency for international technical cooperation in nuclear arms control. The Soviets exploded the myth of a long-term U.S. monopoly on atomic secrets by detonating their own atomic and hydrogen bombs in 1949 and 1950. In turn, Eisenhower returned "atoms for peace" to the United Nations agenda in 1953. In 1957, among the last fields to do so, nuclear experts succeeded in establishing their own UN specialized agency—the International Atomic Energy Agency—to provide technical assistance for the development of civilian nuclear energy and to implement the safeguards provisions of the Nuclear Non-Proliferation Treaty.

It is a tribute to the enduring influence of the committee's ideas that institutions like the International Atomic Energy Agency continue to be central to the resolution of some of global society's most difficult contemporary policy problems. The awarding of the 2005 Nobel Prize to IAEA director Mohamed ElBaradei testifies that the practice of international scientific and technical cooperation remains as highly regarded in world affairs, today, over a half-century later, as it did in 1946–1947. In considerations of global security, the postwar goal of international political control over atomic weapons remains a distant pipe dream. However, the world still pins many of its hopes for a future free from the scourge of nuclear war on international cooperation among nuclear experts.

Atoms for Peace, Scientific Internationalism, and Scientific Intelligence

By John Krige[*]

ABSTRACT

The promotion of the benign atom as an instrument of American foreign policy and hegemonic ambitions was important to scientists and policy makers alike who sought to win "hearts and minds" in the early years of the cold war. The distribution of radioisotopes to friendly nations for research and medicinal purposes in the late 1940s was followed by Eisenhower's far more spectacular Atoms for Peace initiative, announced at the United Nations in December 1953. This chapter describes the polyvalent significance of the diffusion first of radioisotopes, then of reactor technology, notably at the famous conference in Geneva in 1955. It places particular emphasis on the role of scientists and their appeal to scientific internationalism to promote national scientific leadership. It is stressed that openness and security, sharing knowledge or technology and implementing regimes of surveillance, were two sides of the same coin.

INTRODUCTION

On December 8, 1953, U.S. president Dwight D. Eisenhower made a major speech before the General Assembly of the United Nations. He had just returned from a meeting in Bermuda with his British and French allies. There Prime Minister Winston Churchill and his science adviser, Lord Cherwell, had been informed of a new idea that was "designed to ease even by the smallest measure the tensions of today's world." This reduction of tension was not to be achieved by appeasing the Soviets or by lowering the defensive shield. Any atomic attack on the United States, Eisenhower assured the assembled gathering, would lead to swift and resolute retaliation. The consequences of such an engagement, however, would be disastrous. The time had come for the "two atomic colossi" to work together to build a more peaceful world; failing which they were "doomed malevolently to eye each other indefinitely across a trembling world." Stalin was dead. The Korean War was over. The Soviet Union had shown a new willingness to hold a Four Power meeting without the "unacceptable preconditions [regarding disarmament] previously put forward." This was the moment to explore "a new avenue

* School of History, Technology and Society, Georgia Institute of Technology, 685 Cherry St., Atlanta, GA 30309; john.krige@hts.gatech.edu.

I would like to thank Itty Abraham, Gabrielle Hecht, and Bruno Strasser for helpful remarks on an earlier version of this chapter.

of peace" that was coherent with his country's wish to be "constructive, not destructive." The way he chose was "atoms for peace."[1]

A new international atomic energy agency, set up under the auspices of the United Nations, lay at the core of Eisenhower's plan. The major powers, notably the United States and the Soviet Union, would be invited to "make joint contributions from their stockpile of normal uranium and fissionable materials" to the agency. Its most important function would be "to devise methods whereby this fissionable material would be allocated to serve the peaceful pursuits of mankind," especially in the areas of agriculture and medicine. Above all the material would be used "to provide abundant electrical energy in the power-starved areas of the world." In this way, the American president concluded, "the contributing powers would be dedicating some of their strength to serve the needs rather than the fears of mankind." Eisenhower's proposal was greeted with rapturous applause; the president himself was almost moved to tears.[2]

Atoms for Peace was a polyvalent policy initiative. I shall focus on just four of its multiple dimensions. First, it was intended to divert attention from Eisenhower's commitment to the use, expansion, and improvement of increasingly lethal nuclear weapons. Banalizing the bomb, National Security Council directive NSC162/2 redefined America's cold war military posture. It was officially approved on October 30, 1953, and affirmed that, if attacked, the United States would regard nuclear weapons as munitions like any other, to be used if the situation called for them. To satisfy this New Look military doctrine, which shifted the burden of defense from manpower to nuclear power, Eisenhower was engaged in what would be the most massive weapons buildup in U.S. history: in 1952, just before he took office, the United States counted 841 nuclear weapons in its stockpile; by 1960, near the end of his presidency, the number would have grown to 18,638.[3]

Atoms for Peace would displace public attention from the military to the benign atom. It was a weapon of what Eisenhower called "psychological warfare." It would help win the "struggle for the minds and wills of men," the struggle to get them to grasp one fundamental truth: "That truth is that Americans want a world at peace, a world in which all peoples shall have an opportunity for maximum individual development."[4] Eisenhower recruited C. D. Jackson from Time Incorporated to get this truth across. In Jackson's view, Atoms for Peace served as a "direct challenge to the Soviets' near monopoly of 'peace' propaganda," just the thing the United States

[1] Available online at http://www.eisenhower.utexas.edu/atoms.htm. It is also reproduced in full in Joseph F. Pilat, Robert E. Pendley, and Charles K. Ebinger, eds., *Atoms for Peace: An Analysis after Thirty Years* (Boulder, Colo., 1985), appendix C.

[2] Richard M. Holl, "The Peaceful Atom: Lore and Myth," in Pilat et al., *Atoms for Peace* (cit. n. 1), 149–59, on 149.

[3] Robert J. Watson, *History of the Office of the Secretary of Defense,* vol. 4, *Into the Missile Age, 1956–1960* (Washington, D.C., 1997), 457, table 6.

From 1952 to 1954, military spending was 70 percent of federal government spending, reflecting the demands of the Korean War. By the end of Eisenhower's term, it had dropped to 50 percent. For comparison, with demobilization after World War II, the figure from 1948 to 1950 hovered just above 30 percent, far less in both relative and absolute terms.

[4] Dwight D. Eisenhower, "Address by Dwight D. Eisenhower on Psychological Warfare, October 8, 1952," cited by Martin J. Medhurst, "Atoms for Peace and Nuclear Hegemony: The Rhetorical Structure of a Cold War Campaign," *Armed Forces and Society* 23 (1997): 571–93, on 572. Medhurst's paper provides a fine account of the place of "atoms for peace" in Eisenhower's domestic and foreign policy thinking. See also Ira Chernus, *Eisenhower's Atoms for Peace* (College Station, Texas, 2002).

needed to "go on the moral and ideological offensive against the Communists . . . give it a bite and a punch which would really register on both sides of the Iron Curtain."[5]

There was added need for a "punchy" propaganda initiative after the successful test of a fusion weapon that followed soon after the president's UN appeal. The first of the Castle test series, undertaken to evaluate prototypes of utilizable thermonuclear weapons, was conducted on March 1, 1954. It was a technological triumph, and a human and public relations disaster. The bomb Bravo vaporized three islands in the Bikini atoll in the Marshall Islands.[6] Its explosive power was equivalent to that of fifteen megatons of TNT, about a thousand times the destructive force of the atomic bomb dropped on Hiroshima. The yield was two-and-a-half times greater than expected and, coupled with unfavorable winds, dispersed radioactive material over a far wider area than anticipated. Military personnel and equipment had to be rapidly evacuated along with native islanders, some of whom were exiled from their homes for several years. A Japanese tuna fishing boat, the *Lucky Dragon Five,* was about eighty nautical miles east of the Bikini atoll where the bomb was detonated and was caught in the path of Bravo's fallout. For nearly three hours, white radioactive ash rained down on the boat, causing nausea, skin irritation, and hair loss among most of the twenty-three crew members (one of whom died in September 1954). The ensuing domestic and international protest was vociferous. It seemed to confirm that thermonuclear weapons were not only militarily superfluous, as some had said, but also morally repugnant instruments of genocide whose use would undermine the legitimacy and leadership of any power daring to detonate them. At a meeting of the National Security Council (NSC) on May 6, 1954, the day after the fifth of the six Castle tests, Eisenhower worried that the world would "think that we're skunks, saber-rattlers and warmongers."[7] His secretary of state, John Foster Dulles, confirmed that the image of the United States was becoming increasingly tarnished among its European allies "because they are all insisting that we are so militaristic."[8] Three weeks later, on May 27, at a meeting of the NSC, Eisenhower again expressed concern about "a future which contained nothing but more and more bombs."[9] Something had to be done to project a more positive image of the United States abroad, something that would show the world that a country that had mastered the power of the nucleus to unleash unimaginable destruction could contain that power and use it for human betterment.

Atoms for Peace was not simply an instrument of propaganda, an attempt to promote a nonbellicose image of the United States abroad and to allay the fear of the nuclear at home. It was also intended to divert skills and resources from Moscow's military program and to restrict developing nations to purely civil activities.[10] In planning

[5] For Jackson and the quote, see Spencer R. Weart, *Nuclear Fear: A History of Images* (Cambridge, Mass., 1998), 156. The remark is in a letter of Nov. 10, 1953.

[6] Richard G. Hewlett and Jack M. Holl, *Atoms for Peace and War, 1953–1961: Eisenhower and the Atomic Energy Commission* (Berkeley, 1989), 172 ff.

[7] Peter Galison and Barton Bernstein, "In Any Light: Scientists and the Decision to Build the Superbomb, 1952–1954," *Historical Studies in the Physical and Biological Sciences* 19 (1989): 267–347, on 331.

[8] Ibid., 331.

[9] Ibid., 331.

[10] The U.S. administration's attempted use of technological sharing in the space sector to divert resources away from national, military missile programs into international, civilian rocket programs is described in John Krige, "Technology, Foreign Policy, and International Cooperation in Space," in *Critical Issues in Space History,* ed. Steven J. Dick and Roger D. Launius (Washington, D.C., forthcoming).

for Eisenhower's proposal to the UN, his aides suggested that the amount of fissionable material to be donated to the atomic pool should be *X* where "*X* could be fixed at a figure which we could handle from our stockpile, but which it would be difficult for the Soviets to match."[11] The administration reckoned that this approach would contain the proliferation of nuclear weapons by redirecting the materials and the engineers available to develop them, and by establishing a system of safeguards that were enforceable thanks to the transparency provided by an international agency. Atoms for Peace was thus an attempt to maintain U.S. nuclear superiority by ensuring that, as far as possible, other countries, including the Soviet Union itself, devoted their limited nuclear resources to civil programs under international surveillance.[12]

The Atoms for Peace proposal also dovetailed with steps being taken by the Atomic Energy Commission (AEC) to pressure an immensely reluctant private sector to invest in a domestic civilian nuclear power program.[13] Although the idea had been in the air since 1947, its implementation was far from straightforward. As late as the end of 1952, nuclear reactor technology was still a military secret (embodied most notably in the navy's nuclear submarine) and a government monopoly. Little technological information was available in the public domain, the engineering challenges were substantial, there was no one best process for power generation, and economic prospects were dismal. Without financial guarantees, firms such as General Electric and Westinghouse were unwilling to develop the technology. The AEC tried to generate enthusiasm by supporting studies of possible designs by different firms, using information released to select groups of engineers and acquired at various official training programs. Lacking any sound economic rationale, however, another kind of argument was needed to cajole industry into civilian nuclear power. Cold war rivalry and psychological warfare provided that argument: the program was essential to maintain the United States' international prestige and scientific and technological leadership.[14] At a convention of electric utility companies in Chicago in October 1953, AEC commissioner Thomas E. Murray announced that the AEC would build a full-scale 60kW demonstration pressurized water reactor at Shippingport in Pennsylvania to show the way to industry and to stimulate the private sector to invest in civilian nuclear power. Murray was a devout Roman Catholic determined to do all he could to combat atheistic Communism.[15] "For years," he proclaimed, "the splitting atom, packaged in weapons has been our main shield against the Barbarians—now, in addition it is to become a God-given instrument to do the constructive work of mankind." *U.S. News and World Report* was enthusiastic. "An international race for supremacy has started. Britain, with one atomic-powered project, is in the race. Russia is probably starting. Now the U.S. is jumping in."[16]

The development of a domestic civilian nuclear power industry, and the export of

[11] Weart, *Nuclear Fear* (cit. n. 5), 158.

[12] On this policy, and its failure due to the discovery of far more uranium than originally thought, as well as to the ease of producing plutonium in reactors, see James R. Schlesinger, "Atoms for Peace Revisited," Robert R. Bowie, "Eisenhower, Atomic Weapons and Atoms for Peace," Richard G. Hewlett, "From Proposal to Program," and Henry Sokolski, "The Arms Control Connection," in Pilat et al., *Atoms for Peace* (cit. n. 1), 5–15, 17–23, 25–33, and 35–50, respectively.

[13] Hewlett and Holl, *Atoms for Peace and War* (cit. n. 6), chap. 7.

[14] Brian Balogh, *Chain Reaction: Expert Debate and Public Participation in American Commercial Nuclear Power, 1945–1975* (Cambridge, 1991), chap. 3.

[15] Hewlett and Holl, *Atoms for Peace and War* (cit. n. 6), 11.

[16] Murray is quoted and *U.S. News and World Report* is cited in ibid., 194–5.

nuclear technology to foreign markets, required that the extremely tight security restrictions embodied in the 1946 Atomic Energy Act be substantially relaxed. Three weeks before the president made his official proposal at the United Nations, the AEC sent two draft bills to this effect to the U.S. Bureau of the Budget. One broadened the legal base to enable private industry to develop nuclear technology; the other provided for a freer flow of information.[17] After considerable revision and debate, these arrangements were enshrined in a new, less restrictive Atomic Energy Act that the president signed into law on August 30, 1954. A major effort was made to rapidly declassify information for use by private firms, and by February 1956 no fewer than 25,000 technical reports had been reviewed. About a third were declassified completely, and about a quarter were reclassified *L* (limited clearance) and made available to engineers from industry.[18]

The relaxation of security surrounding the civilian aspects of nuclear energy was a condition for the success of the fourth facet of the Atoms for Peace plan: an international scientific conference on the "benign and peaceful uses of atomic energy." AEC chairman Lewis Strauss first mooted the idea in Bermuda in December 1953: he thought "that an international conference might have propaganda value in winning worldwide support among scientists for the President's plan."[19] The original plan was to hold a relatively small meeting in the United States sponsored by the National Science Foundation. In consultation with Isidor I. Rabi, Columbia University physicist, Nobel Prize winner, and chairman of the AEC's General Advisory Committee, it was decided that the meeting should rather eschew overt political and ideological issues and serve as "a real forum for the exchange of information in biology, medicine, basic science and engineering."[20] In subsequent discussions in Europe, notably with Sir John Cockcroft in England, nuclear power reactors emerged as the main focus for the conference. Representatives from Britain, Canada, and the United States would present "papers of real substance on the technical aspects" of reactor construction, and many features of the technology, from the social and economic aspects of nuclear energy, to medical, biological and industrial uses of radioisotopes, would be discussed. The British also suggested that there be an exhibition of nuclear information and equipment to serve as both a trade fair and an explanation of the complexities of reactors and their applications to potential clients and the general public.[21] The meeting would take place under the auspices of the United Nations in Geneva in August 1955. This would give the British and the Americans time to declassify significant amounts of information as permitted by the new Atomic Energy Act.

Historians of science and technology have paid little or no attention to Atoms for Peace. By and large, their interest in postwar American nuclear science, notably physics, has concentrated on the transformations in the discipline and its practitioners required for winning future wars and for winning Nobel Prizes, not for winning hearts and minds. What is more, the propaganda and the popularization, the economics and the engineering might lead one to believe that the meeting in Geneva was of marginal scientific interest anyway and that the leading world scientists who attended in droves found little there to inspire them. "You probably saw the program," one

[17] Ibid., 119 ff.
[18] Ibid., 252.
[19] Ibid, 232.
[20] Ibid., 233.
[21] Ibid., 233–234.

Swiss physicist wrote to Max Delbrück after the meeting, "Nothing very interesting."[22] That said, it must be stressed that, even if the *content* was not scientifically riveting, the event was an outstanding scientific success in other ways. An (anonymous) columnist in the British *New Scientist* remarked at the time that the meeting, "which started out as if it was going to be a dull and almost formal affair, was suddenly brought to life after about three days by the discovery that it was becoming the most momentous scientific occasion the post war world had ever seen."[23] It provided an opportunity for hundreds of scientists, engineers, and technicians to be exposed to, and to learn about, nuclear reactors and their applications in nonmilitary fields. It also lifted the veil of secrecy from the reactor programs in the industrialized countries, above all in the Soviet Union, and it was the first time since the war that Soviet and American researchers had met and exchanged views on such matters relatively openly. As our otherwise disappointed Swiss physicist put it, "Did not see many people who had come for the Atomic Conference. They were so very busy with one another. The few Russian physicists [were] very nice and *open:* pleasant to talk to about everything."[24] It was, said Soviet accelerator expert Vladimir Veksler, "not only the first truly international conference in the field of physics; we can certainly claim, as regards scope and significance, that it was a conference of scientists unique in history."[25]

At Geneva, international scientific exchange flourished, winning hearts and minds and building mutual respect between very different, even rival, scientific communities. However, for scientists (and the states they represented) the occasion was not simply an opportunity to share knowledge and build trust and credibility. It was also a way to probe into the laboratory life of others, to learn about their research techniques, to access their research results, and to assess the quality of what they were doing. The Geneva conference helped scientists situate their work with respect to the (declassified) research frontier. It helped them as well to gauge the strategic significance of what their rivals were doing and the implications that work had for the security of their own countries. It was a site not only of scientific exchange but also of scientific intelligence gathering.

The use of international scientific exchange as an instrument of scientific intelligence gathering was officially promoted and sanctioned in the classified appendix to a report prepared by a panel established by Lloyd Berkner at the request of the State Department. Titled *Science and Foreign Relations,* and partially released in May 1950, the Berkner report insisted that an awareness of foreign scientific developments was crucial to the progress of American science.[26] A classified appendix noted how difficult this was in the postwar era, when scientific capability was identified with

[22] Jean Weigle to Max Delbrück, Aug. 30, 1955, cited in Bruno J. Strasser, *Les Sciences de la Vie a l'Age Atomique: Identités, pratiques, et alliances dans la construction de la biologie moléculaire á Genève (1945–1970)* (Ph.D. diss., Univ. of Geneva, Univ. of Paris, 2002), chap. 1, n. 92.

[23] Geminus, "It Seems to Me," *New Scientist,* Sept. 4, 1955, 742.

[24] Quoted in Strasser, *Les Sciences de la Vie a l'Age Atomique* (cit. n. 22), chap. 1, n. 92.

[25] Vladimir Veksler, USSR International Service, press statement, Sept. 8, 1955, "Atom Conference 'Tremendous Success,'" Folder 7, Box 55, Rabi papers, Library of Congress, Washington, D.C. (hereafter cited as Rabi LoC).

[26] International Science Policy Survey Group, *Science and Foreign Relations: International Flow of Scientific and Technological Information* (U.S. Dept. of State Publication 3860, General Foreign Policy Series 30, released May 1950), 3. For a description of the history of the report, see Allan A. Needell, *Science, Cold War, and the American State: Lloyd V. Berkner and the Balance of Professional Ideals* (Amsterdam, 2000), especially 141–4.

state power, and science was increasingly protected behind walls of secrecy.[27] In this situation, to know what others were doing the United States had to resort to indirect means of intelligence gathering. The "prime target" was, of course, the Soviet Union, but "other areas are also of major importance, first, because research and development results in those countries may contribute to our own scientific and technological advancement, and second, because such discoveries may become known to the Soviet Union and so be of potential use against this country."[28]

Science attachés located in U.S. embassies abroad were supposed to bear the main burden of scientific intelligence gathering. The system had its limits, however, since they generally lacked any scientific credibility.[29] Berkner and his panel stressed that it was highly desirable that qualified American scientists be enrolled in scientific spy work and that they do so informally and without raising suspicions. "The emphasis should be on the free and open discussion of the content, procedures and mechanisms of the science involved," the panel wrote. More specifically, it suggested that the opportunities provided by the international circulation of scientific knowledge be exploited, from scouring publications to capitalizing on personal contacts at meetings of "UNESCO, the international scientific unions, and international scientific congresses and conventions."[30] In short, for Berkner and his panel, in the context of cold war rivalry, scientific internationalism and scientific intelligence were two sides of the same coin. The first pushed back the frontiers of security restrictions and mutual distrust, enabling scientists to build together a shared body of public knowledge. The second exploited that trust to learn what others were doing, to establish the limits of what they could speak about freely, and to assess the dangers that may lurk behind what they left unsaid. International scientific exchange is not just about sharing information. When the science concerned is also an affair of state, of immense importance for national strategic interests, international exchange is at once a window and a probe, an ideology of transparency and, by virtue of that, an instrument of control, a viewpoint from which to look in and watch over. The Atoms for Peace Conference in Geneva in 1955 was such a panopticon.[31]

The Geneva conference was not the first occasion on which the benign atom had served this dual purpose. The precedent had been set as early as 1947, when the AEC agreed, under immense pressure from scientists at home and in Europe, to make select radioisotopes produced in American reactors available to foreign researchers

[27] Needell, *Science, Cold War, and the American State* (cit. n. 26), 145–9, describes this document at some length. I should like to thank him for making a copy available to me. For a general discussion of science and foreign policy, see Ronald E. Doel, "Scientists as Policymakers, Advisors, and Intelligence Agents: Linking Contemporary Diplomatic History with the History of Contemporary Science," in *The Historiography of Contemporary Science and Technology,* ed. Thomas Söderqvist (Amsterdam, 1997), 215–44.

[28] Doel, "Scientists as Policymakers" (cit. n. 27), 5.

[29] Wilton Lexow, "The Science Attache Program," *Studies in Intelligence,* published 4/1/1966, released 7/30/2001, http://www.foia.cia.gov/.

[30] International Science Policy Survey Group, *Science and Foreign Relations* (cit. n. 26), classified appendix, "Scientific Intelligence," 10.

[31] Informal intelligence gathering was common in the cold war before sophisticated technologies such as reconnaissance satellites could peer behind the iron curtain. In the late 1950s, the CIA ran a program code named REDSKIN, in which it recruited nonofficial travelers from the United States, Europe, and "Third World nations," including "tourists, businessmen, journalists, scientists, academics, athletes, chess players, and church leaders." These visitors provided important information about Soviet infrastructure and industrial capabilities (e.g., by buying Soviet merchandise). See Jeffrey Richelson, *American Espionage and the Soviet Target* (New York, 1987), 53–4.

under controlled conditions. By looking briefly at this earlier program, we can grasp better the specificities of Eisenhower's proposal in 1953 and appreciate the significance of the international scientific conference that it included as part of its offensive.

THE BENIGN ATOM IN THE 1940S: THE RADIOISOTOPE PROGRAM

At the Fourth International Cancer Research Congress in St. Louis in September 1947, President Truman let it be known that the United States AEC would make selected radioisotopes available to scientists abroad for research purposes. Truman's scheme was originally restricted to twenty-eight different isotopes of nineteen elements, to be used for research and therapeutic purposes only. Demand increased so sharply that in February 1950 the AEC put into operation "a sort of atomic pharmacy" at Oak Ridge, Tennessee, which "puts radioisotope processing, packaging and shipping on an assembly-line basis, eliminating for the most part the time-consuming method of handling radioisotope shipments manually."[32] In 1951, the U.S. government further expanded the program's scientific scope. Researchers abroad now could use the material in industry and had acess to all domestically available isotopes, except for tritium.[33]

The program owed its immense success to the Manhattan Project. Shortly after the war, the AEC decided that it could use the nuclear piles that had produced plutonium for the bomb as a source of radioisotopes for biomedical research and therapeutic purposes. In the 1930s, particle accelerators had been used to produce "artificially" radioactive substances. Piles rendered the cyclotrons obsolete. The AEC estimated that the reactors at Oak Ridge, for example, could produce 200 millicuries of carbon-14 in a few weeks for about $10,000; it would take 1,000 cyclotrons and operating costs of well over $1 million to do the same.[34] As soon as the word was out that the laboratory was in the business of providing radioisotopes for American scientists, domestic demand soared. By summer 1947, researchers and medical centers in the United States and Hawaii had received more than 1,000 shipments of ninety regularly available radioisotopes.[35] The AEC had also received almost a hundred inquiries from twenty-eight foreign countries, 75 percent of them for radioisotopes for medical research and therapy.[36]

Foreign researchers, notably in Europe, expected their requests to be met without

[32] "Information for the Press and Radio," Feb. 1, 1950, Folder 3, Box 46, RG 326, E67A, AEC Records, National Archives and Records Administration, Washington, D.C. (hereafter cited as AEC NARA). This was also an invitation to press, radio, and periodical representatives to visit the site.

[33] For details of this program, see Angela N. H. Creager, "Tracing the Politics of Changing Postwar Research Practices: The Export of 'American' Radioisotopes to European Biologists," *Studies in History and Philosophy of Biological and Biomedical Sciences* 33 (2002): 367–88; see also idem, "The Industrialization of Radioisotopes by the U.S. Atomic Energy Commission," in *The Science-Industry Nexus: History, Policy, Implications,* ed. Karl Grandin, Nina Wormbs, and Sven Widmalm (Canton, Mass., 2004), 141–67; and John Krige, "The Politics of Phosphorus-32: A Cold War Fable Based on Fact," *Hist. Stud. Phys. Biol. Sci.* 36 (2005): 71–91.

[34] Creager, "Tracing the Politics" (cit. n. 33), 375.

[35] Majority submission to the State Department, "Foreign Distribution of Radioisotopes," undated, but following on the meeting of the Commissioners on Aug. 19, 1947, 1–2, Folder 3, Box 46, RG 326, E67A, AEC NARA.

[36] "Foreign Distribution of Radioisotopes" (cit. n. 35), 2. On Clinton laboratory's biomedical activities, and those of the AEC in general, see Peter J. Westwick, *The National Laboratories: Science in an American System, 1947–1974* (Cambridge, Mass., 2003), chap. 7. On the medical aspect, see C. P. Rhoads, "The Medical Uses of Atomic Energy," *Bulletin of the Atomic Scientists,* Oct. 1, 1946, 22–4.

difficulty. Before the war, it was usual for them to receive isotopes for research from American cyclotron laboratories. The mechanism was formalized in the 1940s, when the cyclotron facility at Massachusetts Institute of Technology was given the task of providing most radioisotopes to people who were not on the bomb project, including scientists abroad. The relocation of radioisotope production from a university cyclotron to a pile in a national laboratory of the Atomic Energy Commission, and the immensely restrictive Atomic Energy Act of 1946, completely changed the situation, however. Foreign scientists found their requests for radioisotopes deflected, pending a policy decision by the commissioners.

The AEC's main preoccupation was, of course, security. It was suggested that "some shipment abroad could fall into the hands of capable persons who wish to develop atomic weapons."[37] And even if the restricted variety and small quantities of isotopes under consideration could never be used to make a bomb, might they not indirectly strengthen the military capability of a foreign power? Surely, Commissioner Lewis Strauss pointed out, the isotopes "would be useful as tools in biological research, metallurgical research, petroleum chemistry, and other areas which are part of the war-making potential of nations."[38] As far as Strauss was concerned in August 1947, without a satisfactory international regime for controlling atomic energy, the United States "could not help scientists who may work for a putative enemy one jot or tittle without displaying naivete and imperiling our own security." For Strauss, "putative" enemies was a broad term. He regarded scientists as politically fickle and willing to "work just as zealously for dictatorships of the Right and Left as they were for democracy." In subsequent skirmishes, he successfully contested the shipment of a small amount of phosphorus-32 to the University of Helsinki on the grounds that it might fall into Soviet hands. He was also deeply concerned by a request from NATO ally Norway for one millicurie of iron-59 for metallurgical research, since it came from a military laboratory, and one member of the research team "could be described as a Communist."

The security roadblocks on the free circulation abroad of small quantities of "civilian" radioisotopes frustrated many European scientists, especially on the continent. They desperately wanted to get back to research after the war. Their cities had been bombed, their laboratories had been destroyed or pillaged, and with the population cold, miserable, and short of basic necessities, their governments had far more important priorities than supporting scientific research. Scientists turned to the United States for material support. There was "a crying, insistent need" to restart the supply of isotopes that had halted since the war, one American scientist wrote in July 1947 after speaking to Niels Bohr: "*even the bottle-washings* we throw away can be used literally for months of research over there."[39] Without even the crumbs from the rich man's table Europeans felt resentful and rebuffed.

The frustration was that much more intense for many on the Continent because the atom held a significance for them that was quite different from the one it held for

[37] Atomic Energy Commission, Minutes of Meeting No. 95 at Bohemian Grove, Aug. 19, 1947, Folder 3, Box 46, RG 326, E67A, AEC NARA.

[38] For Strauss's views here, see Minutes of Meeting No. 95 at Bohemian Grove (cit. n. 37); and Lewis L. Strauss to Carroll L. Wilson, "Foreign Distribution of Isotopes," Aug. 25, 1947, in ibid.

[39] Quoted in Creager, "Tracing the Politics" (cit. n. 33), 374; Albert Stone to U.S. Naval Research Attaché, July 1, 1947, copy in AEC Records, RG326, E67A, Box 46, Folder 3, Foreign Distribution of Radioisotopes, vol. 1 (italics in original).

their American peers. In the United States, "atoms for peace" was haunted—and dwarfed—by atoms for war. As Spencer Weart has pointed out, the nuclear industry "dealt with uranium-235 and plutonium by the ton, while Atoms for Peace imagery relied upon a stock of isotopes that could have been stored in a closet."[40] Not so in much of Europe in the first years after the war. In countries whose scientists and governments had an interest, not in preparing for a third world war but in rebuilding themselves on the ruins of the second, the atom was an opportunity, a symbol of modernity and a better world to come, nuclear power a promise for energy and independence. Hiroshima heralded the dawning of a new age.[41] It was a (ghastly) scientific experiment that showed conclusively that scientists, given the resources and the social authority, could successfully harness the awesome power of the nucleus to constructive ends. To deny a few millicuries of radioactive iodine or phosphorus to European scientists on the grounds that they constituted a security risk was, from this point of view, simply absurd. It showed an abysmal lack of understanding of the hopes and aspirations of European scientists and their peoples to build a better world, an uninformed projection abroad of the meaning that the atom had in America, and a failure to decouple the benign atom from its military uses, to disentangle atoms for peace from atoms for war.

European disenchantment was reinforced by the conviction that American science was becoming compromised by its increasing dependence on military support, which was rendering the freewheeling discussion of research increasingly impossible, as well as fueling paranoia about infiltration and subversion. Charles C. Lauritsen from the California Institute of Technology reported to AEC commissioner Robert F. Bacher that Europeans seemed to have "a somewhat exaggerated idea of the control which the Army and Navy exert over science in this country."[42] Exaggerated or not, many were disturbed by President Truman's Executive Order 9835, signed in March 1947, which required all federal employees to undergo loyalty and security checks, including, of course, scientists working in federal laboratories. In the ensuing campaign to weed out "disloyal and subversive elements," many scientists, notably those of liberal-left political persuasion, were suspected of being unreliable or a security risk, often on the basis of extremely flimsy, not to say irrelevant, evidence.[43] The image of the United States as a democratic society that respected individual liberty and freedom of expression began to crumble. Some foreign scientists went "as far as to class us in somewhat the same light as Russia on scientific and political matters," wrote one American scientist. Wrote another, "certainly not a flattering comparison but one cannot deny many of the facts brought out."[44]

[40] Weart, *Nuclear Fear* (cit. n. 5), 172; Hewlett and Holl, *Atoms for Peace and War* (cit. n. 6), appendix 2, gives financial data for the AEC for 1952 to 1961 inclusive.

[41] For the following, see, notably, Gabrielle Hecht, *The Radiance of France: Nuclear Power and National Identity after World War II* (Cambridge, Mass., 1998); David Pace, "Old Wine—New Bottles: Atomic Energy and the Ideology of Science in Postwar France," *French Historical Studies* 17 (Spring 1991): 38–61; Strasser, *Les Sciences de la Vie a l'Age Atomique* (cit. n. 22), chap. 1. Strasser insists that "the social, political and intellectual history of the Atom in America . . . cannot be transplanted to Europe, whatever many historians seem to think" (33, my translation).

[42] Quoted in Richard G. Hewlett and Francis Duncan, *A History of the United States Atomic Energy Commission,* vol. 2, *Atomic Shield, 1947–1952* (Berkeley, 1990), 97.

[43] This is discussed at length in Jessica Wang, *American Science in an Age of Anxiety: Scientists, Anticommunism, and the Cold War* (Charlotte, N.C., 1999).

[44] For these sentiments, see Creager, "Tracing the Politics" (cit. n. 33), 376, 373. The last two quotations are by scientists Paul Aebersold and Lorin Mullins in letters written early in August 1947.

The refusal to provide "civilian" radioisotopes to foreign researchers was generating hostility toward the United States, was embarrassing American scientists in their dealings with European colleagues, and was breeding suspicion and distrust about U.S. intentions in the nuclear field. In this context of deteriorating U.S.-European scientific relationships, the circulation of select radioisotopes for research and for therapeutic purposes was imperative to win back hearts and minds. American scientists were in favor of it. Four of the five AEC commissioners (Strauss being the exception) were in favor of it. The State Department favored it. Acting Secretary Richard Lovett enthused that "these valuable products of the United States atomic energy plants will now be available in the services of mankind and . . . , to this extent at least, we are able to advance towards the beneficient [sic] use of this new force. This initiative," Lovett added, "should promote harmony and good feeling among nations."[45]

The appeal to scientific internationalism was crucial in promoting this change in policy. It pushed the U.S. administration to redefine the limits of security surrounding the distribution of isotopes abroad. America was "morally" obligated to share material with Europe, said J. Robert Oppenheimer in Congressional Hearings in 1949, because

> [isotopes] were discovered in Europe; they were applied in Europe; they are available in Europe, and the positive arguments for making them available . . . lie in fostering science; in making cordial effective relations with the scientists and technical people in western Europe; in assisting the recovery of western Europe; in doing the decent thing.[46]

That said, international scientific exchange was essential not only to redrawing the boundary between the permissible and the off-limits but also to ensuring that that boundary was *respected.* In doing the "decent" thing, in restoring "the international fraternity of knowledge" (Lilienthal)[47] one was also guaranteeing American scientists access to radioisotope research in foreign laboratories—and access was an insurance against abuse. The transparency that was intrinsic to international scientific exchange was also the means for monitoring what the other was doing, to ensure that security was not being breached.

The AEC put a complex set of procedures in place to ensure that beneficiaries did not abuse the radioisotopes they received. The request had to be made officially through the State Department (rather than directly from one scientist to another), and complete transparency in terms of intended use and results was expected. The client had to provide the commissioners with three copies of a report every six months on the progress of the work, which had to be published, if possible, in the open scientific or technical literature. Recipients also had to agree "that qualified scientists irrespective of nationality will be permitted to visit the institutions where the material will be used and to obtain information freely with respect to the purposes, methods and results of such use, in accordance with well-established scientific tradition."[48]

[45] Richard Lovett to David E. Lilenthial, letter, Aug. 28, 1947, Folder 3, Box 46, RG 326, E67A, AEC NARA.

[46] J. Robert Oppenheimer, testimony, "Investigation into the United States Atomic Energy Project," *Hearings before the U.S. Congress Joint Committee on Atomic Energy,* 81st Cong., part 7, June 13, 1949, 283.

[47] Cited by Creager, "Tracing the Politics" (cit. n. 33), 376.

[48] *Radioisotopes for International Distribution. Catalog and Price List. September 1947.* (Isotopes Branch, United States Atomic Energy Commission, P.O. Box E, Oak Ridge, Tennessee), 15, Folder 3, Box 46, RG 326, E67A, AEC NARA.

Scientific internationalism helped build or rebuild atomic research capabilities in nations that wanted them. It also opened doors and loosened tongues. It enabled "inspectors" to ensure that the radioisotopes sent abroad were not being used for purposes for which they were not intended. Scientific internationalism would benefit American science by contributing to the shared pool of knowledge and, by ensuring that U.S. scientists had access to any major discoveries, enhanced "our national security, which depends on continued progress in the field." Finally, it would strengthen American leadership and supremacy: "With its superior technological potential," the commissioners favoring the policy pointed out, "the United States can expect to profit more quickly and more fully than any other nation from the exploitation of published findings." In short, by trading on the taken-for-granted conventions of scientific internationalism, the foreign isotope program would reinforce, rather than undermine, "the common defense and security of the United States."[49]

A footnote: The AEC's radioisotope program was used not only to win the allegiance of scientists abroad but also to woo the hesitant at home. It was intended to dispel the antipathy that some American scientists were beginning to feel toward nuclear science, encouraging "the wholehearted support of United States scientists and medical doctors for our national program for atomic energy."[50] It could also help wean the general public from its nuclear fear. In 1949, the first American Museum of Atomic Energy was opened at Oak Ridge National Laboratories, in the shadow of the massive uranium enrichment plants built there during the war. Its mission was "to serve as an exhibition and education center for advocating the peaceful uses of atomic energy."[51] The "atomic pharmacy" opened in February 1950 (see above) reinforced the message. It was not only a stockpile of irradiated materials for scientific use; it was also accessible to the press and the public. The production and distribution of radioisotopes for research, medicine, and soon, industry, was the vector that carried the image of the United State as the promoter of the benign atom into hearts and minds at home and abroad.[52]

By the time Eisenhower made Atoms for Peace an official plank of U.S. foreign policy, with the full weight of presidential authority behind it, the idea that nuclear science could be advanced, and foreign-policy objectives could be promoted, without threatening, but actually enhancing, U.S. national security, was already well established. This is not to belittle the significance of Eisenhower's initiative: on the contrary, it was a huge step forward in three notable respects.

First, it went beyond merely sharing isotopes to promoting the proliferation of the

[49] These arguments are found in Majority submission to the State Department, "Foreign Distribution of Radioisotopes," (cit. n. 35); and Atomic Energy Commission, Minutes of Meeting No. 95 (cit. n. 37).

[50] Atomic Energy Commission, Minutes of Meeting No. 95 (cit. n. 37).

[51] Art Molella, "Exhibiting Atomic Culture: The View from Oak Ridge," *History and Technology* 19 (2003): 211–26. There is a photograph of the museum on 215. One of its main attractions seems to have been irradiated dimes people could keep; the gift shop sold small pieces of uranium ore. One could also buy an American Museum of Nuclear Energy ashtray. The net effect was surely to familiarize and banalize the nuclear, making it seem less threatening and dangerous by reducing it to the familiar and everyday. There are Internet sites now that discuss the likelihood that these materials were in fact radioactive.

[52] Krige, "The Politics of Phosphorus-32" (cit. n. 33), is an extended essay on what capturing hearts and minds with the isotope program might mean.

technology needed to produce them: nuclear reactors. One important reason for this was that the United States no longer had a monopoly on reactor technology or on the radioisotopes that were one of its by-products. Britain, Canada, France, and the Soviet Union all had reactors in various stages of development. Already in 1951, the first two were making radioisotopes available on far less restrictive and security-conscious terms than was Washington. Atoms for Peace needed a new technological platform, and research reactors provided it for both superpowers.

Second, reactors were also important to win the allegiance of new states gaining their independence and sovereignty. Some twenty new nations came into being between 1945 and 1955; another thirty were established in the next decade.[53] The U.S. administration "anticipated" that the Soviet Union would use atomic energy "not only for military and industrial purposes, but also as political and psychological measures to gain the allegiance of the uncommitted areas of the world."[54] If America wanted to seize the initiative and to retain its advantage in what was becoming "a critical sector of the cold war struggle," it had to be present in these countries. The growing pressure inside the AEC to place the development of nuclear power in the hands of private industry provided an additional economic and ideological rationale for proliferation abroad. In Medhurst's purple prose, "the reactor program functioned as a form of industrial imperialism whereby an advanced technology could be embedded in a culture not yet ready to exploit its full potential as a means of getting both a technological and economic foothold."[55]

Reactors had a third purpose: they could be used as bargaining chips with friendly governments to ensure the smooth expansion of America's nuclear stockpile in line with the New Look and the nuclearization of NATO.[56] Access to worldwide deposits of uranium and thorium had to be assured. NATO members who had little experience with nuclear science, and few local skills for handling dangerous nuclear materials, had to be familiarized with the techniques. Foreign bases had to be secured, and an icon around which to rally pro-American sentiment had to be paraded.

To secure these diverse U.S. interests abroad, beginning in June 1955 the Eisenhower administration began to sign bilateral agreements with selected countries all over the world, undertaking to supply nuclear reactors for research and sometimes for power generation. Typically, these research bilaterals provided the U.S. partner with unclassified information on the design, construction, and experimental operation of nuclear reactors, as well as up to 6kg at a time of uranium enriched to 20 percent uranium-235. NATO member Turkey became the first country to sign a bilateral agreement; two more NATO members, Greece and Portugal, soon followed. Argentina, Belgium, and Brazil, all major suppliers of uranium, were also among the earliest beneficiaries of the scheme. Franco's Spain (which signed a mutual military assistance agreement with the United States in September 1953) and apartheid South Africa

[53] Akira Iriye, *Cultural Internationalism and World Order* (Washington, D.C., 2001).

[54] NSC 5507/2, Peaceful Uses of Atomic Energy, approved by Eisenhower on March 12, 1955, cited by Medhurst, "Atoms for Peace and Nuclear Hegemony" (cit. n. 4), 588.

[55] Medhurst, "Atoms for Peace and Nuclear Hegemony" (cit. n. 4), 588.

[56] Ibid., 581–6, stresses this point and notes that, immediately after the passage of the relaxed Atomic Energy Act of 1954, a new series of treaties for mutual defense was signed with NATO countries, loosening restrictions on armaments and nuclear facilities and allowing West Germany to engage in atomic energy plans. Eisenhower saw these measures as essential to help NATO "evolve more effective defense plans concerning the use of atomic weapons than have heretofore been achieved" (586–7).

(whose gold mines were rich in uranium ore) were not forgotten.[57] The United States' addiction to nuclear raw materials and its determination to use research reactors as an instrument of foreign policy quashed any scruples about the political and ideological standing of the governments Americans dealt with. By August 1955, the AEC had negotiated two dozen research bilaterals; by 1961, the number had reached thirty-nine.[58]

The Soviet Union, which had been completely excluded from the earlier foreign radioisotope program, was regarded as an essential partner in Atoms for Peace in 1953–1954. In part, this was a sign of the new, more relaxed relationship between Washington and Moscow. It was also a recognition that "the secret was out." The Soviet Union had exploded its first fission bomb in August 1949 and conducted its first significant thermonuclear test (dubbed Joe-4 in the West) in August 1953. Security assumed a new meaning in this context: it made no sense to try to seal all military knowledge, and those who had it, behind an impenetrable wall of secrecy. A greater degree of openness and exchange in the name of scientific internationalism would provide U.S. scientists with a better idea of Soviet capabilities and help them to assess more realistically the extent of the Soviet threat. The scientific sessions and the informal coffee breaks and walks by the lake in Geneva were likely to be a boon in this respect.

THE 1955 ATOMS FOR PEACE CONFERENCE IN GENEVA

The first International Conference on the Peaceful Uses of Atomic Energy opened at the United Nations' Palais des Nations in Geneva on August 8, 1955, a decade almost to the day after the first use of a nuclear weapon. The distinguished Indian nuclear physicist Homi J. Bhabha presided over the twelve-day meeting. It was attended by more than 1,400 delegates, from seventy-three countries, by almost as many observers, and by more than 900 journalists.[59] Welcoming messages arrived from the heads of state of Britain, France, India, Switzerland, and the United States. Eisenhower reaffirmed his pledge "to help find ways by which the miraculous inventiveness of man shall not be dedicated to his death, but consecrated to his life."[60] The United States delegation was the largest: 259 people of whom 183 were scientists. They appointed Laura Fermi, Enrico Fermi's widow, to write an official account of the American contribution to the planning and proceedings of the meeting.[61] The British came second in terms of sheer size, followed by the Soviet Union: 78 official representatives, including physicists, engineers, students, government officials, and "the usual KGB staffers."[62]

The conference was organized around three major themes: physics and atomic

[57] For a discussion of the Spanish case, see Javier Ordoñez and José M. Sánchez Ron, "Nuclear Energy in Spain: From Hiroshima to the Sixties," in *National Military Establishments and the Advancement of Science and Technology,* ed. Paul Forman and José M. Sánchez Ron (Amsterdam, 1996), 185–213.

[58] There is a list in Hewlett and Holl, *Atoms for Peace and War* (cit. n. 6), 581, appendix 6.

[59] For an analysis of Bhabha's opening speech, see Itty Abraham, *The Making of the Indian Atomic Bomb: Science, Secrecy, and the Postcolonial State* (London, 1998), 98–102.

[60] "Eisenhower's Message to Nuclear Parley," *New York Times,* Aug. 9, 1955.

[61] Laura Fermi, *Atoms for the World: United States Participation in the Conference on the Peaceful Uses of Atomic Energy* (Chicago, 1957). Notwithstanding its uncritical admiration and its irritating sexism (the heroic achievements of individual, identified men are described alongside the service activities [e.g., guides, translators] of countless "pretty girls"), Fermi's book gives one a good idea of the planning, organization, and evolution of the conference from the U.S. point of view.

[62] Paul Josephson, *Red Atom: Russia's Nuclear Power Program from Stalin to Today* (New York, 2000), 174.

piles, chemistry, metallurgy and technology, and medicine, biology, and radioactive isotopes. The United States delegation made a major effort to disseminate information about its nuclear reactors and their uses in biomedicine and agriculture. Of the 3,000 scientific and technical papers published in the proceedings of the meeting, more than 550 were from the United States (selected from more than 1,000 submissions), and many of these were presented orally. Notwithstanding the limits imposed by security, full engineering details were provided on nuclear plants already operating or under construction in the country.[63]

The scientific papers were complemented by technical exhibits that were reserved for delegates until 4 PM (when they were opened to the general public) and by a trade fair in downtown Geneva.[64] The centerpiece of the U.S. exhibit was a working swimming pool fission reactor of the type designed and built at the Oak Ridge National Laboratory and operated by the Union Carbide and Carbon Corporation for the U.S. AEC. It was flown in from America in June and installed in a wooden chalet in the grounds of the Palais des Nations, where it was seen by 50,000 visitors. It was sold to Switzerland for $180,000 after the meeting.[65]

The presentation of the U.S. reactor in Geneva was a masterpiece of marketing. It was intended to demystify nuclear power and to show that anyone and any nation could exploit it safely and to social advantage. It was designed to operate at a continuous power level of 10kW. This reproduced as closely as possible the kind of reactor that could be built under the Atoms for Peace plan. It was specifically designed to be fuelled with 18kg of uranium of which 20 percent was uranium-235, a composition "identical to that in the United States' allocation of 200 kilograms of fissionable material for the proposed International pool."[66] Operation was achieved with just three control rods, one raised more slowly than the others to fine-tune the chain reaction. The procedure was deliberately simplified so that "technically qualified visitors" did not simply have the opportunity to observe a functioning reactor but actually to operate it themselves (as President Eisenhower did on a visit to the reactor a few days before the show opened). A Union Carbide document explained that this was meant "to show that an efficient working reactor can be designed, constructed, and operated in complete safety without elaborate preparations or complicated facilities." An accompanying panel explained how personnel could easily be protected from the damaging effects of radiation by combining sensitive detectors with shielding and special disposable equipment, such as gloves and jackets.

The Geneva reactor was primarily a research tool. Provision was made to place capsules in the neutron flux either manually, manipulating them with special long tongs, or with compressed air that propelled them down tubes into or outside the core. A

[63] Hewlett and Holl, *Atoms for Peace and War* (cit. n. 6), 250.

[64] This Atom Fair provided a forum for firms in ten countries to exhibit and sell their nuclear wares. The British were the stars. No fewer than fifty companies from Britain (compared with sixteen from the United States) aggressively promoted nuclear technology. They reputedly received serious inquiries from thirty-three countries within days of the conference's opening. "Britain Exploits Atomic Market," *New York Times,* Aug. 8, 1955.

[65] Report, undated and unsigned, "Background of the Geneva Conference," 6–7, Folder 7, Box 55, Rabi LoC. On the sale, see "Achat d'un réacteur nucléaire américain," May 31, 1955, Documents Diplomatiques Suisses, (Swiss) Département Politique Fédéral, Berne, http://www.dodis.ch/, DoDiS 10835.

[66] IACF papers, Regenstein Library, Chicago, Box 259, Series II, Folder 4, Memo, "Background Data: The United States Exhibit Reactor," prepared by the Union Carbide and Carbon Corporation, undated; Fermi, *Atoms for the World* (cit. n. 61), 94.

panel told visitors in four languages that "An Enriched Uranium-235 reactor has many uses." It itemized education in nuclear science, nuclear physics research, reactor design, radioisotope production, activation analysis, radiation effects, and biomedical research. The United States delegation also produced seven technical films, and installed a research library, thus providing additional educational material for all those who wished to enter the realm of the peaceful atom.[67]

The description of the conference in the U.S. media emphasized the benign wonders of the atom. Watson Davis, for the CBS Radio Network, announced before the meeting that "the accent at the Geneva Atomic Conference which I shall report is upon the good, the true and the beautiful, about atomic energy. This is as it should be, in my opinion," Davis went on, remarking on the absence in the program of "bad and naughty atomic words" such as "bomb," "radiation," "fallout," and "secrecy."[68] The *New York Times* reported that, on the day the meeting opened, Lewis Strauss announced that the Ford family had established a Ford Fund (independent of the Ford Foundation) of $1 million. Annual awards of $75,000 and a medal would be given to individuals or groups, regardless of nationality or political opinion, who made important contributions to the development of the peaceful atom. The family was responding to the president's plea to private business to provide incentives "for finding new ways of using atomic power for the benefit of mankind."[69] During the second week of the conference, the *Times* reported on the "veritable wonderland of advanced designs of atomic power plants" that had been revealed at the conference, with the United States leading the way.[70] Scientists predicted that the atom provided a "virtually unlimited supply of energy." *Times* science reporter William L. Laurence enthused, "[M]an is on the eve of his most extensive industrial, social and economic revolution."[71]

The Soviet Union came to Geneva equally determined to capitalize on the benign atom for propaganda purposes and divert attention from its military program: "Let the atom be a worker, not a soldier,"[72] as the slogan had it. They, too, were determined to demonstrate the success of their system, winning hearts and minds for the Communist road. As Paul Josephson has put it, for the Soviet authorities

> the peaceful atom showed that a nation whose citizens had been illiterate and agrarian less than forty years earlier, had become a leading scientific and industrial power. The achievements of science and technology, with nuclear energy at its summit, were symbols of the legitimacy of the regime both to Soviet citizens and to citizens of the world. The peaceful atom also allowed the USSR to score points with the conquered countries of Eastern Europe . . . each of whom had a nuclear program based on Soviet isotopes, technology, and training programs and, in part, its largesse.[73]

Moscow found it somewhat galling to see the spectacular demonstration of a working American research reactor in the gardens of the United Nations building, particu-

[67] Fermi, *Atoms for the World* (cit. n. 61), 33–4.
[68] Smithsonian Institution Archives, Washington, D.C., RU7091, Box 402, Folder 62, '8/6/55 CBS Geneva Atomic Conference', "Adventures in Science, Saturday, August 6, 1955," CBS Radio Network, 5:00–5:15 p.m. EDT. I thank Teasel Muir-Harmony for finding this document for me.
[69] "Ford Sets Award for Peace Atom," *New York Times,* Aug. 9, 1955.
[70] William L. Laurence, "Parley in Geneva Unveils Advance in Atomic Plants," *New York Times,* Aug. 14, 1955.
[71] "Limitless Supply of Energy Seen by Experts," *New York Times,* Aug. 15, 1955.
[72] Josephson, *Red Atom* (cit. n. 62), 3.
[73] Ibid., 174.

larly when the Soviet Union had brought the first working power reactor on-stream the year before.[74] A model of the Soviet power plant was shown in Geneva: it obviously lacked the impact of the U.S. exhibit. Not to be upstaged, the Soviet authorities made a heavy-handed attempt to hold their own international conference just before the Geneva meeting. In June 1955 they sent invitations to scientists from forty-one nations, including the United States, to a meeting in Moscow from July 1 to 5 on the peaceful uses of atomic energy, just a few weeks before the Geneva gathering. The members of the scientific academies in Britain and the United States politely declined the invitation on grounds of timing. Those scientists who did attend were treated to a visit to the new power station.[75]

Notwithstanding superpower rivalry and mutual "psychological warfare," the conference did apparently contribute to reducing public fear and political tension. It was widely reported in the press: ten of the twelve days of the conference made the first page of the *New York Times*. Indeed the picture of scientists from rival power blocs and from different nations discussing (civilian) nuclear affairs must surely have impressed those who previously saw nuclear science and nuclear scientists as major threats to world peace.[76] Vladimir Veksler, leading Soviet accelerator physicist, described the meeting as having "moved public opinion" and as having "strengthened the atmosphere of mutual understanding and good will born in every country following the Four Power Conference in Geneva" that had ended just a few weeks before the scientific gathering.[77] No formal mention was made of the international control of atomic weapons, but it was generally felt that the scientific cooperation and openness it had fostered would help remove political barriers to such controls. Certainly U.S. commentators felt it had enhanced Eisenhower's efforts to promote the international control of nuclear material through a new agency. An American report written after the meeting claimed that "as a focal point of nuclear cooperation, the International Atomic Energy Agency was given an enormous boost. Delegates began to realize that international cooperation through the agency could now be placed on solid grounds."[78] The need for an agency to "serve regulatory and developmental purposes" was also strengthened by the realization that more than thirty countries were actually embarking on nuclear programs.

For the scientists, the conference provided access not only to new knowledge but also to nuclear researchers on the other side of the iron curtain. As one commentator put it, "Many scientists from the East and West met for the first time. There were many

[74] The Soviet Union built the world's first power reactor, which produced 5,000kW for the national grid in 1954. Paul R. Josephson, "Atomic-Powered Communism: Nuclear Culture in the Postwar USSR," *Slavic Review* 55 (1996): 297–324, 305.

[75] Anon., "The Atomic Energy Conference at Moscow," *Science and Culture (Calcutta)* 21(2) (1955): 76–83. A translation of the proceedings of the *Conference of the Academy of Sciences of the USSR on the Peaceful Uses of Atomic Energy, July 1–5, 1955,* was published by the U.S. AEC in 1956, Report AEC-TR-2435. See also Fermi, *Atoms for the World* (cit. n. 61), 21, 39–40; and Abraham, *Making of the Indian Atomic Bomb* (cit. n. 59), 88.

[76] "It was not an unusual sight to see small groups having spirited conversations as they walked along the Rhone River at dusk." Report, "Background of the Geneva Conference" (cit. n. 65), 7.

[77] Press statement by Veksler, "Atom Conference 'Tremendous Success'" (cit. n. 25). The Four Power conference (Britain, France, the Soviet Union, the United States) opened in Geneva on July 18 and ended on July 23, 1955. The final communiqué encouraged the hope of international détente. Within weeks, and in the midst of the Atoms for Peace meeting, the Soviet Union announced that it would reduce its armed forces by 640,000 by the end of the year.

[78] Report, "Background of the Geneva Conference" (cit. n. 65), 7.

luncheons, dinners, and serious discussions over coffee at the Palais des Nations . . . lasting friendships were formed among these scientists." Soviet scientists were equally enthusiastic. Veksler "noted with satisfaction that the scientists of the world easily found a common language; the significance of this fact is inestimable."[79] In Geneva scientific internationalism blossomed.

The openness, familiarity, and trust fostered at the meeting also provided an opportunity for scientific intelligence gathering, which was given added urgency by the closed character of the Soviet program. The 1955 Atoms for Peace meeting opened eyes in the West to what the Soviets had achieved, and it alerted Moscow to the extent of the nuclear programs in the United States and its allies. It created a declassified space in which one could learn what others were doing, judge their competences, and assess their priorities—and take the necessary steps to ensure that one did not lag behind friends or foes in the civilian nuclear field.

In 1955, much of the work of Soviet scientists had been virtually inaccessible to outsiders for almost a decade. The brief spring of international rapport, encouraged by Stalin's determination to "catch up with West," had come to an abrupt halt in 1947.[80] Very few Soviet scientific periodicals had been translated into English. International meetings were often canceled at the last minute when invited Soviet colleagues were ordered by Moscow to boycott them. When they did attend it was difficult to glean much information from them.[81] Secrecy and evasion fueled fear and paranoia. Many in the West had little regard for Soviet scientific and technological capabilities anyway, thanks to Soviet and Western propaganda, naïve assumptions about the incompatibility between science and Communism, and the belief that Soviet scientific and technological achievements owed much to foreign help and to spies. Edward Teller was convinced that Klaus Fuchs's treachery had advanced the Soviet atomic bomb project by ten years: a ludicrous exaggeration.[82] At Geneva, write Hewlett and Holl, the U.S. delegation was "surprised" by "the highly technical competence of Russian scientists and engineers generally, and the large numbers of students in training in universities and technical schools."[83] Until the Soviets launched Sputnik, General John B. Medaris remembered afterward, it was fashionable to think of the Russians as "retarded folk who depended mainly on a few captured German scientists for their achievements, if any. And since the cream of the German planners had surrendered to the Americans, so the argument ran, there was nothing to worry about."[84] Hence Veksler's pride: on returning home he reported that before the meeting many in the United States "had believed that, with the isolation of the USSR, the development of science and technology there would be prevented. The Geneva Conference showed, however, that the USSR is very successfully advancing along its own

[79] Report, "Background of the Geneva Conference" (cit. n. 65), 6–7; press statement by Veksler "Atom Conference 'Tremendous Success'" (cit. n. 25).

[80] Nikolai Kremenstov, *Stalinist Science* (Princeton, 1997).

[81] The scientific consequences ensuing on Soviet secrecy and the mistrust of their work in the West are nicely described by Ronald E. Doel, "Evaluating Soviet Lunar Science in Cold War America," *Osiris,* 2nd ser., 7 (1992): 44–70. See also Jacob Darwin Hamblin, "Science in Isolation: American Marine Geophysics Research, 1950–1968," *Physics in Perspective* 2 (2000): 293–312.

[82] See David Holloway, *Stalin and the Bomb: The Soviet Union and Atomic Energy, 1939–1956* (New Haven, 1994), 222–3.

[83] Hewlett and Holl, *Atoms for Peace and War* (cit. n. 6), 250.

[84] Quoted in Clarence Lasby, *Project Paperclip: German Scientists and the Cold War* (New York, 1971), 6.

road and has achieved great results both in science and technology." He was particularly happy with the impression that his own work had made. Veksler proudly told the press, "In the course of our conversations my foreign colleagues repeatedly declared how impressed they were by the new data concerning the construction in the USSR of a vast new accelerator of charged particles which is nearing completion and is intended for the production of protons of 10-billion-electron-volt energy."[85]

Veksler's "impressed" western colleagues were not just stunned, but panicked, by Soviet achievements. They demanded that the United States immediately take steps to ensure that they did not lose their lead over their Communist rivals. Melvyn Price, the chairman of the subcommittee of the U.S. Congress's Joint Committee on Atomic Energy responsible for research and development, drew the conclusion that America was not producing enough qualified scientists and engineers for both the peaceful and military atomic programs. "When the Committee attended the Geneva conference last summer," wrote price in March 1956, "it gained a firsthand impression of this alarming fact." Immediate and strenuous measures were needed to resolve the situation: "at stake," said Price, "is not only our national defense and well-being but our ability to compete with the Soviets in the struggle for men's minds throughout the free world."[86]

Fred Seitz, a solid state physicist at the University of Illinois, was particularly disturbed by Veksler's description of the 10BeV accelerator. For Seitz, Veksler's revelations were a call to arms. "High energy nuclear physics," he wrote in April 1956, "is the principle frontier area of research in the physical sciences at the present time." It produced new scientific results and new devices; "many of the most important innovations in modern engineering have found their origin in the nuclear laboratory," he claimed. It also produced what Seitz called "uniquely trained manpower," intellectually imaginative and daring nuclear physicists who had played a key role in the development of a whole range of weapons during the war and who had "also demonstrated great aptitude in the planning of *weapons systems.*" During the past five years, "the Soviet has challenged our leadership through the establishment of several institutes devoted to high-energy physics." The 10 BeV accelerator would give them additional leverage. Pleading for a Department of Defense program in high-energy physics, Seitz insisted it was "essential that the United States retain its leadership in all essential parts of the field and that the Department of Defense profit as much as is conceivably possible from the development."[87]

We do not know how Soviet scientists and their administration used the information they got from the British and the Americans in Geneva to strengthen and reorient their national nuclear programs: Veksler noted how impressed he was with Ernest Lawrence's account of accelerators and how much his colleagues liked Walter Zinn's report on the boiling water reactor. But just as the state sought legitimacy and credibility, just as the Soviet Union sought to be respected as a modern, scientifically and technologically capable society, so too did their scientists and their engineers. The respect in which they were held in the United States after the meeting built their self-esteem and their self-confidence. Scientific credibility, legitimacy, and cognitive

[85] Press Statement by Veksler, "Atom Conference 'Tremendous Success'" (cit. n. 25).
[86] Melvyn Price to Henry Barton, March 29, 1956, letter, Folder 8, Box 58, Barton files, American Institute of Physics, Center for History of Physics, College Park, Md.
[87] "Proposal for Department of Defense Program in High Energy Physics," Oct. 31, 1955, memo attached to letter from George D. Lukes to I. I. Rabi, April 6, 1955, Folder 8, Box 25, Rabi LoC (italics in the original).

authority are social accomplishments. They are constructed and consolidated in interactions between peers whose assessment of the plausibility of another's truth claims is interwoven with an assessment of their competence. American scientists arrived in Geneva with a view of their Soviet colleagues as trapped in a closed, backward Communist society that had little respect for science. They went home chastened. Soviet scientists arrived in Geneva awed by the achievements of their American counterparts and uncertain of the reception their work would be given. They went home reassured.

CONCLUSION

In a paper published more than thirty years ago, Paul Forman showed how scientific internationalism in Weimar Germany was an expression of deeply felt nationalistic sentiments. He remarked that historically these two apparently contradictory allegiances were reconciled "through the eminently simple formula that the [international] fame and honor which the scientist wins accrues also to his nation and patron." Forman goes on:

> According to this classical conception—largely due to and propagandized by the scientists themselves—the contribution of science to national prestige is an automatic and inevitable byproduct of scientific achievement. It does not require a choice on the scientist's part between serving the interests of science and serving the interests of his nation, between behaving like a good scientist and behaving like a good patriot.[88]

This paper has shown how the fusion between the invocation of internationalism and the pursuit of national interest was achieved in the nuclear field in the early cold war. For the United States (and the Soviet Union) a demonstration of scientific and technological generosity and prowess on the international stage was intended to win hearts and minds and to confirm the legitimacy or even the superiority of rival politico-economic systems. To this vague and rather general cultural agenda were added more specific and tangible scientific and intelligence-gathering goals. Sharing knowledge and techniques would both advance understanding and scientific authority and provide a window into the scientific life-worlds of allies and enemies alike. International scientific exchange deftly reconciled the universalistic appeal to the pursuit of truth with the particularist needs of national security. By weaving surveillance and security into the fabric of openness and internationalism in the Atoms for Peace program one could be both a good scientist and a good patriot. This double movement was indeed constitutive of scientists' behavior. One chord was struck when they spoke to their colleagues abroad, the other when they spoke to their patrons in Washington or Moscow.

The same logic informed the diffusion of nuclear technology to the less industrialized countries. The United States was prepared, through the IAEA, to share knowledge and nuclear technology with them.[89] However, the reactor on display in Geneva,

[88] Paul Forman, "Scientific Internationalism and the Weimar Physicists: The Ideology and Its Manipulation in Germany after World War I," *Isis* 64 (1973): 151–80. See also Daniel J. Kevles, "'Into Hostile Political Camps': The Reorganization of International Science in World War I," *Isis* 62 (1971): 47–60.

[89] The United States authorities explained to European scientists in December 1954 that, as far they were concerned, in nuclear matters a distinction had to be drawn between developed countries, like those in Western Europe, less developed countries like Turkey and Iran, and what a Swiss report called "des territoires primitifs." As far as the United States was concerned, it was the second-tier countries that would benefit from U.S. economic aid and nuclear reactors using fuel made available through the

the model that the United States wanted to diffuse to win hearts and minds, used enriched uranium-235. As such it propelled the recipient up the "barometer of nuclearity" and inevitably embedded the "benefactor" in a regime of surveillance implemented through the imposition of "safeguards."[90] Indeed, it might be argued that the United States' insistence on promoting a technology which contained fissionable material (while Britain, Canada, and France marketed power reactors that used natural uranium as a fuel) was intended to combine sharing with surveillance. "Autonomy" in the nuclear field was interwoven with dependence, a dependence reinforced for the less developed countries by coupling U.S. economic aid with the acquisition of reactors using enriched uranium fuel and built by firms such as General Electric and Westinghouse.[91]

IAEA: "Note pour le Chef du Département. Entretien avec le prof. Scherrer sur l'énergie atomique," Jan. 27, 1955, Documents Diplomatiques Suisses, http://www.dodis.ch/, DoDiS-9598.

[90] Gabrielle Hecht, "Negotiating Global Nuclearities: Apartheid, Decolonization, and the Cold War in the making of the IAEA" (this volume).

[91] Thus regarding "the Spanish case and the matter of American control: every significant element of choice" in the Spanish nuclear energy program between 1955 and 1958 "was preempted by the Americans' requirement that the fissionable material loaned could be used only as fuel in a 'swimming pool' type reactor," such as the one shown at Geneva: Ordoñez and Sánchez Ron, "Nuclear Energy in Spain" (cit. n. 57), 200. For a discussion of the dilemmas and contradictions faced by a "self-reliant" India offered U.S. reactor technology, see Abraham, *Making of the Indian Atomic Bomb* (cit. n. 59), 91–8.

Catalysts of Change:
Scientists as Transnational Arms Control Advocates in the 1980s

By Kai-Henrik Barth[*]

ABSTRACT

In 1986, the Natural Resources Defense Council (NRDC), a U.S. nongovernmental organization, began an unprecedented collaboration with the Soviet Academy of Sciences to establish three U.S. nuclear monitoring stations in the vicinity of the Soviet nuclear test site in Semipalatinsk. This transnational collaboration of scientists accomplished what U.S. administrations from Eisenhower to Carter had tried and failed to do, namely, to place U.S. scientists and their seismic monitoring equipment on Soviet soil. This chapter analyzes how, why, and to what extent this coalition of U.S. and Soviet scientists succeeded in shaping Soviet foreign policy and at the same time altering the nuclear test ban debate in the United States. The chapter places this project in a framework of transnational relations and highlights the importance of trust between the involved scientists, in particular between the project leaders, NRDC physicist Thomas Cochran and Soviet Academy physicist and vice-president Yevgeny Velikhov.

INTRODUCTION

> Our objective is to get some movement in the arms control arena. We want to show that American and Soviet scientists can set aside their differences and work together to prove that verification is not an obstacle to a nuclear test moratorium or test ban treaty.[1]
>
> Thomas B. Cochran, NRDC, 1986

On May 28, 1986, the Natural Resources Defense Council (NRDC), a U.S. nongovernmental organization, signed an agreement with the Soviet Academy of Sciences to establish three U.S. nuclear monitoring stations near the principal Soviet nuclear test site at Semipalatinsk, Kazakhstan.[2] At the same time, the two sides agreed

* Security Studies Program, Edmund A. Walsh School of Foreign Service, Georgetown University, Mortara Building, 3600 N Street NW, Washington, D.C. 20057; khb3@georgetown.edu.

I would like to thank John Krige, Natalie Goldring, Sunil Dasgupta, Thomas B. Cochran, and two anonymous reviewers for comments on earlier drafts. I owe special thanks to the scientists I interviewed for this project, most prominently Cochran and Yevgeny Velikhov. I am particularly grateful to Cochran, who has given me access to the unprocessed NRDC Files, Natural Resources Defense Council, Washington, D.C. (hereafter cited as NRDC Files), and has provided me with additional documents.

[1] Thomas Cochran, quoted in Fred Kaplan, "Soviets to Allow US Scientists to Build Stations to Monitor Nuclear Tests," *Boston Globe,* June 4, 1986, 9.

[2] The NRDC–Soviet Academy collaboration is discussed in a number of publications, most prominently in Philip G. Schrag, *Listening for the Bomb: A Study in Nuclear Arms Control Verification*

to prepare Soviet monitoring stations in the vicinity of the U.S. Nevada Test Site. Only weeks later, NRDC scientists arrived in Kazakhstan to set up three seismic monitoring stations in an area usually off-limits to foreigners. In collaboration with their Soviet colleagues, they operated these stations successfully from July 1986 to the end of 1988.[3]

NRDC scientists accomplished what U.S. administrations from Eisenhower to Carter had attempted and failed to do, namely, to put U.S. scientists and their monitoring equipment on Soviet soil to take seismic measurements of the test site and ultimately to control Soviet nuclear testing activities.[4] At the same time, the NRDC–Soviet Academy project brought a fresh impetus to the national and international debate about a comprehensive nuclear test ban treaty. Soviet leader Mikhail Gorbachev had announced a nuclear test moratorium in August 1985 and sought to convince U.S. president Ronald Reagan to join. For Gorbachev and his informal science adviser, Yevgeny Velikhov, a physicist and vice-president of the Soviet Academy, the collaboration with the NRDC offered an opportunity to demonstrate to U.S. policy makers that the new Soviet leadership was serious about arms control. In fact, they were willing to change the traditional Soviet position on nuclear test ban verification and on-site measurements. Suddenly, it appeared that the Soviets were offering a way out of the arms control stalemate.

This chapter analyzes how, why, and to what extent this coalition of U.S. and Soviet scientists succeeded in shaping Soviet foreign policy and altering the nuclear test ban debate in the United States. It seeks to understand why scientists were able to change the terms of the national and international arms control debate at a time when diplomats on both sides had failed to make progress in their negotiations. Furthermore, it discusses how and why this "private diplomacy" effort succeeded despite strong opposition from parts of the U.S. and the Soviet administrations. Why did the Soviets change their nearly thirty-year policy on on-site verification, and to what extent did the transnational scientists contribute to this policy change? Was this agreement simply a Soviet propaganda coup, as some critics suggested, in support

---

Policy (Boulder, Colo., 1989); Matthew Evangelista, *Unarmed Forces: The Transnational Movement to End the Cold War* (Ithaca, 1999), 279–85. For Cochran's own analysis, see Thomas B. Cochran, "The NRDC/Soviet Academy of Sciences Joint Nuclear Test Ban Verification Project," *Physics and Society* 16 (July 1987): 5–8. The project was widely covered in the news media, from which I draw for this article. Other important analyses of this project include Michèle A. Flournoy, "A Controversial Excursion into Private Diplomacy: The NRDC/SAS Verification Project," in *Annual Review of Peace Activism,* ed. John Tirman (Boston, 1989), 11–24; Leslie Y. Lin, "Gaining Ground Zero," *Sierra Journal* (Jan./Feb. 1987): 33–6; for a hostile view, see Rael Jean Isaac, "The Nuclear Test Ban Hoax: Brought to You by the NRDC and Its Soviet Friends," *American Spectator* (May 1987): 21–6.

[3] Flournoy, "A Controversial Excursion into Private Diplomacy" (cit. n. 2), 20.

[4] For overviews of nuclear arms control negotiations, see George Bunn, *Arms Control by Committee: Managing Negotiations with the Russians,* Studies in International Security and Arms Control (Stanford, 1992), 18–58; David B. Thomson, *A Guide to the Nuclear Arms Control Treaties* (Los Alamos, 2001), 31–41; G. Allen Greb, "Survey of Past Nuclear Test Ban Negotiations," in *Nuclear Weapon Tests: Prohibition or Limitation?* ed. Jozef Goldblat and David Cox (Oxford, 1988), 95–117; National Academy of Sciences, *Nuclear Arms Control: Background and Issues* (Washington, D.C., 1985), especially 187–223. For major works on the early test ban negotiations, see Harold Karan Jacobson and Eric Stein, *Diplomats, Scientists, and Politicians: The United States and the Nuclear Test Ban Negotiations* (Ann Arbor, Mich., 1966); Glenn T. Seaborg and Benjamin S. Loeb, *Kennedy, Khrushchev, and the Test Ban* (Berkeley, 1981); Glenn T. Seaborg with Benjamin S. Loeb, *Stemming the Tide: Arms Control in the Johnson Years* (Lexington, Mass., 1987); Herbert F. York, *Making Weapons, Talking Peace: A Physicist's Odyssey from Hiroshima to Geneva* (New York, 1987).

of Gorbachev's arms control policies?[5] In other words, did Soviet officials use the NRDC? Alternatively, did the NRDC use Velikhov and even Gorbachev to change the U.S. test ban debate?

Political scientists and commentators have suggested a variety of factors that led to this unprecedented project and its success. The first political science concept of relevance to this paper concerns the notion of "epistemic communities."[6] Peter M. Haas's original definition characterized an epistemic community as "a network of professionals with recognized expertise and competence in a particular domain and an authoritative claim to policy-relevant knowledge within that domain or issue-area."[7] For the purpose of this chapter, we can regard the participating scientists from East and West as an epistemic community, in which each member shared four core beliefs: First, they believed that scientists could, and even should, move the international arms control agenda outside of official government-to-government contacts. Second, they believed that a comprehensive nuclear test ban treaty was an important step toward more substantial arms control agreements and détente in general. Third, they believed that such a treaty could be adequately verified by seismic means. Finally, they believed that their partners from the other side of the cold war divide shared the first three beliefs.[8]

The second political science concept of relevance to this paper is the notion of "transnational actors."[9] Transnational actors are entities that engage in transnational relations, that is, "regular interactions across national boundaries when at least one actor is a nonstate agent or does not operate on behalf of a national government or an intergovernmental organization."[10] As political scientists have shown, transnational actors—such as the approximately 7,000 multinational corporations and 5,000 international nongovernmental organizations (NGOs), including Greenpeace and Amnesty International—have often played a significant role in international affairs and the behavior of states.[11] The concept can be extended to collaborations between scientists from various countries, in which the scientists don't represent their nations and

[5] See, e.g., Frank Gaffney, "Test Ban: The 'Quick Fix' Won't Work," *Washington Post,* Aug. 29, 1986, A15.

[6] On the concept of epistemic communities in political science, see Peter M. Haas, ed., *Knowledge, Power, and International Policy Coordination,* which appeared as a special issue of *International Organization* 46 (Winter 1992). See, especially, Peter M. Haas, "Introduction: Epistemic Communities and International Policy Coordination," ibid., 1–35; Emanuel Adler, "The Emergence of Cooperation: National Epistemic Communities and the International Evolution of the Idea of Nuclear Arms Control," ibid., 101–45. For a critical discussion of the concept, see Sheila Jasanoff, "Science and Norms in Global Environmental Regimes," in *Earthly Goods: Environmental Change and Social Justice,* ed. Fen Osler Hampson and Judith Reppy (Ithaca, 1996), 173–97.

[7] Haas, "Introduction" (cit. n. 6), 3.

[8] This is based on my interviews with scientists involved in transnational arms control networks, including Yevgeny Velikhov, Thomas Cochran, Roald Sagdeev, Stanislav Rodionov, Jeremy Stone, Frank von Hippel, Charles Archambeau, and Jack Evernden.

[9] For a good overview, see Thomas Risse, "Transnational Actors and World Politics," in *Handbook of International Relations,* ed. Walter Carlsnaes, Thomas Risse, and Beth A. Simmons (London, 2002), 255–74.

[10] Thomas Risse-Kappen, "Bringing Transnational Relations Back in: Introduction," in *Bringing Transnational Relations Back in: Non-State Actors, Domestic Structures, and International Institutions,* ed. Thomas Risse-Kappen (Cambridge, 1995), 3.

[11] Numbers based on ibid., 3. For the influence of transnational networks of civil society groups on governments, see Ann M. Florini, ed., *The Third Force: The Rise of Transnational Civil Society* (Tokyo, 2000).

therefore national agendas (as in scientific exchanges between nations), but rather nonstate or transnational agendas.

Arguably, the most relevant study of scientists as transnational actors for the purpose of this chapter is Matthew Evangelista's book *Unarmed Forces: The Transnational Movement to End the Cold War.*[12] Evangelista, a political scientist at Cornell University, concluded that transnational networks of scientists played a significant role in arms control affairs of the 1980s and in the end of the cold war. However, he focused on the importance of Soviet domestic structure to explain the scientists' successes and did not analyze further why scientists accomplished these successes. I agree with Evangelista that the very hierarchical Soviet domestic structure provided a window of opportunity for transnational groups to shape Soviet foreign policy, especially when they gained direct access to the top leadership, such as Soviet general secretary Mikhail Gorbachev and his foreign secretary, Eduard Shevardnadze.

However, I argue here that Soviet domestic structure is only one factor in the explanation of the effectiveness of these groups and that two factors have been overlooked. First, the NRDC–Soviet Academy collaboration rested on the political activism of the involved American and Soviet scientists, in particular the leaders of the two delegations, the NRDC physicist and senior scientist, Thomas Cochran, and Gorbachev's informal science adviser, Yevgeny Velikhov, a physicist and vice-president of the Soviet Academy. It took the daring and tenacity of both Cochran and Velikhov to overcome opposition in their own countries. Second, it took trust between these scientists, a trust that depended on the shared belief that arms control was desirable and manageable by a combination of technical and political approaches. In other words, they belonged to the same epistemic community of arms control supporters and activists, one that regarded it as crucial to work toward a relaxation of superpower tensions and, most importantly, to prevent the ultimate catastrophe—nuclear war. This trust stands in marked contrast to the deep suspicions both governments harbored.

This chapter analyzes the role of these scientists as transnational arms control activists and unofficial diplomats and raises some broader questions about the role of scientists in international affairs. What have scientists contributed to conflict prevention and resolution? To what extent has their contribution been distinctive? In other words, to what extent have scientists been able to address international problems that diplomats, business people, artists, or athletes have not, and what kind of groups or individual scientists has it taken to effectively shape international affairs? In addition, to what extent has technical expertise been necessary for scientists to become influential? Have other skills and qualities been significant as well?[13]

This chapter does not suggest that scientists have changed or can change international affairs simply by the force of their knowledge claims. They operate within an international system and a domestic structure that constrains every actor. However, the chapter suggests that input from transnational coalitions of scientists has shaped international affairs. In the Soviet Union in the mid-1980s, a political system in flux, small-scale changes could have dramatic effects, and scientists became catalysts of change.

[12] Evangelista, *Unarmed Forces* (cit. n. 2).

[13] Similar questions have been raised by Allison L. C. de Cerreño and Alex Keynan, *Scientific Cooperation, State Conflict: The Roles of Scientists in Mitigating International Discord,* Annals of the New York Academy of Sciences, vol. 866 (New York, 1998).

The chapter is divided into three sections and an epilogue. The first section highlights the two trends out of which the NRDC–Soviet Academy cooperation developed: first, the rich tradition of scientists as transnational arms control activists during the 1960s and 1970s; second, the efforts of governments to seek a complete, or comprehensive, nuclear test ban treaty. Then I turn to the immediate origins of the project and introduce the principal players: on the one side, Thomas Cochran and the Natural Resources Defense Council, and on the other, Yevgeny Velikhov and the Soviet Academy of Sciences. The third section discusses the monitoring agreement and the actual establishment of stations in both countries. The epilogue returns to the core question: How, why, and to what extent were these scientists from East and West able to do what official diplomatic efforts had failed to do?

SCIENTISTS AS TRANSNATIONAL ARMS CONTROL ADVOCATES

During the cold war, scientists from East and West repeatedly worked together to discuss arms control measures and to seek solutions for complex technical-political problems of arms control verification.[14] The NRDC–Soviet Academy project, arguably the most visible case of such cooperation during the Gorbachev era, can be seen in this tradition.

The most important transnational effort of scientists during the Khrushchev and Brezhnev eras was the Pugwash movement, which brought together scientists from the Soviet Union, the United States, and other countries to discuss fundamental problems of arms control and the survival of humanity.[15] Pugwash's underlying assumption was that scientists would share a common language and therefore could find common ground where diplomats could not. Pugwash conferences, which began in 1957 and continue to this day, provided an influential and open communication forum, especially during times of tensions between the superpowers. They led to informal contacts between U.S. and Soviet scientists and to the generation of new ideas that have shaped foreign policy decisions, most notably in the cases of the 1963 Limited Test Ban Treaty (LTBT) and the 1968 Nuclear Non-Proliferation Treaty.[16] The Norwegian Nobel Committee awarded the 1995 Nobel Peace Prize to Pugwash secretary Joseph Rotblat and the Pugwash movement "for their efforts to diminish the part played by nuclear arms in international politics and, in the longer run, to eliminate such arms."[17]

A second example of such transnational cooperation is the Soviet-American Disarmament Study Group, which met from 1961 to 1975.[18] This group developed partly in reaction to the changing focus and participation of Pugwash conferences. While early Pugwash participants were predominantly nuclear physicists, who sought technical solutions to nuclear arms control matters, the conferences soon broadened to

[14] Evangelista, *Unarmed Forces* (cit. n. 2), is the best source. See also Joseph Rotblat, "Movements of Scientists against the Arms Race," in *Scientists, the Arms Race, and Disarmament: A Unesco/Pugwash Symposium,* ed. Joseph Rotblat (London, 1982), 115–57.

[15] No scholarly history of the Pugwash conferences exists. A starting point is Joseph Rotblat, *Scientists in the Quest for Peace: A History of the Pugwash Conferences* (Cambridge, Mass., 1972).

[16] Rotblat, "Movements of Scientists against the Arms Race" (cit. n. 14), especially 138–44.

[17] Cited in http://nobelprize.org/peace/laureates/1995/ (accessed Nov. 20, 2005).

[18] Bernd W. Kubbig, *Communicators in the Cold War: The Pugwash Conferences, the U.S.-Soviet Study Group, and the ABM Treaty; Natural Scientists as Political Actors: Historical Success and Lessons for the Future,* Peace Research Institute Frankfurt Report no. 44 (Frankfurt am Main, 1996).

include more social scientists and topics, such as world hunger. In addition, the conferences had swollen to more than a hundred participants and had become, in the eyes of some critics, too unwieldy and unfocused. The new group returned to the origins of Pugwash and limited the membership to only a handful of politically well-connected experts on both sides, who focused on policy-relevant aspects of arms control. U.S participants were affiliated with the Cambridge-based American Academy of Arts and Sciences; members of the Soviet Academy of Sciences dominated the Soviet delegation. Among the Americans we find Harvard chemist Paul Doty and Harvard political scientist Henry Kissinger, who became a dominant figure in U.S. foreign policy, first as head of the National Security Council (1969–1973) and then as secretary of state (1973–1977). Leaders on the Soviet side included influential academicians Mikhail Millionshchikov, a vice-president of the academy, and Lev Artsimovich, both physicists and committed supporters of arms control measures.[19] Bernd W. Kubbig has shown how this group succeeded in convincing Soviet officials to abandon their missile defense posture and to accept instead the intellectual foundations of the Anti-Ballistic Missile Treaty in 1972.[20]

The early 1980s posed a particular challenge for transnational scientific collaborations, since relations between the United States and the Soviet Union had reached a dangerous low point.[21] The Soviet occupation of Afghanistan in 1979, the election of Ronald Reagan in 1980, and the tensions about deployment of Pershing II missiles in Western Europe led to a breakdown of official diplomatic channels between the superpowers from 1983 to 1986. Concerned U.S. citizens sought contacts with Soviet groups to reduce tensions and to keep communication channels open.[22] U.S. scientists reacted to these international tensions by discussing arms control again. For example, in 1980 alone, the American Association for the Advancement of Science established an arms control group, the American Physical Society held a symposium on arms control, and Physicians for Social Responsibility revived their arms control activity.[23]

Some groups of U.S. scientists sought direct contact with Soviet scientists. Three such efforts stand out: first, the close collaboration between the Federation of American Scientists (FAS) leadership, that is, Jeremy Stone and Frank von Hippel, and Soviet physicists, such as the ubiquitous Velikhov and his colleague, academician Roald Sagdeev, the former director of the Soviet Space Research Institute;[24] second, the formal meetings between the U.S. National Academy's Committee on International

[19] On Millionshchikov and Artsimovich, see ibid., especially 26–7, 33–4; and Evangelista, *Unarmed Forces* (cit. n. 2), 25–140.

[20] Kubbig, *Communicators in the Cold War* (cit. n. 18).

[21] Raymond L. Garthoff, *The Great Transition: American-Soviet Relations and the End of the Cold War* (Washington, D.C., 1994); Don Oberdorfer, *From the Cold War to a New Era: The United States and the Soviet Union, 1983–1991,* updated ed. (Baltimore, Md., 1998).

[22] For efforts by other groups, especially the Dartmouth Conferences, see David D. Newsom, ed., *Private Diplomacy with the Soviet Union* (Lanham, Md., 1987).

[23] Michael Heylin, "NAS, Others Take on Arms Control Issues," *Chemical and Engineering News,* May 19, 1980, 43–4.

[24] The FAS collaborated with Velikhov and Sagdeev on issues such as nuclear winter and the critical response to President Reagan's Strategic Defense Initiative. See *F.A.S. Public Interest Report* 36 (Dec. 1983) and 37 (Jan. 1984), available online at http://www.fas.org/faspir/archive.htm (accessed Nov. 13, 2005). For von Hippel, see Joel Primack and Frank von Hippel, *Advice and Dissent: Scientists in the Political Arena* (New York, 1974); Frank von Hippel, *Citizen Scientist* (New York, 1991). For Stone, see Jeremy Stone, *"Every Man Should Try": Adventures of a Public Interest Activist* (New York, 1999). For Sagdeev, see Roald Z. Sagdeev, *The Making of a Soviet Scientist* (New York, 1994).

Security and Arms Control (CISAC) with its Soviet counterpart delegation from the Soviet Academy, chaired for the better part of the 1980s by Velikhov and Sagdeev;[25] and third, the NRDC–Soviet Academy interaction, which is analyzed below in more detail.

THE TEST BAN DEBATES IN THE EARLY 1980s

The second root of the NRDC–Soviet Academy project lies in the nuclear test ban debates of the early 1980s. After many years of negotiations, the United States, the United Kingdom, and the Soviet Union signed the Limited Test Ban Treaty in 1963.[26] The treaty banned tests in the atmosphere, underwater, and in outer space, but not underground, since the United States felt that seismic monitoring capabilities were not sufficiently developed to detect small Soviet underground explosions.[27] With the signing of the LTBT, the test ban issue largely faded from public interest and returned only in the final weeks of the Nixon presidency.

In 1974, President Nixon, under pressure at the height of the Watergate affair, met with Leonid Brezhnev in Moscow. To make the summit worthwhile, Nixon needed a bilateral arms control agreement of some significance. After a few days of negotiation, both sides agreed to limit their underground tests to a maximum yield of 150 kilotons. Two years later both sides extended this Threshold Test Ban Treaty (TTBT) to cover the so-called peaceful nuclear explosions (PNEs) as well, resulting in the Peaceful Nuclear Explosions Treaty (PNET).[28] Negotiations for a comprehensive test ban treaty (CTBT) among the United States, the United Kingdom, and the Soviet Union began in 1977, accompanied by a raging test ban debate in Washington.[29] The negotiations recessed in 1980, at least in part as a consequence of the Soviet invasion of Afghanistan. The incoming Reagan administration explicitly rejected the CTBT and formally withdrew from the trilateral talks in 1982, citing the need for continued testing and the lack of adequate verification measures.[30] As Eugene V. Rostow, director of the Arms Control and Disarmament Agency, put it, "[G]iven the uncertainty of the nuclear situation, the nuclear balance and the need for new weapons and modernization . . . we are going to need testing and perhaps even testing above the 150-kiloton [TTBT] limit for a long time to come." In addition, he voiced "grave doubts" about existing test ban verification measures.[31]

While the United States had signed both the TTBT and the PNET in 1974 and 1976, respectively, both treaties were left hanging in the Senate Foreign Relations Committee and languished in the no-man's-land of nonratified treaties. With Gorbachev's push for a comprehensive nuclear test ban, the Reagan administration, determined to continue testing, used the ratification issue of the TTBT and PNET as a stumbling block for a comprehensive nuclear test ban treaty. Since 1982, Reagan officials had

[25] These meetings continue in the present.

[26] For major works on the early test ban negotiations, see Jacobson and Stein, *Diplomats, Scientists, and Politicians;* Seaborg and Loeb, *Kennedy, Khrushchev, and the Test Ban.* (Both cit. n. 4.)

[27] Jacobson and Stein, *Diplomats, Scientists, and Politicians* (cit. n. 4).

[28] Thomson, *A Guide to the Nuclear Arms Control Treaties* (cit. n. 4), 38–41.

[29] Greb, "Survey of Past Nuclear Test Ban Negotiations" (cit. n. 4), 105–9.

[30] Ibid.

[31] Eugene V. Rostow, quoted in Murrey Marder, "Two Dormant Treaties Awaken Dispute over A-Test Inspection," *Washington Post,* July 26, 1982, A2.

argued that the TTBT and PNET had to be ratified first before serious consideration of a CTBT could begin.[32] However, ratification of the two treaties was again held up by the alleged uncertainty of seismic verification: To what extent was the United States able to detect small Soviet underground explosions?[33] Reagan administration officials assumed that Soviet leaders would never allow on-site inspections of their nuclear testing facilities or the direct on-site monitoring of nuclear explosions. Without on-site monitoring, testing proponents continued to claim that a nuclear test ban was not verifiable.

Test ban supporters realized that access to Soviet test grounds would eliminate this argument. But would the Soviets let U.S. monitoring experts set up seismic stations around their test site at Semipalatinsk? It took the daring of Velikhov to overcome the traditional suspicions of Soviet hardliners. He outmaneuvered internal dissent of leading policy makers and presented Gorbachev essentially with an accomplished fact of U.S. seismic experts on the ground in Kazakhstan.

By 1985, the two superpowers disagreed sharply about the value and usefulness of a CTBT. Leading defense officials in the Reagan administration rejected a CTBT for a variety of reasons. Not only was such a ban not verifiable, they argued, but further nuclear tests were necessary to guarantee the safety and reliability of the existing stockpile. Critics argued that the Reagan administration wanted to keep the testing option open, including testing of new nuclear weapons designs for the Strategic Defense Initiative. Reagan and his security advisers had repeatedly refused to join Gorbachev's test moratorium, some dismissing it as a propaganda gimmick.[34] For supporters of continued testing, the verification issue provided a convenient and powerful argument against a test ban. They argued that a test ban without adequate verification measures would be against U.S. interests, since the United States could never be sure whether the Soviets would test nuclear devices clandestinely and thereby improve the quality of their nuclear arsenal and even possibly shift the balance of power.

Test ban supporters, in contrast, highlighted that a continued nuclear arms race could have only one outcome. In their view, a test ban was verifiable, and they challenged the notion that further testing for the safety and reliability of warheads was, in fact, needed. Some academic seismologists in particular challenged official government statements that the state-of-art in seismology did not permit a viable monitoring system.[35]

With no movement in sight, a number of scientists thought about initiatives to break the impasse and push for a test ban. It took a very unconventional approach of test ban supporters in East and West to change the debate.

[32] Judith Miller, "U.S. Confirms a Plan to Halt Talks on a Nuclear Test Ban," *New York Times,* July 21, 1982, A1.

[33] Some seismologists felt strongly that the United States was already able to verify a CTBT. For a good overview of their arguments, see Gregory E. van der Vink, "The Role of Seismologists in Debates over the Comprehensive Test Ban Treaty," in de Cerreño and Keynan, *Scientific Cooperation, State Conflict* (cit. n. 13), 84–113. "Small explosion" in this context meant nuclear yields of about one kiloton. In comparison, the yields of the Hiroshima and Nagasaki bombs were thirteen kilotons and twenty-two kilotons, respectively. See Richard L. Garwin and Georges Charpak, *Megawatts and Megatons: A Turning Point in the Nuclear Age?* (New York, 2001), 59–60.

[34] Institute for Defense and Disarmament Studies (IDDS), *The Arms Control Reporter* (Brookline, Mass., 1982), sec. 608.B.66 ("Comprehensive Test Ban, Conference on Disarmament, Chronology 1985").

[35] Van der Vink, "Role of Seismologists" (cit. n. 33).

THE ORIGIN OF THE NRDC–SOVIET ACADEMY COLLABORATION

In the sections above, I highlighted two roots of the NRDC–Soviet Academy project: first, scientists as transnational arms control advocates; second, efforts to negotiate a comprehensive nuclear test ban treaty. In this section, we will focus first on the two organizations involved in the project, the NRDC and the Soviet Academy of Sciences, and the project leaders, Thomas Cochran and Yevgeny Velikhov. Then I will discuss the origins of their collaboration.

Thomas Cochran and the NRDC

In the mid-1980s, the Natural Resources Defense Council could look back to more than a decade of public interest activism, with a focus on environmental, public health, and arms control causes. Founded in 1970 "by 50 lawyers and scientists," NRDC claimed more than 60,000 members by 1986.[36] Headquartered in New York, with branch offices in Washington, D.C., and San Francisco, NRDC had the staff, the scientists, the determination, and the connections of a "heavyweight" nongovernmental organization.[37] The NRDC–Soviet Academy seismic verification project, the "largest privately-funded scientific exchange ever undertaken with the Soviet Union,"[38] became NRDC's largest project, with an estimated cost of $2.4 million for the first year alone, a large sum for an NGO with an annual budget of approximately $7 million.[39]

In 1973, NRDC hired physicist Thomas Cochran as a senior scientist to work on litigation related to the Atomic Energy Commission's plutonium breeder reactor program. Cochran, born in 1940, had served in the navy, had taught mathematics and physics at the U.S. Naval Postgraduate School, and had received his Ph.D. in physics at Vanderbilt University.[40] In 1974, he published an analysis of the Clinch River Breeder Reactor Project and demonstrated that the government's assumptions about the economic and environmental consequences of the reactor were faulty.[41] Later he recalled that "the Atomic Energy Commission people were fudging the information to make the breeder look good, when really it looked terrible."[42] Cochran and other

[36] Philip Taubman, "New Yorkers Sign Soviet Test Pact," *New York Times,* May 29, 1986, A3.

[37] Besides Cochran, other NRDC officials played a major role in this project. NRDC board chairman Adrian DeWind signed the original agreement and aided in the project's implementation. In addition, NRDC executive director John A. Adams helped with the fund-raising. Cochran shared responsibility for the project with NRDC senior staff attorney S. Jacob Scherr; Cochran, "The NRDC/Soviet Academy of Sciences Joint Nuclear Test Ban Verification Project" (cit. n. 2), 5–8.

[38] NRDC, Status Report, May 1987, 2, NRDC Files.

[39] NRDC, Status Report, Nov. 1986, 4, NRDC Files. Major donors included the "John D. and Catherine T. MacArthur Foundation, the Carnegie Corporation of New York, Ploughshares Fund, the W. Alton Jones Foundation, J. M. Kaplan Fund, the Columbia Foundation, and an anonymous donor." Jonathan Berger, James N. Brune, Paul A. Bodin et al., "A New U.S.-U.S.S.R. Seismological Program," *EOS: Transactions of the American Geophysical Union* 68, Feb. 24, 1987, 105 and 110–1; see the acknowledgements. According to Thomas Cochran, the anonymous donor was the philanthropist Joan Kroc, widow of McDonald's Corporation founder Ray A. Kroc. Cochran, personal communication, Oct. 11, 2005.

[40] On Cochran, see Tim Beardsley, "Rebottling the Nuclear Genie," *Scientific American* 278 (May 1998): 34-5; Schrag, *Listening for the Bomb* (cit. n. 2), 11; René T. Riley, "A Physicist Blasts Away at Verification 'Myths,'" *National Journal,* Jan. 31, 1987, 273.

[41] Thomas B. Cochran, *The Liquid Metal Fast Breeder Reactor: An Environmental and Economic Critique* (Washington, D.C., 1974).

[42] Cochran, quoted in Riley, "Physicist Blasts Away at Verification 'Myths'" (cit. n. 40), 273.

antinuclear activists fought the project and succeeded, with Congress canceling the project in 1983.[43]

At NRDC, Cochran and his small group of nuclear experts began to analyze nuclear weapons research and production, nuclear waste, arms control, proliferation, and nuclear energy policy. In particular, the group collected information about the environmental effects of nuclear weapons production, testing, and possible use, which led to a series of volumes, the *Nuclear Weapons Databooks.* The five *Databooks* provided the most comprehensive unclassified information about U.S., Soviet, French, British, and Chinese nuclear weapons and their production facilities.[44]

As part of the *Databook* work, NRDC analysts collected data about each nuclear test, including its yield and purpose. Earlier U.S. administrations had announced essentially all nuclear tests, but in 1982 the Reagan administration decided to announce only selected tests, arguably to downplay the magnitude of the U.S. nuclear testing program. Cochran and his colleagues decided to produce a list of these unannounced U.S. nuclear tests, but information, especially on the small tests, was extremely difficult to obtain. Underground nuclear tests at the Nevada Test Site (NTS) an hour north of Las Vegas would generate seismic waves similar to that of an earthquake and would be recorded by seismic stations of the U.S. Geological Survey (USGS). However, small explosions in the one- to two-kiloton range would evade detection by USGS stations. Yet Cochran learned about a number of unannounced small explosions at NTS from an analysis of an unclassified paper by Lawrence Livermore National Laboratory physicist Ray Kidder.[45]

In early 1986, Cochran and William M. Arkin, a coauthor of the *Nuclear Weapons Databook* series, half-jokingly discussed setting up seismic stations around the Nevada Test Site to guarantee that there would be no more secret tests at NTS. They dismissed the idea because it was unlikely to get political support.[46] Based on a suggestion by a staff reporter of the *Wall Street Journal,* John J. Fialka, Cochran then developed a plan to set up seismic stations around the Soviet nuclear test site.[47] NRDC analysts agreed it was very unlikely that the Soviets would ever agree to such a plan, since they had never permitted on-site monitoring despite decades of U.S. insistence. At the same time, the plan offered some rewarding policy options: if the United States rejected the plan, it would demonstrate that the Reagan administration was not interested in seriously testing the Soviets' position on verification. This, in turn, would demonstrate that the real reasons behind the Reagan administration's opposition to a nuclear test ban was not the verification problem but the determination to develop and test the next generation of nuclear weapons. Such a position was politically costly in the mid-1980s in the light of the popular Nuclear Weapons Freeze Campaign.[48] The

[43] Beardsley, "Rebottling the Nuclear Genie" (cit. n. 40).

[44] Thomas B. Cochran, William M. Arkin, Robert S. Norris et al., *Nuclear Weapons Databook,* vols. 1–4 (Cambridge, Mass., 1984–1989); Robert S. Norris, Andrew S. Burrows, and Richard W. Fieldhouse, *Nuclear Weapons Databook,* vol. 5 (Cambridge, Mass., 1994).

[45] Cochran, phone interview with author, May 14, 2001. Thomas B. Cochran, Robert S. Norris, William M. Arkin, and Milton Hoenig, "Unannounced U.S. Nuclear Weapons Tests, 1980–1984," Nuclear Weapons Databook Working Paper NWD 86–1, NRDC, 1986.

[46] Schrag, *Listening for the Bomb* (cit. n. 2), 12.

[47] Cochran, personal communication, Oct. 11, 2005.

[48] About the Freeze movement, see Lawrence S. Wittner, *The Struggle against the Bomb,* vol. 3, *Toward Nuclear Abolition: A History of the World Nuclear Disarmament Movement: 1971 to the Present* (Stanford, 2003), 169–97.

NRDC plan would also call the Soviets' bluff: the Soviets had repeatedly claimed that the technical means of verification were sufficient to monitor a nuclear test ban. If they rejected the NRDC's proposal, it would demonstrate to the world that the Soviets, too, were unwilling to seriously consider a test ban. Both the U.S. and the Soviet administrations relied on the posture of the other side to justify why they rejected a test ban, and a movement by one side would call the bluff of the other side.

Cochran began to shop this concept around. He met with Vitalii Zhurkin, a leading defense expert at the Soviet embassy in Washington, D.C.[49] Cochran "outlined the idea of writing Reagan and Gorbachev requesting permission for NGO scientists to jointly establish seismic stations in each country. [Zhurkin] seemed very interested in the idea."[50] However, when the two met again, Zhurkin insisted that such a letter to Gorbachev and Reagan had to include a call to the U.S. president to join the moratorium. In Cochran's view, such an approach would be fruitless.[51]

After this initial disappointment, Cochran sought other ways to present his proposal to leading Soviet officials. Jeremy Stone, then director of the Federation of American Scientists, suggested that Cochran go through the Soviet Academy of Sciences instead of the Soviet Embassy.[52] Stone invited Cochran to present his idea to a group of Soviet scientists visiting the FAS in April 1986. The initial response of the Soviet delegation, led by Soviet defense expert Andrei Kokoshin, encouraged Cochran.[53] At the meeting, he raised his idea with FAS chairman Frank von Hippel, who had developed a close working relationship with Velikhov. Von Hippel discussed Cochran's idea with Velikhov, who in turn agreed to holding a workshop on verification methods in Moscow in May 1986.[54]

Cochran also contacted officials from U.S. nuclear weapons laboratories, who apparently saw no problems for national security if Soviet scientists were to set up monitoring stations in the vicinity of the Nevada Test Site.[55] To get feedback from official government representatives, NRDC chairman Adrian DeWind arranged a meeting with U.S. deputy secretary of state John Whitehead and Paul Nitze, special adviser to the president for arms control and one of the nation's most experienced arms control negotiators. Whitehead replied in a letter to DeWind on March 4, "[T]here is potential that private citizens could draw conclusions about Soviet compliance with testing constraints that differ from the judgments of U.S. government officials with access to more sources of information. There is obvious potential here for confusion." Furthermore, he concluded that "the establishment of such seismic stations would not resolve the current uncertainty in measuring the yield of nuclear tests, nor would it alter our national security need for on-site measurements of nuclear tests."[56] However, while

[49] In 1986, Vitalii Zhurkin was a deputy director of the influential Institute of the USA and Canada, ISKAN.

[50] Cochran, "Notes on History of the NRDC Verification Project," n.d. I thank Thomas Cochran for making these notes available to me.

[51] Cochran, personal communication, Oct. 11, 2005.

[52] Cochran, "Notes on History of the NRDC Verification Project" (cit. n. 50).

[53] Ibid.

[54] Frank von Hippel, interview with author, Princeton University, Feb. 20, 2004.

[55] Cochran to Velikhov, Jan. 23, 1987. I thank Thomas Cochran for making this document available to me.

[56] Whitehead, quoted in Flournoy, "A Controversial Excursion into Private Diplomacy" (cit. n. 2), 13–4.

Whitehead did not endorse the NRDC proposal, he also did not actively oppose it. Based on this meeting, Cochran felt safe that the administration would not attack the NRDC.[57]

Yevgeny Velikhov and the Soviet Academy of Sciences

Yevgeny Velikhov, born in 1935, graduated from Moscow State University in 1958 and began a long career as a physicist at the Kurchatov Institute of Atomic Energy.[58] In 1971, he joined the CPSU, and during the same year, he became deputy director of the Kurchatov Institute. In 1974, at the unusually young age of thirty-nine, he was elected a full member of the Soviet Academy. Three years later, he became the academy's vice-president for applied physics and mathematics. Velikhov's work on nuclear fusion brought him into direct contact with Soviet military organizations. By the late 1970s, he was involved in internal discussions on space-based missile defenses and laser weapons, which prepared him for his later role in challenging Reagan's Strategic Defense Initiative.[59]

How did Velikhov get involved in arms control issues? In 1982, he was one of the Soviet delegates to the Pontifical Academy of Sciences. The pope had invited an international group of scientists to discuss the dangers of nuclear war. At the academy, he worked with Manhattan Project veteran Victor Weisskopf on a declaration, which included "some of the ideas we now call the new thinking—the need for a nuclear-free world, the impossibility of nuclear superiority or of a defense against nuclear weapons."[60] The declaration appealed to national leaders, scientists, religious leaders, and "people everywhere" to make nuclear war impossible. It urged scientists specifically "to apply their ingenuity in exploring means of avoiding nuclear war and developing practical methods of arms control."[61] It was this conviction that Velikhov shared with arms control physicists from East and West.

In 1982 as well, academician Nikolai Inozemtsev invited Velikhov to join a high-level Soviet Academy delegation, which met regularly with a delegation of the U.S. National Academy of Sciences, the above-mentioned Committee on International Security and Arms Control, to discuss arms control matters. When Inozemtsev passed away unexpectedly in August 1982, Velikhov became the Soviet delegation's chairman and remained in this position until 1986.[62]

In 1983, in response to Reagan's "Star Wars" speech, Velikhov, Roald Sagdeev, Andrei Kokoshin and other arms control scientists and political scientists founded the Committee of Soviet Scientists against the Nuclear Threat (CSS). They modeled CSS

[57] Cochran, personal communication, Oct. 11, 2005.

[58] As of this writing (Nov. 2005), Velikhov continues to be president of the Kurchatov Institute, which was renamed the Russian Research Centre Kurchatov Institute in 1992. See http://www.kiae.ru/eng/str/direct/vep.htm (accessed Nov. 5, 2005).

[59] Yevgeny Velikhov, interview with author, Kurchatov Institute, Moscow, May 25, 2001.

[60] Yevgeny Velikhov, "Chernobyl Remains on Our Mind," in *Voices of Glasnost: Interviews with Gorbachev's Reformers,* ed. Stephen F. Cohen and Katrina vanden Heuvel (New York, 1989), 160. For the "Declaration on Prevention of Nuclear War," see Yevgeny Velikhov, ed., *The Night After . . . : Climatic and Biological Consequences of a Nuclear War* (Moscow, 1985), 155–8. For the context of Velikhov's involvement in arms control issues, see Velikhov interview (cit. n. 59).

[61] Velikhov, ed., *The Night After* (cit. n. 60), 157–8.

[62] Velikhov interview (cit. n. 59); Notes of Meeting Records, Committee on International Security and Arms Control, National Academy of Sciences, Washington, D.C. (hereafter cited as CISAC files).

after the Federation of American Scientists[63] and saw themselves as contributing to an emerging discipline, *politicheskaya fisika* (political physics), that is, technical-political analyses of military projects.[64] For example, CSS chairman Velikhov played a leading role in the Soviet response to Reagan's Strategic Defense Initiative.[65] Out of this initiative, he developed a close working relationship with the FAS's chairman Frank von Hippel, which culminated in joint research projects between the FAS and the CSS. This collaboration led to a new journal, *Science and Global Security,* which became a unique publication on scientific-technical aspects of arms control.[66]

By the mid-1980s, Velikhov was an accomplished nuclear fusion physicist with extensive international contacts, a vice-president of the Soviet Academy, and arguably the most politically influential Soviet scientist of his generation, his position comparable to that his teacher and mentor Lev Artsimovich had occupied in the 1960s.[67] Velikhov had close working relationships with many leading U.S. arms control scientists, including Frank von Hippel, Wolfgang Panofsky, and Richard Garwin.[68] He worked with Senator Ted Kennedy and Carl Sagan on the possible catastrophic climate changes after a nuclear war, which became popularized as "Nuclear Winter."[69] In short, by 1985 he had an established record of unconventional approaches to arms control and had won the trust of Western arms control scientists.

As an academician and vice-president of the academy, Velikhov had access to high-level policy makers. His defense-related work gave him access to top military leaders such as Marshal Sergei Akhromeev, first deputy chief of the general staff.[70] In addition, he had developed a relationship with Mikhail Gorbachev, whom he first met in 1978.[71] When Gorbachev became general secretary in 1985, Velikhov moved into Gorbachev's inner circle and advised him predominantly on "nuclear issues, strategic weapons, and computers."[72] According to Velikhov, the relationship was so close that he could call Gorbachev with an idea about arms control.[73] He accompanied Gor-

[63] On the Committee of Soviet Scientists against the Nuclear Threat, see Frank von Hippel, "The Committee of Soviet Scientists against the Nuclear Threat," *F.A.S. Public Interest Report* 37 (Jan. 1984): 1–4. For CSS's membership list, see ibid., 8.

[64] According to Stanislav Rodionov, a Soviet physicist, space weapon expert, and CSS member, CSS actually used the term *politicheskaya fisika.* Rodionov, interview with author, Moscow, May 22, 2001.

[65] Frank von Hippel, "Arms Control Physics: The New Soviet Connection," *Physics Today* (Nov. 1989): 39–46. Yevgeny Velikhov, Roald Z. Sagdeev, and Andrei A. Kokoshin, *Weaponry in Space: The Dilemma of Security* (Moscow, 1986).

[66] The outcome of this collaboration is captured in Frank von Hippel and Roald Z. Sagdeev, *Reversing the Arms Race: How to Achieve and Verify Deep Reductions in the Nuclear Arsenals* (New York, 1990). For *Science and Global Security,* see http://www.princeton.edu/~globsec/publications/SciGloSec.shtml (accessed Nov. 5, 2005).

[67] Relevant books and articles by Velikhov include *The Night After* (cit. n. 60); "Science and Scientists for a Nuclear-Weapon-Free World," *Phys. Today* (Nov. 1989): 32–6; and, for autobiographical information, "Chernobyl Remains on Our Mind" (cit. n. 60), 157–73; see also Velikhov interview (cit. n. 59).

[68] For the collaboration with Frank von Hippel, see von Hippel, "Arms Control Physics" (cit. n. 65). As chairman of the Soviet counterpart to the U.S. National Academy's Committee on International Security and Arms Control from 1982 to 1986, Velikhov worked with leading U.S. scientists such as Panofsky and Garwin on arms control issues. See Notes of Meeting Records, CISAC Files.

[69] *F.A.S. Public Interest Report* 37 (Jan. 1984); Velikhov, introduction to *The Night After* (cit. n. 60), 1–33, especially 4–18. On Sagan and Nuclear Winter, see Lawrence Badash, "Nuclear Winter: Scientists in the Political Arena," *Physics in Perspective* 3 (2001): 76–105.

[70] Velikhov interview (cit. n. 59).

[71] Velikhov, "Chernobyl Remains on Our Mind" (cit. n. 60), 160; Velikhov interview (cit. n. 59).

[72] Velikhov, "Chernobyl Remains on Our Mind" (cit. n. 60), 161.

[73] Velikhov interview (cit. n. 59).

bachev on important foreign travel and, in particular, to summit meetings with the American president.[74] Most notably, Gorbachev trusted Velikhov with organizing the immediate response to the April 1986 Chernobyl nuclear disaster, and Velikhov became well known for his daring efforts to curb the reactor's fire.[75]

Velikhov was the prototype of the scientist as political entrepreneur, a political operator who could move things quickly in an otherwise inflexible bureaucracy. To what extent was he, or any Soviet Academy leader for that matter, able to influence Soviet foreign policy decisions?[76] To answer this question, we have to look briefly into the Soviet foreign policy-making process. Soviet foreign policy-making, and more narrowly arms control decision-making in the early 1980s, was primarily the domain of five institutional actors: the general secretary and his close advisers; the Politburo; the relevant branches of the Central Committee of the Communist Party of the Soviet Union (CPSU);[77] the Ministry of Defense; and the Ministry of State.[78] A detailed analysis of the Soviet arms control process is beyond the scope of this paper. Here it suffices to emphasize that direct access to the leading policy makers gave transnational actors an opportunity to shape arms control policies, if, and only if, the policy makers decided to use their information and agenda for their own purposes. Political scientists have pointed out to what extent Gorbachev, in contrast to some of his predecessors, was willing to engage new ideas and to learn from others, including western scientists and physicians, about the dangers of a nuclear arms race and the consequences of a nuclear war.[79]

The following picture emerged: Gorbachev was open to learning, and he was willing to listen when his trusted science adviser Velikhov had an unconventional idea.

[74] Velikhov accompanied Gorbachev on his trip to Great Britain in December 1984. Archie Brown, *The Gorbachev Factor* (Oxford, 1996), 76. Velikhov was also part of Gorbachev's team at the 1986 Reykjavik summit meeting. Anatoly Chernyaev, *My Six Years with Gorbachev,* trans. and ed. Robert D. English and Elizabeth Tucker (University Park, Pa., 2000), 84.

[75] Velikhov, "Chernobyl Remains on Our Mind" (cit. n. 60), 162.

[76] The influence of the Soviet Academy of Sciences on Soviet foreign policy-making during the 1980s was clearly limited. However, the Soviet Academy of Sciences included political science departments that acted as think tanks on issues ranging from international economic developments to arms control. Most important for international affairs, in particular arms control matters, were two Soviet Academy branches, the Institute for Studies of Canada and the United States, ISKAN, run by the influential adviser Georgi Arbatov, and the Institute of World Economy and International Relations, IMEMO, directed by academician Inozemtsev.

[77] The most relevant departments are the International Department and the Department for Relations with Communist and Workers' Parties of Socialist Countries.

[78] Archie Brown, "The Foreign Policy-Making Process," in *The Soviet State: The Domestic Roots of Soviet Foreign Policy,* ed. Curtis Keeble (Boulder, Colo., 1985), 191–216; Aleksandr' G. Savel'yev and Nikolay N. Detinov, *The Big Five: Arms Control Decision-Making in the Soviet Union,* trans. Dmitriy Trenin and ed. Gregory Varhall (Westport, Conn., 1995); for the role of the Central Committee's International Department, see Mark Kramer, "The Role of the CPSU International Department in Soviet Foreign Relations and National Security Policy," *Soviet Studies* 42 (July 1990): 429–46; on the role of Shevardnadze, see John Van Oudenaren, *The Role of Shevardnadze and the Ministry of Foreign Affairs in the Making of Soviet Defense and Arms Control Policy* (Santa Monica, 1990). Shevardnadze struggled to shape Soviet foreign policy in the first year or two of his term. Later he built up his own think tank in the Ministry of Foreign Affairs and became more influential. No documentary evidence suggests that Shevardnadze was involved in the decision to allow American scientists on Soviet soil. On foreign policy-making under Gorbachev, see Brown, *The Gorbachev Factor* (cit. n. 74), especially 212–51.

[79] For Gorbachev's willingness to listen to new ideas, see Brown, *The Gorbachev Factor* (cit. n. 74); Sarah E. Mendelson, *Changing Course: Ideas, Politics, and the Soviet Withdrawal from Afghanistan* (Princeton, 1998); Janice Gross Stein, "Political Learning by Doing: Gorbachev as Uncommitted Thinker and Motivated Learner," *International Organization* 48 (Spring 1994): 155–83.

The placing of U.S. scientists near the Semipalatinsk test site was clearly unconventional and would have been impossible under Gorbachev's predecessors. Gorbachev, deeply worried about nuclear weapons, especially after the radioactive catastrophe of Chernobyl, agreed with Velikhov that the NRDC–Soviet Academy collaboration offered an opportunity to push nuclear arms control forward. He adopted the idea developed by Cochran, Velikhov, and others and made it his own, thereby allowing the project to succeed against resistance in the Politburo and the military.[80]

The 1986 Moscow Meeting and the Monitoring Agreement

The NRDC–Soviet Academy project developed out of a flurry of initiatives aimed at reducing the danger of nuclear war.[81] First, in conversations with FAS chairman Frank von Hippel in the fall of 1985 Velikhov suggested that it might be possible to get foreign seismologists on Soviet soil.[82] Von Hippel then talked with officials of the Parliamentarians for Global Action, an interest group of about 600 representatives, who according to von Hippel "already had the idea of setting up something like that."[83] In 1985, the Parliamentarians for Global Action developed a Five Continent Peace Initiative, in which six nonaligned countries—Argentina, Greece, India, Mexico, Sweden, and Tanzania—proposed a joint monitoring of a nuclear test moratorium.[84] As part of the proposal, the six nations suggested placing seismic monitors near the American and Soviet test sites. A group of U.S. scientists, including seismic verification specialists Charles Archambeau, Jack Evernden, and Lynn Sykes, provided technical guidance about how such a limited moratorium might be monitored.[85] However, the proposal required approval of both the U.S. and Soviet governments, and since the Reagan administration was openly hostile to a test ban, the proposal was "dead on arrival," as Cochran put it.[86] In addition, USGS seismologist Jack Evernden, a nuclear test detection specialist and longtime critic of test ban opponents, had suggested setting up "an experimental network of seismic monitoring stations at up to 18 sites within the Soviet Union."[87] Compared with the NRDC's plan, Evernden's proposal was significantly more expensive and would take longer to implement.

These various initiatives came together at a Moscow workshop (May 22–23, 1986)

[80] Velikhov interview (cit. n. 59). On the role of ideas in foreign policy-making, see Jeff Checkel, "Ideas, Institutions, and the Gorbachev Foreign Policy Revolution," *World Politics* 45 (1993): 271–300; idem, *Ideas and International Political Change: Soviet/Russian Behavior and the End of the Cold War* (New Haven, 1997); Mendelson, *Changing Course* (cit. n. 79).

[81] Von Hippel, "Arms Control Physics" (cit. n. 65).

[82] Len Ackland, "Testing—Who Is Cheating Whom?" *Bulletin of the Atomic Scientists* (Oct. 1986): 9–11, 11.

[83] Ibid., 11.

[84] See Natural Resources Defense Council, "Proposal to Place Seismic Monitoring Stations Near the Soviet Nuclear Test Site," June 26, 1986, NRDC Files.

[85] These advisers included leading geophysicists Charles Archambeau, a professor of geophysics at the University of Colorado; Jack F. Evernden, a senior scientist at the USGS; and Lynn Sykes, a professor of geophysics at Columbia University. Also included were Frank Barnaby, the former director of SIPRI; Princeton physicist and former FAS president Frank von Hippel; Carson Mark, retired director of the theory division of Los Alamos; and Robert Socolow, the director of Princeton's Center for Energy and Environmental Studies. "Technical Experts Assisting in the Development of the New Delhi–Six Moratorium Monitoring Proposal," NRDC Files; Charles Archambeau, Lynn Sykes, and Jack Evernden, "Monitoring of a Nuclear Test Moratorium by Seismic Methods," NRDC Files.

[86] Cochran phone interview (cit. n. 45).

[87] Jack Evernden, interview with author, Golden, Colo., June 16, 1998; R. Jeffrey Smith, "Soviets Agree to Broad Seismic Test," *Science* 233, Aug. 1, 1986, 511–2.

on test ban verification, organized by von Hippel and Velikhov. At the meeting, Archambeau presented the technical basis of the Five Continents Peace Initiative, Evernden advocated his verification plan, and Cochran presented the NRDC plan. The participants agreed "that the current state of geophysical knowledge gives reasonable confidence in the detectability, using practical seismic networks, of nuclear weapons tests down to yields at, or below, one kiloton."[88] During the meeting, it became clear that Velikhov was looking for a quick political statement and that he favored the NRDC proposal,[89] which suggested installing three seismic stations in Semipalatinsk and three in the United States. Cochran later recalled, "[W]hen we made the proposal to the Soviets [at the May 1986 meeting], in a sense we were shooting from the hip. We didn't have any seismologists organized. We didn't have any money. We just challenged them."[90]

Velikhov took up the challenge. At this time, he still enjoyed easy access to Gorbachev and informed the general secretary of the various arms control initiatives presented at the workshop. According to Velikhov, Gorbachev gave his science adviser "freedom of action," and Velikhov decided in favor of Cochran's plan, arguably the most political of the three major proposals.[91]

The tireless Velikhov and von Hippel followed the May workshop with the larger International Forum of Scientists for Stopping Nuclear Tests (July 11–13, 1986), in Moscow. The meeting brought together about 150 scientists from thirty-four nations, who opposed nuclear testing and lobbied for Gorbachev to extend the moratorium, thereby putting pressure on the Reagan administration to join. On the day after the meeting, a delegation of western and Soviet scientists, including von Hippel, Velikhov, and Cochran, met with Gorbachev.[92] In an open conversation, some scientists, including von Hippel, Cochran, and Velikhov, praised the unilateral Soviet moratorium and appealed to Gorbachev to extend the moratorium yet again, beyond the August 6 expiration date. Gorbachev, in return, praised the scientists and declared, "Convincing arguments saying that the monitoring of halting nuclear tests is possible rang out both in the [scientists'] declaration and here at our meeting. This is of enormous significance as it reflects the view of people who know what they are talking about."[93] In short, transnational scientists had access to Gorbachev and were able to put their views of arms control directly to him.

In sum, the forging of this transnational collaboration depended on three factors: Frank von Hippel's close working relationship with Velikhov; Velikhov's easy access to Gorbachev and the secretary's particular leadership style; and Cochran's plan that promised rewards for the Soviet leadership in demonstrating glasnost and its seriousness about a nuclear test ban.

A few days after the Moscow meeting, on May 28, 1986, the NRDC's chairman Adrian DeWind and Velikhov signed a simple two-page agreement, a marked contrast to official treaty documents. The agreement, vague on technical details, specified that the Soviet Academy of Sciences and NRDC would set up three seismic stations

[88] "Agreement between NRDC and Soviet Academy of Sciences," May 28, 1986, NRDC Files.
[89] According to Cochran, at this point Evernden dropped out. Cochran phone interview (cit. n. 45).
[90] Cochran, quoted in Flournoy, "A Controversial Excursion into Private Diplomacy" (cit. n. 2), 14.
[91] Velikhov interview (cit. n. 59).
[92] TASS, "Gorbachev, Scientists Discuss Issues at Forum," July 15, 1986, USSR International Affairs, Arms Control and Disarmament, Foreign Broadcasting Information Service (FBIS), AA 1-10.
[93] Ibid., AA 7.

adjacent to the Semipalatinsk and Nevada test sites. The six stations would be manned jointly by the two organizations, and the equipment would be obtained by the NRDC. The agreement specified that the project should begin before the end of June 1986.[94] After returning from Moscow, DeWind and Cochran walked into a meeting of the foundation community in New York and began their fund-raising for the project.[95]

Cochran gave numerous talks about the "Verification Breakthrough" with the Soviet Academy and placed a number of op-ed pieces in newspapers and journals.[96] Cochran and NRDC senior staff attorney S. Jacob Scherr informed leading newspaper editors, including Nicholas Wade of the *New York Times* and Stephen Rosenfeld of the *Washington Post,* about this "very significant citizen arms control initiative and a major breakthrough in American-Soviet relations."[97] Consequently, the project was covered prominently in major newspapers[98] and magazines[99] as well as on radio and TV shows.[100] As Scherr put it, "Certainly, we were buoyed by the tremendous response of the general public. We seem to have struck a chord in a population tired of waiting for arms control progress."[101]

The publicity clearly was an embarrassment for the Reagan administration. Tom Wicker blasted President Reagan on the editorial pages of the *New York Times* and criticized the president's unwillingness to engage in serious arms control negotiations with the Soviets. Wicker noted that while Reagan did not believe Gorbachev's offer to accept on-site monitoring, the NRDC project was underway, raising the question of whether Reagan was serious about arms control.[102]

From the beginning of the project, Cochran sought to involve U.S. government officials. He regarded the NRDC "as an 'honest broker' to stimulate agreement between the U.S. and the Soviet Union."[103] Cochran consulted with various State Department officials before initiating the project. In particular, he had met with Deputy Secretary of State John C. Whitehead before the trip to Moscow and again after his return.[104] Whitehead remained skeptical about the use of private groups to conduct foreign policy initiatives but noted that State had not decided how to respond to the NRDC initiative.

By the time the Department of Defense (DoD) officials learned about the project, it was too late to derail it. When Richard Perle, then assistant secretary of defense, learned in press reports about the NRDC–Soviet Academy interaction, he was in-

[94] "Agreement between NRDC and Soviet Academy of Sciences" (cit. n. 88).

[95] Cochran, personal communication, Oct. 11, 2005; "$500,000 Is Needed to Verify A-Tests," *New York Times,* June 4, 1986, A13.

[96] See, e.g., flier announcing talk by Cochran on Oct. 1, 1986, at the University of Chicago, titled "Clearing the Way for a Nuclear Test Ban: NRDC's Verification Breakthrough," Folder 2, Box 1, NRDC Files. For an op-ed piece, see Robert S. Norris and Thomas B. Cochran, "Why Not a Test Ban?" *Washington Post,* Sept. 9, 1986, A25.

[97] Cochran and Scherr, June 17, 1986, form letter to various editors, with list of editors attached, Folder 2 "Verification Project: Agreement and Draft Propos," Box 1, NRDC Files.

[98] The original agreement of May 28, 1986, was reported in the *New York Times* and many other national newspapers. See Taubman, "New Yorkers Sign Soviet Test Pact" (cit. n. 36).

[99] See, e.g., R. Jeffrey Smith, "Soviet, U.S. Scientists Reach Seismic Agreement," *Science* 232, June 13, 1986, 1338; Joseph Palca, "Private Diplomacy Emergent," *Nature* 321, June 12, 1986, 638.

[100] NRDC, Status Report, November 1986, 3, NRDC Files.

[101] Scherr cited in Lin, "Gaining Ground Zero" (cit. n. 2), 35.

[102] Tom Wicker, "A Dark New Identity," *New York Times,* June 6, 1986, A31.

[103] NRDC, Nuclear Test Ban Verification Project, Status Report, Nov. 1986, 2, NRDC Files.

[104] "$500,000 Is Needed to Verify A-Tests" (cit. n. 95); Schrag, *Listening for the Bomb* (cit. n. 2).

credulous and called it "an absurd private excursion."[105] From his point of view, nongovernmental groups had no business conducting such initiatives. He retaliated by preventing Soviet scientists from coming to set up seismic stations in the United States.[106] Other DoD nuclear test supporters rejected the project as an NRDC publicity stunt and Soviet propaganda. For example, Frank J. Gaffney Jr., a deputy assistant secretary of defense for nuclear forces and arms control policy, defended the administration's nuclear testing policy and criticized the NRDC project in a widely cited conclusion: "We are under no illusion as to the mischievous and counterproductive purpose that Soviet authorities hope the NRDC experiment will serve: to confuse the domestic debate about the need for American nuclear testing and the reasons why we oppose the effort of the Soviet Union and others to promote an inequitable and unverifiable ban on nuclear testing."[107] In general, however, government officials treated the project in various ways, ranging from assistance (as in the case of the expedited export controls) to indifference to open hostility.[108] The media savvy NRDC essentially outplayed its critics in the DoD.

Cochran and others at NRDC were able to raise more than $500,000 dollars in a matter of weeks, primarily from foundations such as Carnegie, Ford, and MacArthur. In addition, NRDC succeeded in getting the necessary export licenses in a record time of four business days.[109] Besides fund-raising and applying for permits, Cochran had to assemble a team of U.S. seismologists to set up the stations in Kazakhstan. He left the job of formulating a coherent seismic detection and research program to Charles Archambeau, an accomplished seismologist from the University of Colorado–Boulder, who had advised the DoD on nuclear test detection issues since the 1960s.[110] In the 1970s, Archambeau published articles in geophysical journals that highlighted new seismic detection techniques, and he became an outspoken test ban supporter.[111] Archambeau took over the technical direction of the project and recruited James N. Brune and Jonathan Berger, two accomplished geophysicists from the Scripps Institution of Oceanography in La Jolla, California.[112] Brune was uniquely qualified since he had operated a network of portable seismic stations in the Soviet Union in cooperation with Soviet scientists from 1974 to 1980.[113] Berger, an instrumentation specialist, essentially ran the NRDC field operation, with the help of University of Nevada, Reno (UNR), and University of California, San Diego (UCSD) graduate students.[114]

[105] David J. Lynch, "SALT Fallout Rains on Reagan," *Defense Week,* June 9, 1986, 8.

[106] See Schrag, *Listening for the Bomb* (cit. n. 2).

[107] Gaffney, "Test Ban" (cit. n. 5). Gaffney's statement is cited, for example, in Edgar Ulsamer, "Nuclear Test Delusions," *Air Force Magazine* (Oct. 1986): 16.

[108] Schrag, *Listening for the Bomb* (cit. n. 2).

[109] For an analysis of why the Department of Commerce granted the export licenses in an expedited way, see ibid.

[110] Charles Archambeau, interview with author, Boulder, Colo., June 18, 1998.

[111] On Archambeau, see Jeannine Malmsbury, "Surface Tension: Nuclear Verification No Longer an Underground Issue," *Summit Magazine* (Winter 1987): 22–3.

[112] Both Brune and Berger worked at the Institute of Geophysics and Planetary Physics of the Scripps Institution of Oceanography at the University of California in San Diego. See Berger and Brune's research proposal from June 10, 1986, for "Seismic Studies of Nuclear Test Sites: Part I—Semipalatinsk," in which the two scientists apply for $416,279, Folder 2 "Verification Project: Agreement and Draft Propos," Box 1, NRDC Files.

[113] Ibid.

[114] Courtney Brenn, "Soviets in Reno to Discuss Data from Nuke Tests," *Reno Gazette-Journal,* Dec. 2, 1987, 1.

The participating Soviet geophysicists came from the Soviet Academy's Institute of Physics of the Earth (IPE) in Moscow, directed by Mikhail A. Sadovsky, who was one of the leading Soviet nuclear test detection experts.[115] Mikhail Gokhberg, IPE's deputy director, was responsible for the overall management, and Igor Nersesov, IPE's chief seismologist, headed the Soviet field team.[116]

In early July, the seismologists arrived in Moscow, but at that time they had no formal permission beyond Velikhov's word to install the stations. However, Velikhov's good relationship with Gorbachev and his ties to local party bosses paid off. In a meeting on July 3, 1986, in which Politburo members discussed the consequences of Chernobyl, Velikhov brought up the permission for the NRDC.[117] To his dismay, former ambassador to the United States Anatoly Dobrynin and Central Committee member in charge of defense industries Lev Zaikov, who had assured him of their support, voted against the NRDC plan and requested more discussion of the matter.[118] However, since the American scientists were already in Moscow, Velikhov asked Gorbachev directly about what to do. Again, Gorbachev gave Velikhov the freedom to decide, and Velikhov essentially accepted responsibility for the project.[119] This unusual permission did earn him enemies, and he remembers that a Politburo member called him a traitor for allowing "U.S. intelligence equipment" close to the Soviet test site.[120]

As Michèle Flournoy has pointed out, the debate within the Soviet administration focused on two questions: whether the NRDC project would lead to an extension of the Soviet moratorium, and whether NRDC scientists would be permitted to take data in case the Soviets returned to testing.[121] In a memorandum signed by Cochran and Velikhov on July 5, both agreed that "the recording of the tests of nuclear weapons is not necessary to the success of the joint research being undertaken."[122] In addition, both sides limited the original agreement for the duration of one year. In other words, this additional memorandum addressed the two concerns expressed by critics in the Soviet government. After reaching agreement on July 5, the NRDC was allowed to continue to Kazakhstan.

AMERICANS IN KAZAKHSTAN

On July 9, 1986, Tom Cochran, James Brune, and their Soviet colleagues[123] flew in small biplanes to Karkaralinsk, Kazakhstan, about 1,800 miles southeast of Moscow. In the following weeks, the NRDC team and their Soviet colleagues set up three seismic stations, each about 120 miles from the Soviet test site. They established the first

[115] On Sadovsky, see Kai-Henrik Barth, "Detecting the Cold War: Seismology and Nuclear Weapons Testing, 1945–1970" (Ph.D. diss., Univ. of Minnesota, July 2000), 136.

[116] Cochran, "The NRDC/Soviet Academy of Sciences Joint Nuclear Test Ban Verification Project" (cit. n. 2), 5.

[117] Velikhov interview (cit. n. 59). While Velikhov did not remember the exact date of the meeting, his description of the meeting's content fits the analysis given by Anatoly Chernyaev, Gorbachev's foreign policy adviser. See Chernyaev, *My Six Years with Gorbachev* (cit. n. 74), 66.

[118] Velikhov interview (cit. n. 59).

[119] Ibid.

[120] Ibid.

[121] Flournoy, "A Controversial Excursion into Private Diplomacy" (cit. n. 2), 15.

[122] The memorandum is quoted ibid., 15.

[123] The Soviet team included Igor Nersesov, corresponding member of the Armenian Academy of Sciences, who was in charge of the experiment on the Soviet side; Mikhail Gokhberg, acting director of the Institute of Physics of the Earth of the Soviet Academy; and Institute of Physics of the Earth research associates Sergei Daragan, Oleg Stolyarov, Evgeny Sutlov, and Nikolai Tarassov.

station near Karkaralinsk, where the Americans stayed in trailers, nicknamed "Soviet Winnebagos."[124] The other two stations, Bayanaul and Karasu, were operational by the end of August 1986 and recorded earthquake signals, which allowed a calibration of the geologic setting.[125] This calibration, in turn, was necessary to determine the yield of nuclear explosions in this particular geologic formation. With these measurements, NRDC and the Soviet Academy claimed to "[obtain] valuable information about the geological characteristics of the Soviet nuclear test site, data previously unavailable in the West."[126] The seismic data seemed to confirm the geologic differences between the Nevada and the Semipalatinsk test sites, which led to very different seismic amplitudes for the same nuclear yields. As University of Nevada geophysicist Keith Priestley put it, "If two tests, one at Nevada and one at Semipalatinsk, are identical in size, the amplitude of the Kazakhstan test will show up twice as large."[127]

These results had direct consequences for the controversy about Soviet compliance with the 150-kiloton yield limit of the Threshold Test Ban Treaty. U.S. officials had charged that the Soviets had repeatedly violated the threshold and had tested weapons of higher yields, a violation with potential military significance. However, accusations of TTBT violations depended on the assumption that the geology around the Semipalatinsk test site was well enough understood to allow seismologists to determine the yield of a nuclear test based on the seismic waves generated by such an explosion. The NRDC measurements confirmed, as Thomas Cochran pointed out, that "preliminary analysis of our data, incidentally, is consistent with Soviet compliance [with the Threshold Test Ban Treaty]."[128]

The scientists recorded seismic signals from small earthquakes and mining explosions at a strip mine sixteen miles from Karkaralinsk[129] as well as signals from nuclear explosions at the Nevada Test Site, thousands of miles away. However, at first they did not record any Soviet nuclear tests, since the Soviet moratorium continued until February 26, 1987.[130]

Why did the Soviets end the moratorium they had begun on August 6, 1985, the fortieth anniversary of Hiroshima? According to Soviet Foreign Ministry spokesman Gennady I. Gerasimov, speaking on ABC News, Soviet nuclear tests would resume "because we waited, waited and waited, we waited for 560 days and 25 of your explosions. So our military people are saying to our politicians: 'Look, we must do something because otherwise we'll be behind.' It's that simple."[131]

In a letter to Velikhov on January 23, 1987, Cochran wrote: "I cannot overemphasis [sic] how disastrous it would be from a political point of view if we are required to turn off our equipment during a Soviet test."[132] He continued that

[124] Bob Guldin, "Glasnost Prevails among Pioneers of Nuclear Monitoring," *Guardian,* Sept. 23, 1987, 10–1. American and Soviet seismologists stayed in Shakhtyor, a small miners' resort four miles from Karkaralinsk.

[125] NRDC, Status Report, May 1987, 1, NRDC Files.

[126] NRDC, Nuclear Test Ban Verification Project, Status Report, Feb. 1987, 1, NRDC Files.

[127] Keith Priestley, cited in Guldin, "Glasnost Prevails among Pioneers" (cit. n. 124), 10–1.

[128] Thomas Cochran, cited ibid.

[129] Alexander Grigoryev, "U.S. Seismologists in Kazakhstan," *Soviet Life* (Oct. 1986): 2–3.

[130] On Feb. 5, 1987, the Soviets announced that the moratorium had officially ended, since the United States had continued to test. "Comprehensive Test Ban, Conference on Disarmament, Chronology 1987," *Arms Control Reporter,* sec. 608.B.128.

[131] Gerasimov, quoted in "Soviets Confirm Move for Nuclear Tests," *Los Angeles Times,* Feb. 23, 1987, 5.

[132] Cochran to Velikhov, Jan. 23, 1987, Folder 2, Box 1, NRDC Files.

> if we are not permitted to listen to tests the public perception will be that 1) the project won't work; 2) the Soviet Union has something to hide; 3) the project was all a propaganda stunt; 4) now that you are resuming testing, it's back to business as usual; and 5) the Soviets have given up on a [Comprehensive Test Ban] and are abandoning those of us who are continuing to fight in Congress this year for a moratorium on U.S. nuclear testing.[133]

Cochran warned that essentially the credibility of perestroika was at stake: "It will be asserted by Administration officials and others that oppose a CTB that the Soviets were never really serious about verification and were merely using the American scientists." He then appealed to his counterpart that "if we are not permitted to monitor tests, our chances of getting a moratorium adopted by Congress would be jeopardized."[134] Finally, he pointed out that future support from U.S. foundations and individuals was at stake. Cochran urged Velikhov to make arrangements so that Cochran and Archambeau "could meet with appropriate people in your government in Moscow on February 12 or 13 to discuss this issue further."[135] However, on February 21, 1987, Soviet officials ordered the NRDC instruments to be shut off.[136] According to NRDC attorney Jacob Scherr, the scientists were told that they "had to stop work for three days and possibly longer."[137] On February 26, the Soviet Union tested its first device since July 25, 1985.[138]

Cochran appealed directly to Gorbachev at a Kremlin reception for participants of the Moscow "International Forum for a Nuclear-Free World, for the Survival of Humanity" and met with former Soviet ambassador to the United States Anatoly Dobrynin, who was now the Communist Party secretary for foreign affairs.[139] Cochran was not successful. From February to April, the Soviets exploded eleven nuclear devices and ordered the NRDC stations to be shut down a few days before each test or test series.[140] This setback destroyed hopes that the NRDC would be allowed to measure a Soviet nuclear explosion yield at a close distance and disappointed U.S. participants.[141] The participating Soviet seismologists were not happy about this restriction either, since they had to turn off their equipment as well.[142] NRDC officials argued, however, that "the scientific objectives of the Project are not compromised by this limitation, which we have, nonetheless, pressed Soviet officials to remove. The primary purpose of the Project is to demonstrate technology to verify the *absence* of nuclear tests."[143]

Despite this setback, in June 1987 the NRDC signed a second agreement with the Soviet Academy, extending the project for another fourteen months. This agreement determined that the three stations in Kazakhstan would operate until December 1987,

[133] Ibid.

[134] Ibid.

[135] Ibid.

[136] Bill Keller, "Soviet Advance Notice of A-Blast Likely," *New York Times,* Feb. 22, 1987, A20.

[137] "Russia Planning Nuclear Test, U.S. Group Says," *San Francisco Examiner,* Feb. 21, 1987, 1.

[138] For information about the Soviet nuclear test program and a list of Soviet nuclear tests, see Pavel Podvig, ed., *Russian Strategic Nuclear Forces* (Cambridge, Mass., 2001), 439–566.

[139] Keller, "Soviet Advance Notice of A-Blast Likely" (cit. n. 136), A20.

[140] NRDC received orders on Feb. 21, March 9, March 25, and April 12, 1987. See Warren Strobel and James M. Dorsey, "Soviets Expected to Conduct Nuclear Test by End of Week," *Washington Times,* April 15, 1987, 3A. The Soviets conducted tests at Semipalatinsk on Feb. 26, March 12, April 3, and April 17. See Podvig, *Russian Strategic Nuclear Forces* (cit. n. 138), 558–9.

[141] S. Jacob Scherr, "Karkaralinsk Diary," *Amicus Journal* (Spring 1987): 4–7.

[142] Guldin, "Glasnost Prevails among Pioneers" (cit. n. 124), 10–1.

[143] NRDC, Status Report, May 1987, 1, NRDC Files (emphasis in original).

then relocate to greater distances from Semipalatinsk (from 150 miles to 600 miles). After the relocation, the stations were permitted to record Soviet nuclear tests, beginning in 1988. Subsequently, on February 6, 1988, American scientists took their first measurements of Soviet nuclear tests on Soviet soil with U.S.-made seismic instruments.

The NRDC experience in Kazakhstan highlighted that Soviet officials were divided in their views about the project: while the Soviet Academy and in particular Velikhov regarded the project as a unique chance to put pressure on the Reagan administration, opponents of glasnost in the military objected and emphasized security concerns. While Gorbachev and Velikhov won the battle and made the implementation of American stations possible, hardliners put up roadblocks where they could. For example, Soviet officials would not release the exact coordinates of the three monitoring stations, which the U.S. seismologists needed to accurately measure seismic wave properties from the test site.[144] Again, as in the case of equipment shut down during actual nuclear tests, security precautions won over glasnost.

SOVIETS IN NEVADA

Restrictions placed on NRDC scientists in the Soviet Union were minor compared with what awaited Soviet Academy scientists in the United States. The NRDC–Soviet Academy agreement was reciprocal, and both sides understood that Soviet scientists would set up similar seismic monitoring stations in the United States. Despite significant opposition from within the U.S. administration, a Soviet delegation of scientists made it to Nevada, and they eventually established monitoring sites around the Nevada Test Site. Philip G. Schrag has analyzed in detail how the bureaucratic fights within the U.S. administration, especially between officials in the Department of Defense and the State Department, led to significant delays in the granting of visas, permits, and licenses for the Soviets and their equipment.[145] Decisions often depended on second- or third-tier agency officials, who had the power to expedite, hinder, or simply ignore NRDC's requests for permits and licenses necessary to establish the Soviet stations. Here I am concerned not with the reasons for these delays but with the actual establishment of Soviet monitoring stations on U.S. soil and the collaborative effort of the transnational group of scientists.

In November 1986, seismologists from the Institute of the Physics of the Earth of the Soviet Academy were scheduled to arrive in the United States to select sites for monitoring stations around the Nevada Test Site. The Soviets arrived in the United States, but visa restrictions prevented them from visiting potential monitoring sites. NRDC circumvented this roadblock by providing the Soviets with data and rock samples from these sites. The Reagan administration placed unusual restrictions on the visas for the Soviet scientists. They could choose between two options: either observe a nuclear test at Nevada and a demonstration of a new nuclear yield measurement favored by the Reagan administration, and be allowed to go to the potential monitoring sites; or be restricted to a visit to Scripps Institution and receive no permission to visit potential monitoring sites. As the NRDC pointed out, "In effect, the Administration is trying to compel the Soviets to accept their top agenda item on nuclear testing, which

[144] Guldin, "Glasnost Prevails among Pioneers" (cit. n. 124), 10–1.
[145] Schrag, *Listening for the Bomb* (cit. n. 2).

involves an agreement on improved verification as a basis for ratification of the 1974 Threshold Test Ban Treaty."[146] The Soviets objected and consequently received no permission to visit the potential monitoring sites in November 1986.[147]

In October 1987, scientists from the University of Nevada, Reno, and the Scripps Institution of Oceanography began to install state-of-the-art seismic equipment for the Soviet monitoring stations. The three stations were located in the remote, "dry, boulder-strewn foothills of the Last Chance Range" near Deep Springs Valley, California, and two equally remote places in Nevada, at Nelson Valley south of Las Vegas, and Railroad Valley, 150 miles east of Tonopah.[148] In early December 1987, four Soviet scientists arrived in Reno, Nevada, greeted by thirty cheering peace activists. Felix Tregub, Mikhail Gokhberg, Nicolai Yukhnin, and Igor Nersesov met with seismologists Keith Priestley and James Brune of the UNR to discuss the seismic data from U.S. stations in Kazakhstan and the Soviet stations in Nevada and California.[149] However, the Soviets were still not permitted to visit the monitoring sites. By January 1988, the permanent stations were operational, and in April the Soviet seismologists detonated three chemical explosions to measure the propagation and attenuation of seismic signals in the vicinity of the Nevada Test Site.[150] These measurements demonstrated that the attenuation near the NTS was significantly larger than the attenuation near Semipalatinsk, an important result to counter charges of Soviet Threshold Treaty violations.

Overall, the Reagan administration placed visa restrictions on the Soviet scientists and prevented them for months from visiting the NRDC–Soviet Academy stations in California and Nevada. This led to the paradoxical situation that American scientists had greater degrees of freedom traveling in the Soviet Union than Soviet scientists did traveling in the United States.

EPILOGUE: SCIENTISTS AS TRANSNATIONAL ARMS CONTROL ADVOCATES

NRDC was active in the seismic verification business only from May 1986 to April 1989. Cochran and his colleagues never intended to become seismic monitoring specialists. Rather, they wanted to change the test ban debate, then hand over the stations to another agency. Ultimately, the five new stations in the Soviet Union continued to operate as part of a worldwide seismic detection network managed by the Incorporated Research Institutions for Seismology (IRIS), a nongovernmental consortium of research universities. The Moscow Institute of Physics of the Earth owned the stations and operated them with Scripps Institution.[151] While Cochran and his group left the

[146] NRDC, Status Report, Nov. 1986, 3, Folder 4, Box 1, NRDC Files.

[147] Ibid.

[148] Kristine Moe, "Scripps Scientists Install Devices to Monitor U.S. N-tests for Soviets," *San Diego Union,* Oct. 20, 1987, A3. For the Nevada stations, see Courtney Brenn, "Soviets in Reno to Discuss Data from Nuke Tests" (cit. n. 114), 1.

[149] Brenn, "Soviets in Reno to Discuss Data from Nuke Tests" (cit. n. 114), 1.

[150] NRDC, Update Sept. 1988, 2, Folder 4, Box 1, NRDC Files. The explosions included two ten-ton and one fourteen-ton chemical explosion. Cochran, NRDC/Soviet Academy of Sciences, Nuclear Test Ban Verification Project, 1, Folder 3, Box 1, NRDC Files.

[151] NRDC ended its involvement with the original NRDC–Soviet Academy stations in late 1988. See Flournoy, "A Controversial Excursion into Private Diplomacy" (cit. n. 2), 20. NRDC "extended its agreement with the Academy until 1 April 1989 to provide for the in-country seismic measurements of the two CORRTEX related nuclear tests." Cochran, "NRDC/Soviet Academy of Sciences, Nuclear Test Ban Verification Project," 2, Folder 3, Box 1, NRDC Files.

seismology project in 1989, they continued, often in collaboration with Velikhov, to conduct equally provocative scientific-political experiments, including unprecedented nuclear warhead detection measurements on a Soviet cruiser[152] and the surprising permission to visit one of the most secret Soviet installations, the Krasnojarsk radar.[153]

The NRDC project accomplished what U.S. governments from Eisenhower to Carter had attempted, but failed, to do: to put American scientists on Soviet soil to monitor Soviet explosions and gain data of the particular geologic environment of the test site. These data were useful in determining if the Soviets had, in fact, violated the TTBT, as the Reagan administration had alleged. To get seismic data from a Soviet nuclear test area was unprecedented.

The transnational effort between scientists from East and West challenged the positions of hardliners within the Soviet Union as well as the U.S. administration: for Soviet hardliners, especially in the Ministry of Defense, any suggestion that American scientists be allowed to take measurements close to the Soviet nuclear test site was utterly unacceptable;[154] for U.S. hardliners, who were convinced that nuclear testing had to continue to develop the next generation of nuclear warheads, any initiative that strengthened arguments for a comprehensive nuclear test ban were equally abhorrent.[155]

The project turned out to be a public relations coup for the NRDC as well as for the Soviets. For the Reagan administration, it proved to be a serious embarrassment, since the NRDC, with a $7 million annual budget, essentially outmaneuvered the Department of Defense on a question of importance for national security. This transnational collaboration between scientists from East and West was widely covered by the media, discussed, supported, and often attacked by policy makers in the United States and the Soviet Union. It changed Soviet foreign policy and U.S. domestic debates about a comprehensive nuclear test ban, because it demonstrated that test ban verification was possible, that Soviet leaders had changed their minds about on-site inspections, and that U.S. administration arguments against a nuclear test ban were disingenuous.

How did the involved scientists expect to shape international relations? What did they hope science and scientists could contribute to the arms control debate? Velikhov, for example, highlighted the good personal relationships he had developed with U.S. plasma physicists since the 1960s, with whom he shared a professional interest in nuclear fusion and magnetohydrodynamics. The professional relationships often led to trust, which then made frank political discussions possible.[156] Such open exchanges, in turn, could lead to new ideas, such as the seismic verification experiment, which some of the politically well-connected physicists could then present to leading policy makers.

FAS director Frank von Hippel, a leading public interest scientist, argued that physicists have an ability to reach agreements, even when they come from very different political perspectives.[157] Velikhov's group, the NRDC, and the FAS came, as I argued,

[152] See Steve Fetter, Thomas B. Cochran, Lee Grodzins et al., "Gamma-Ray Measurements of a Soviet Cruise-Missile Warhead," *Science* 248 (1990): 828–34.

[153] On the Krasnojarsk radar visit, see William J. Broad, "Inside a Key Russian Radar Site: Tour Raises Questions on Treaty," *New York Times,* Sept. 7, 1987, 1; Evangelista, *Unarmed Forces* (cit. n. 2), 325–8.

[154] Velikhov interview (cit. n. 59).

[155] Gaffney, "Test Ban" (cit. n. 5).

[156] Velikhov interview (cit. n. 59).

[157] Frank von Hippel, interview with author, Princeton Univ., Feb. 20, 2004.

from quite comparable perspectives: as members of the same epistemic community of transnational antinuclear scientists they shared a strong sense of social responsibility to develop technical-political solutions for arms control problems.

Returning to the original question that motivated this study: Why were nongovernmental scientists successful in shaping Soviet foreign policy? What kind of factors contributed to their success? Arguably, the most important reason for the scientists' success was the structure of Soviet society and the emergence of a new leader, one willing to listen to new ideas. As Matthew Evangelista has shown, access to Gorbachev turned out to be the way to shape Soviet foreign policy. Velikhov worked directly with Gorbachev in the early years of his restructuring of the Soviet political system. Velikhov accompanied Gorbachev as adviser to the 1986 Reykjavik summit meeting with President Reagan. When political professionals began to restrict access to Gorbachev about 1987 and Velikhov's influence receded, the window of opportunity closed.[158]

Domestic structure, however, is a necessary factor, not a sufficient one. Other factors played a significant role as well. I argued that we have to understand the communication strategies of the scientists who sought to decrease the risk of nuclear war by opening backchannel communications with their counterparts. The point of this paper is that we find a whole spectrum of approaches among the scientists. Both Cochran and Velikhov were at heart political players with a keen sense of timing, masters of public relations, and pragmatists. Both were driven by concerns about the dangers of an extended nuclear arms race.

Why were these scientists able to reduce the dangers of nuclear war, when diplomats had failed for so long? Did scientists succeed because they found a common language that transcended national and ideological boundaries? I argued that these scientists shared the belief in the responsibility of scientists and the desirability and technical capability of technical solutions to arms control problems. I also argued that the most significant factor for the scientists' success was the collaboration of activist-minded scientists such as Cochran and Velikhov with scientists such as von Hippel and Archambeau. Cochran had a brilliant idea and the stamina to pull it through, but of course, he depended on the scientific expertise of geophysicists such as Charles Archambeau, James Brune, and Jonathan Berger.

In sum, it took more than a few well-trained scientists to shape Soviet foreign policy. Other factors included the particular Soviet domestic structure; the existence of a large peace movement in Europe and the freeze movement in the United States; the Gorbachev factor; the existence of professional full-time public interest physicists such as von Hippel and Cochran; and finally, the media savvy scientists who became political players, such as Velikhov. Beyond all these necessary factors, I have argued, we need to take the coherence of the transnational network of arms control scientists into account: scientists such as Cochran, Velikhov, Sagdeev, Archambeau, and von Hippel shared the belief that communication channels between scientists could lead to results when government-to-government negotiations had failed; they believed that a comprehensive test ban treaty was not only desirable but also technically verifiable. These scientists, constituting an epistemic community of transnational arms control advocates, shaped Soviet foreign policy and thereby became catalysts of change.

[158] According to von Hippel, Roald Sagdeev complained in July 1987 that professionals from the Soviet Ministry of Foreign Affairs increasingly controlled access to Gorbachev and forced Sagdeev and Velikhov out of the inner circle. Von Hippel interview (cit. n. 157).

SCIENCE, TECHNOLOGY, AND GLOBALIZATION

Hallowed Lords of the Sea:
Scientific Authority and Radioactive Waste in the United States, Britain, and France

*By Jacob Darwin Hamblin**

ABSTRACT

In 1959, oceanographers and atomic energy officials met at an international conference in Monaco to discuss the scientific aspects of dumping radioactive waste into the ocean. The result was a broad consensus among oceanographers that there was not enough scientific knowledge of the oceans to merit large-scale dumping. Because nuclear nations already had been dumping for years, the new consensus challenged existing practices. This paper focuses on the conflicts between oceanographers and the atomic energy establishments of the United States, Britain, and France. It reveals the perception, shared by atomic energy officials in all three countries, that oceanographers manipulated public and international opinion to seize authority, influence, and potential patronage for research on oceanography. While historians often debate the consequences of government (usually military) funding on scientists' agendas and practices, few address the impacts that international consensus and scientists' patronage strategies have had upon the policies and status of patrons. This paper reveals a siege mentality within governments and shows the birth of international collaboration between atomic energy establishments as a means of combating the negative publicity caused, in their view, by oceanographers seeking influence and financial support. The international debates about radioactive waste disposal, seen as a contest for scientific authority, highlight the reciprocal influences of patronage practices during the cold war era.

INTRODUCTION

The Wise Man says
That only those who bear the nation's shame
Are fit to be its hallowed lords
—John Isaacs (1959), quoting Lao Tzu[1]

* Department of History, Clemson University, 126 Hardin Hall, Box 340527, Clemson, SC 29634; jdhamb@yahoo.com.

Special thanks to Odile Frossard for her help in the archives of the Commissariat à l'Énergie Atomique and to Janice Goldblum for her help in the National Academies Archives. This article is based on archival research financially supported by the American Philosophical Society through its Franklin Research Grant. Unless otherwise noted, translations into English from French are the author's.

[1] John D. Isaacs to Pacific Coast Committee, Dec. 21, 1959, Folder "ES: Com on Ocean: Subc on Disposal of Low-Level Rad Waste into Pacific Coastal Waters: Report: Review, 1959," National Academies Archives, Washington, D.C. (hereafter cited as NAS Archives).

Perhaps sensing a heavy burden of responsibility, U.S. oceanographer John Isaacs reached deep into the cultural past for inspiration, drawing on ancient Chinese philosophy to conjure images of wise men, national shame, and hallowed lords. At the time, he chaired a National Academy of Sciences (NAS) panel to report on the biological effects of dumping radioactive waste into the Pacific. He knew that his panel's conclusions, warning about the dangers of existing practices in radioactive waste disposal, would be at odds with the statements of atomic energy establishments in the United States, Britain, and France. They also would cast doubt on what leading oceanographers themselves had been saying for most of the decade—that the sea could be considered a giant sewer.[2] The Lao Tzu passage itself, only one of several such quotations Isaacs sent to his fellow panelists, stands as a provocative encapsulation of a political attitude: those who acknowledge a nation's past sins have a moral claim to authority and leadership. In the present essay, we can extend this lordship to the sea, to examine the international issue of contested scientific authority between oceanographers and atomic energy establishments.

Isaacs's panel had its roots in a conflict about ocean dumping between the federal government and the state of California, but its conclusions spoke to questions of worldwide importance: Was it safe to dump radioactive waste at sea? If so, where and how much? By the mid-1950s, the nuclear powers had been dumping radioactive waste into sewers, rivers, and oceans for years without much international conflict. But the 1958 United Nations Conference on the Law of the Sea had declared itself against ocean pollution. It charged the newly created International Atomic Energy Agency with the task of developing appropriate regulations for radioactive waste, and international scientific bodies seemed poised to play a leading role in authoring them.[3] In late 1959, delegations made up of oceanographers and atomic energy officials met in Monaco to discuss the important scientific problems related to radioactive waste disposal. Although it was not supposed to be a diplomatic meeting, it resulted in a broad acceptance of many of the same conclusions being made by Isaacs and his committee, with a consensus among the oceanographers that more research was needed before nuclear powers could dump radioactive waste at sea on a large scale. Because such dumping policies were already in effect, the Monaco meeting implicitly challenged existing practices. Scientists, apparently, had become diplomats.

Historians of science have taken up the story of scientists engaged in international affairs, revealing some of the ways in which they tried to balance the professional demands of science with their need to act as part of the cold war national security state.[4] At the international level, scientific communities advised diplomats, helped to negotiate regulations, and established scientific problems requiring international cooperation. Certainly, historians agree that many international activities could, in fact, be

[2] *Disposal of Low-Level Radioactive Waste into Pacific Coastal Waters* (Washington, D.C., 1962).

[3] For overviews of the IAEA, see David Fisher, *History of the International Atomic Energy Agency: The First Forty Years* (Vienna, 1997); and Lawrence Scheinman, *The International Atomic Energy Agency and World Nuclear Order* (Baltimore, 1987).

[4] Ronald Doel and Allan Needell have written extensively about these connections, exploring the ways in which scientists were incorporated into diplomacy and national security. Ronald E. Doel, "Scientists as Policymakers, Advisors, and Intelligence Agents: Linking Contemporary Diplomatic History with the History of Contemporary Science," in *The Historiography of Contemporary Science and Technology,* ed. Thomas Söderqvist (Amsterdam, 1993), 215–44; Allan A. Needell, *Science, Cold War, and the American State: Lloyd V. Berkner and the Balance of Professional Ideals* (Amsterdam, 2000).

construed as scientists acting in the service of their respective states.[5] However, scientists were not always simply co-opted into state goals, so we must also examine the array of other motivations at the international level. Like any other group tied to governmental politics during the cold war era, scientists might have had entirely separate interests to pursue.[6]

One of these interests undoubtedly was patronage, especially given the power of international scientific consensus to justify or even compel it. An ongoing dispute among historians is the question of how military patronage altered the practice of science and the research agendas of scientists.[7] With our focus on the consequences of military funding, however, we risk seeing the patron as the sole source of pressure; it is easy to forget that scientists often courted their patrons, and they did so vigorously. Oceanographers, in particular, were hugely successful in convincing a variety of patrons that research on the oceans was worth a large grant or two.[8] In many ways, their pursuit of funding fits our conception of science in the service of the cold war state; for the military, it often was a relatively simple relationship in which one side wanted science and technology, and the other provided the necessary expertise.[9] But once we look at the variety of other situations, in which expertise was claimed by both the patron and the potential recipient of funds, a more complex portrait of national and international science emerges. In the case of radioactive waste, the international consensus

[5] See, e.g., the growing literature on the International Geophysical Year (1957–1958), in which historians increasingly are pointing out the diplomatic, national, and strategic motivations behind cooperative activities. Aant Elzinga, "Antarctica: The Construction of a Continent by and for Science," in *Denationalizing Science: The Contexts of International Scientific Practice,* ed. Elisabeth Crawford, Terry Shinn, and Sverker Sörlin (Dordrecht, 1992), 73–106; Ronald E. Doel, "Constituting the Postwar Earth Sciences: The Military's Influence on the Environmental Sciences in the USA after 1945," *Social Studies of Science* 33 (2003): 635–66; Jacob Darwin Hamblin, "Mastery of Landscapes and Seascapes: Science at the Strategic Poles during the International Geophysical Year," in *Extremes: Oceanography's Adventures at the Poles,* ed. Helen M. Rozwadowski (Cambridge, forthcoming).

[6] Kristine Harper has pointed this out in her discussion of the competing national and international goals in meteorology. Kristine C. Harper, "Research from the Boundary Layer: Civilian Leadership, Military Funding, and the Development of Numerical Weather Prediction (1946–55)," *Soc. Stud. Sci.* 33 (2003): 667–96.

[7] For representative publications taking different sides on this issues, see Paul Forman, "Behind Quantum Electronics: National Security as Basis for Physical Research in the United States, 1940–1960," *Historical Studies in the Physical and Biological Sciences* 18 (1987): 149–229; Daniel J. Kevles, "Cold War and Hot Physics: Science, Security, and the American State, 1945–1956," *Hist. Stud. Phys. Biol. Sci.* 20 (1990): 239–64.

[8] Oceanographers' changing patronage strategies are discussed in Jacob Darwin Hamblin, *Oceanographers and the Cold War: Disciples of Marine Science* (Seattle, 2005). The military (naval) patronage for oceanography in the United States is discussed at length in Gary E. Weir, *An Ocean in Common: American Naval Officers, Scientists, and the Ocean Environment* (College Station, Texas, 2001). Efforts to court naval patronage at the Scripps Institution of Oceanography are discussed in Ronald Rainger, "Patronage and Science: Roger Revelle, the U.S. Navy, and Oceanography at the Scripps Institution," *Earth Sciences History* 19 (2000): 58–89. Naomi Oreskes has pointed out that some ostensibly nonmilitary projects have had their military origins erased or forgotten, as in the case of the *Alvin* submersible craft; rather than painting their projects blue, that is, making basic research seem like applied military research to attract funding, oceanographers often painted their projects white, obscuring the military origins of their research. See Naomi Oreskes, "A Context for Motivation: U.S. Navy Oceanographic Research and the Discovery of Sea-Floor Hydrothermal Vents," *Soc. Stud. Sci.* 33 (2003): 697–742.

[9] Much of the discussion about naval patronage of oceanography has centered on the use of scientists as experts. Chandra Mukerji, for example, went so far as to call oceanographers a reserve labor force, ready to provide the military with much-needed expertise at any time. Mukerji, *A Fragile Power: Scientists and the State* (Princeton, 1990).

of oceanographers at Monaco went *against* the wishes of the authorities in their own countries. Emphasizing the need for research (certainly a typical patronage strategy) called into question existing policies. It also implied that oceanographers were better suited as authoritative experts on marine radioactivity than the scientists already working for atomic energy establishments.

Because we often see the problem of scientific authority as a way of understanding the combative relationship between environmentalists and bureaucracies such as the AEC, we might overlook other turf wars played out on an international scale.[10] One of these was waged between established scientists in government and (what they perceived as) the opportunistic nongovernmental scientists using public opinion to grasp power and money by asserting scientific authority. The relationship between these groups was often collaborative, to be sure, but it could be confrontational as well—perhaps surprisingly so, given the desire of one to solicit funding from the other. The present chapter centers on the dilemmas of three of the four major atomic energy establishments of the late 1950s: the U.S. Atomic Energy Commission (AEC), the UK Atomic Energy Authority (AEA), and the French Commissariat à l'Énergie Atomique (CEA). Leaving out the Soviet Union (except in passing) allows us to focus on the crucial link, or battleground, between oceanographers and the atomic energy establishments: democratically elected public officials and the lay public at large. It also helps to reveal the emergence of an international community separate from the one created by academic oceanographers, in the form of interagency liaison between major figures in all three establishments, across national lines. They created such links, formally or informally, to contest oceanographers' claims to authority and to find ways of bringing oceanographers back into line with established assumptions about the role of the ocean in waste disposal. Even when the three establishments were not acting in concert, their concerns were strikingly homogeneous, centered on the perception of oceanographers not as accomplices but as opportunistic adversaries. Were the oceanographers acting in the interests of the state? Certainly, atomic energy officials did not think so; instead, they saw the oceanographers as acting in their own, self-aggrandizing interest. Atomic energy officials perceived a struggle: for scientific authority about how the ocean could or could not be used—or, to continue with Isaacs's imagery, for hallowed lordship of the sea.

THE SEARCH FOR WASTE SITES

The first serious assessments by oceanographers of radioactive waste disposal at sea came about because of the fallout debates in the mid-1950s. To address public concern, the National Academy of Sciences conducted a major study of the biological effects of atomic radiation (the BEAR study) in 1956. One of the panels, chaired by Scripps Institution of Oceanography director Roger Revelle, was devoted to the oceans. Although it acknowledged the uncertainty of knowledge of the seas, this group confidently agreed with existing policies that the ocean could be used to some degree as a repository for nuclear waste.[11] The United Nations made a similar assess-

[10] Stephen Bocking recently has emphasized the importance to environmentalists of the uncertainty of environmental effects, because it allows them to call into question the scientific authority of those who have already made policies. See Bocking, *Nature's Experts: Science, Politics, and the Environment* (New Brunswick, N.J., 2004).

[11] See *The Effects of Atomic Radiation on Oceanography and Fisheries* (Washington, D.C., 1957).

ment through its Scientific Committee on the Effects of Atomic Radiation (UNSCEAR). The UNSCEAR report was more cautious than the American report, but it echoed the findings of American and British oceanographers who supported their countries' decisions to dump at sea. The search for atomic graveyards in the deep sea became a major goal of international cooperation in the late 1950s, particularly the International Geophysical Year (IGY), an eighteen-month cooperative project beginning in the summer of 1957.[12]

The IGY proved disappointing to those hoping to use the sea for the disposal of high-level wastes, because most of the scientific work cast doubt on widely held assumptions about the existence of deep stagnant water. Early IGY evidence of deep circulation came from Japanese and French scientists, who used the French-made bathyscaphe, a recently invented manned submersible craft, in their joint study. About 120 miles off the coast of Japan, descending to a depth of more than 9,000 feet, the two-man *FNRS III* measured slow currents. A Japanese scientist who went down with the bathyscaphe, Tadayoshi Sasaki of the Tokyo Fishery College, reported water movement at less than an inch per second. He concluded that the movement was probably caused by ice melting at the poles. To Sasaki, the implications were crystal clear. "Considering the length of half-life of radioactive waste," he said, "this sluggish flow of deep sea water would make the sea unsafe as a dumping place for atomic reactor waste."[13]

Soviet scientists came to similar conclusions. "Huge canyons in the oceans, far deeper than the Grand Canyon, are not good ash cans for the nuclear age, a Soviet scientist said today." Thus the *New York Times* heralded the Soviet findings of the IGY in August 1958. Quoting Lev Zenkevich, a marine biologist, the newspaper noted that the water was not stagnant in the deepest parts of the ocean, as some oceanographers had hoped. Instead, the water circulated, meaning that radioactive wastes could poison sea life and, ultimately, man.[14] The Soviet newspaper *Tass* reiterated this point in November 1958, when pointing out that the Soviet oceanographic vessel *Vityaz* had measured the deepest parts of the ocean during the IGY. Even in these deep areas, there were no stagnant waters, in direct contradiction to what American oceanographers had suggested.[15]

National delegates met for the United Nations Conference on the Law of the Sea (UNCLOS) from February through April 1958. Meeting while the IGY was in progress added a renewed urgency to the problem. Scientists needed to find answers quickly in order to inform future agreements. International law now stated that United Nations signatories would have to regulate against pollution of the seas from the exploitation of offshore oil and other resources. The law also required states to regulate the dumping of radioactive waste at sea, according to principles to be set forth by a competent international organization. Because that organization would be the International Atomic Energy Agency, officially created just months before, the handwriting was on the wall: oceanographers and other scientists would have a limited amount of time to gather data and bolster their professional judgments about international regulation of

[12] An overview of the IGY can be found in Walter Sullivan, *Assault on the Unknown: The International Geophysical Year* (New York, 1961).
[13] "Current is Found Far Down in Sea," *New York Times,* June 29, 1958, 45.
[14] "Sea Canyons Held Poor Atom Ash Cans," *New York Times,* Aug. 3, 1958, 8.
[15] "Oceanographers Split," *New York Times,* Nov. 22, 1958, 11. The *Tass* story is reported here. See also Walter Sullivan, "Sea Depths Yield Secrets in IGY," *New York Times,* Jan. 5, 1959, 4.

dumping because the IAEA would see the formulation of specific regulations as one of its principal mandates.[16]

While physical oceanographers were studying the deep sea, biological oceanographers were studying the concentration of isotopes in organisms. Perhaps, the biological argument went, radioactive materials could reach humans through the food chain, despite the physical mixing and chemical dilution of radioactivity in the sea. If true, this might limit even the amount of low-level wastes put into the sea. In the United States, the National Academy of Sciences established the Committee on Oceanography (NASCO) to formulate recommendations to the government on areas needing policy recommendations, including these biological implications of radioactive waste disposal.[17] The new status acquired by oceanographers in the United States raised some hackles in the AEC because it opened a possible avenue for oceanographers to appropriate a large chunk of the AEC's policy authority. Moreover, as historian Ronald Rainger has argued, NASCO scientific assessments of risk became more than policy statements; they became moral pronouncements on what risks were acceptable.[18] As we shall see, this was happening on an international scale, and it irritated scientists in atomic energy establishments enormously. Oceanographers' efforts to claim such policy territory presumed to second-guess the decisions already made by experts within the AEC, the AEA, and the CEA.

While the IGY studies were in progress, the AEC sought the NAS's advice about using the Atlantic coast as a major dumping area. The AEC wanted to allow the dumping of low-level radioactive waste closer to shore than usual—that is, closer than 100 miles out. Actually, the AEC *already* had been allowing a commercial firm to dump its low-level radioactive wastes in shallow water (fifty fathoms) less than 15 miles from shore.[19] The AEC had made the decision: AEC sanitary engineer Arnold Joseph mentioned to the Academy that "the Atomic Energy Commission feels that as many as 4 or 5 disposal areas can be established along the Atlantic Seaboard," all conveniently located near ports but not the most densely populated ones. As its own experts had arrived at this conclusion already, asking the oceanographers to study the problem can only be interpreted as the AEC's effort to consolidate the legitimacy of its decision. Certainly, it would have expected no dissent: the panel's chairman, Dayton Carritt of the Chesapeake Bay Institute, was handpicked by the AEC.[20]

Although the Carritt panel did precisely what the AEC wanted, the experience soured Academy oceanographers on toeing the AEC line. After mulling over the possibility of dumping off the Atlantic seaboard, Carritt's panel chose twenty-eight sites

[16] For a contemporary discussion of UNCLOS, see Charles Swan and James Ueberhorst, "The Conference on the Law of the Sea: A Report," *Michigan Law Review* 56 (1958): 1132–41.

[17] Harrison Brown to Detlev Bronk, Jan. 8, 1958; Detlev Bronk to Harrison Brown, Feb. 3, 1958, Folder "ADM: ORG: NAS: Coms on BEAR: Oceanography and Fisheries: Cooperation with NRC Com on Oceanography," NAS Archives.

[18] Ronald Rainger, "'A Wonderful Oceanographic Tool': The Atomic Bomb, Radioactivity, and the Development of American Oceanography," in *The Machine in Neptune's Garden: Historical Perspectives on Technology and the Marine Environment,* ed. Helen M. Rozwadowski and David K. van Keuren (Sagamore Beach, Mass., 2004), 93–131, 115.

[19] Arnold Joseph to Richard C. Vetter, memo, Jan. 10, 1958, Folder "ES: Com on Ocean: Subcommittee on Radioactive Waste Disposal into Atlantic and Gulf Coastal Waters, General, 1958," NAS Archives.

[20] Ibid.

that might be suitable. When the locations were revealed to the public, in a map on the pages of the *New York Times,* city and state officials all along the coast were horrified. Virginia congressman Thomas N. Downing was shocked to discover that three of the sites were off Virginia, and he wrote to AEC chairman John A. McCone to register a firm protest. While he had great confidence in NASCO's scientific abilities, he wrote, "I cannot see where it is either necessary or practical to dispose of this radioactive material in waters so close to our shore." Aside from the possible physical harm, such disposal would be bad for business. The sites, it seemed, were very close to resort areas, so "there would also arise a psychological factor which could possibly be harmful to the economy of this area."[21]

The outcry by public officials embarrassed the National Academy of Sciences, which terminated Carritt's working group immediately. The experience helped to drive a wedge between academy scientists and the AEC. With its foregone conclusions dictated by the AEC, the report threatened the Academy's status as a reliable, independent, authoritative voice. Academy president Detlev Bronk was very annoyed at all of the negative publicity "because of Carritt." He wondered if the NAS ought to make a formal statement to rectify Carritt's missteps, adding, "I was never impressed by his scientific quality as are some."[22]

In the United States, the Carritt study marked a point of departure for oceanographers and the AEC, with the former trying to assert a separate scientific authority and to attune themselves to the needs of public officials and the lay public. This would be interpreted by the AEC as opportunism. The best example of this was the committee under Scripps Institution of Oceanography scientist John Isaacs, which tried to do (or not to do) for the Pacific what Carritt had done for the Atlantic. Because depth dramatically increased much closer to shore in the Pacific than it did in the Atlantic, the AEC did not anticipate the kinds of commercial gripes that plagued them on the East Coast. Taking ships out to deep water would not be as costly or time consuming. Nonetheless, the California State Legislature formally objected to the AEC's methods of dumping radioactive waste offshore. It petitioned the federal government and the armed forces to extend the depth requirement to 2,000 fathoms and to ensure that dumping areas were at least sixty miles away from known seamounts. Moreover, California politicians seemed to want to make a statement: as Joseph summarized it—"to go on record that they are opposed to the philosophy of unsafe bulk disposal of radioactive wastes."[23]

It was in the decidedly unpleasant aftermath of the Atlantic report that Isaacs's working group made its own report on the Pacific. The crucial difference was that the Pacific group knew beforehand that public officials represented potential adversaries. Should the group make an effort to placate the Californians or simply to back up the AEC's position? Here was where Isaacs hoped a quote from Lao Tzu would aid his fellow scientists in shaping the final report:

[21] Thomas N. Downing to John A. McCone, June 25, 1959, Folder "ES: Com on Ocean: Subc On Rad Waste Disposal into Atlantic and Gulf Coastal Waters, 1959," NAS Archives.

[22] Harrison Brown to Dayton Carritt, July 25, 1959; Detlev Bronk to S. Douglas Cornell, July 31, 1959, Folder "ES: Com on Ocean: Subc On Rad Waste Disposal into Atlantic and Gulf Coastal Waters, 1959," NAS Archives.

[23] Arnold B. Joseph to Richard C. Vetter, June 19, 1956, Folder "ES: Com on Ocean: Subc on Rad Waste Disposal into Atlantic and Gulf Coastal Waters, General, 1958," NAS Archives.

> The Wise Man says
> That only those who bear the nation's shame
> Are fit to be its hallowed lords;
> That only one who takes upon himself
> The evils of the world may be its king.

In his reflective way, Isaacs appeared to be committed to doing good science, possibly suggesting that the sea should be used for dumping—while asserting that oceanographers should shoulder more responsibility for such decisions.[24]

Isaacs attempted to avoid the problems that plagued Carritt's Atlantic report by talking to local organizations about the issues. He later told the *Houston Post* that he consulted more than thirty bodies, including sportsman's clubs, commercial fishermen's groups, and antipollution leagues. The report targeted not scientists but rather the lay person, because Isaacs recognized that the biggest difficulties would come not from scientists or government, but from the public at large. "Thus there are no vital steps of erudition that an audience must take on faith, but, rather, each step in our picture can be considered and criticized by any intelligent 'natural naturalist,' such as a crab fisherman, as well as by the formal scientist."[25]

When Isaacs handed over his committee's draft, AEC scientists were taken aback. The draft emphasized the biological uncertainties connected to isotope concentration and implied that continued radioactive waste disposal was going to limit man's other uses of the sea. "If true," Arnold Joseph complained, "AEC perhaps should have curtailed sea disposal some time ago." The biological argument was made in such a way that "it appears to be a fact, whereas in reality this is still largely hypothesis."[26] What really surprised Joseph, however, was the implication that the oceanographers had more of a right to speak for the public than the AEC did. "We, too, are very sensitive to reactions by public, civic, political and business interest groups." The report was poised, he believed, to paint a negative portrait of atomic energy establishments everywhere. Isaacs's group seemed to suggest widespread complacency and an unwarranted confidence that problems would solve themselves. Joseph countered: "Has AEC exhibited 'complacent confidence'? Does not the fact that this study was requested mean anything?"[27]

Joseph understood the international stakes. After all, sea disposal already was routine, especially by Britain. Joseph thought that the British would be offended by the report. "As worded," Joseph criticized, "this is a real slap at the British in their pipeline disposal only a few miles off shore." The oceanographers, he claimed, were giving the impression that atomic energy establishments everywhere could not be trusted. The report "calls the AEC, collectively from the Commissioners to the janitors, a bunch of untrustworthy people."[28]

Because Isaacs's group called for more research, AEC officials sensed opportunism.

[24] Isaacs to Pacific Coast Committee, Dec. 21, 1959 letter (cit. n. 1).

[25] John D. Isaacs to Ralph S. O'Leary, March 15, 1961, Folder "ES: Com on Ocean: Subc on Disposal of Low-Level Rad Waste into Pacific Coastal Waters: Rept: Review: 1961," NAS Archives.

[26] Arnold B. Joseph to John D. Isaacs, Sept. 22, 1960, Folder "ES: Com on Ocean: Subc on Disposal of Low-Level Rad Waste into Pacific Coastal Waters: Rept: Review: General: 1960," NAS Archives.

[27] Arnold B. Joseph to Roger Revelle, July 20, 1961, Folder "ES: Com on Ocean: Subc on Disposal of Low-Level Rad Waste into Pacific Coastal Waters: Rept: Review: 1961," NAS Archives.

[28] Arnold B. Joseph, "Comments on Report of Pacific Coast Subcommittee on Sea Disposal of Low Level Radioactive Waste," revised April 1961, Folder "ES: Com on Ocean: Subc on Disposal of Low-Level Rad Waste into Pacific Coastal Waters: Rept: Review: 1961," NAS Archives.

"Contrary to the opinions held by some," Joseph complained, "AEC is limited in the funds it can spend for research." The oceanographers had not spent much time asking the AEC for its expertise, but they had spent a disproportionately high amount of time gauging the views of the public. The oceanographers were making their own conclusions about what the public could or could not handle, while clamoring for more research money. What did they mean when they used terms like "unacceptable levels"? This was not a scientific viewpoint, Joseph held. "There will always be 'unacceptable' levels of radioactive substances to some people."[29] In his view, the oceanographers were pandering to public perceptions to augment their own authority by second-guessing experts who had been studying these issues for years.

The sense of outrage within the AEC was perhaps best captured by a comparison, used by Joseph himself, to racial integration. After World War II, President Harry Truman had ordered the armed forces to integrate their units, and in 1954, the U.S. Supreme Court ruled that racial segregation in public schools was unconstitutional. But there was widespread, often violent, resistance to government-mandated integration. Surely this was a clear case, Joseph argued, of experts needing to stay their course and do what was right, rather than accommodating public opinion. The widespread fear and visceral opposition to racial integration had to be overcome for the common good. The same was true of radioactive waste. Joseph challenged the oceanographers to resist equating majority views with right ones, and more importantly, to avoid appealing to the emotions of laypersons. Issues leading to feelings of "'[r]epugnance or apprehension' like race integration problems will probably be with us for a long time," Joseph argued. "Is it proper for a scientific community to sway in its scientific judgment because these states of mind 'might cause' rejection?" The only way to quiet people's fears was to inform and to educate, not to finesse the findings and present them according to what the public might find palatable. Joseph's allusion to racial conflict helps to highlight AEC officials' indignation at oceanographers who not only second-guessed their scientific decisions but also did so by exploiting negative public attitudes.[30]

AN ATOMIC TWIST: INTERNATIONAL SCIENCE AND THE MONACO CONFERENCE

Joseph was right that Isaacs's report would rattle the UK Atomic Energy Authority (AEA), which had worked hard over the years to convince other British government offices to authorize ocean dumping.[31] Although the report was not officially published until 1962, largely because of repeated draft criticisms by the AEC, its conclusions were known to oceanographers in 1959 and would inform the Monaco conference. When finally released, the report would indicate that packaged waste should only be dumped into very deep water, on the order of 1,200 fathoms.[32] If accepted, the British would have to abandon their most convenient dumping ground, the Hurd Deep (in the English Channel). AEA health physicist H. J. Dunster wrote to Roger Revelle, Isaacs's boss at the Scripps Institution of Oceanography, that the report ignored the physical

[29] Ibid.
[30] Ibid.
[31] The "authorization" process in Britain is discussed, along with other pertinent waste disposal issues, in Frans Berkhout, *Radioactive Waste: Politics and Technology* (London, 1991).
[32] *Disposal of Low-Level Radioactive Waste into Pacific Coastal Waters* (cit. n. 2).

and chemical properties of the ocean—the power of dilution—in favor of a biological approach. This led to some pretty restrictive recommendations, ones the British were not prepared to accept.[33]

British atomic energy officials shared the opinion of their American counterparts that leading scientists were exploiting public opinion to bolster their own positions and to ask for money. They expected questions of uncertainty to arise at the IAEA's international scientific conference, held in November 1959 in Monaco, and they saw the main international problem as one involving Western scientists, not cold war diplomacy. Although the Soviet Union's increasingly shrill outcries against ocean disposal loomed as a sticky political issue, the AEA was far more concerned about the effects of an international group of oceanographers entering into atomic matters.[34] AEA officials assumed an authoritative stance to ward off any impression that a scientific negotiation was taking place between them and nongovernmental scientists. Two of the British representatives, health physicists H. J. Dunster and A. H. K. Slater, agreed that their main object would be "to give oceanographers and geologists an idea of what was involved in waste disposal problems" and to discuss the general difficulties, "but not their detailed solution." In other words, the AEA was to instruct the oceanographers, not vice versa. They saw the nongovernmental scientists as special interests, and it would be crucial "to make sure that these other interests did not seek to divert money from atomic energy projects for their own particular problems." The British delegates to Monaco were thoroughly warned to beware of oceanographers and geologists who wanted to give their research "an atomic twist merely in order to divert funds to them for their purposes."[35] Dunster, the head of the delegation, belittled the "marked tendency" of oceanographers to "batten on to waste disposal" as a way to obtain funds, without any genuine interest in solving the waste disposal problem.[36]

With such attitudes in mind, the absence of an oceanographer on the British delegation to this scientific conference should come as no surprise. AEA scientists and officials went to the conference to defend their policies, and the last thing they wanted was an elaborate research program (for which they would have to pay) that might question those policies. Instead, the AEA sent people knowledgeable about disposal practices near the coast and the deep sea, people whose primary purpose would be to speak about the "evidence that these are non-hazardous."[37] Still, in the interest of a balanced delegation, Dunster suggested asking someone from Britain's National Institute of Oceanography to come to Monaco, too. But the AEA offered no travel funds, and the institute politely declined.[38]

Few oceanographers saw the Monaco conference in quite the same way the British AEA did. They were there not to be instructed but to discuss the science and, in doing so, to help the IAEA develop international regulations. The IAEA's director-general,

[33] H. J. Dunster to Roger Revelle, July 17, 1961, AB 54/16, United Kingdom National Archives, Kew, U.K. (hereafter cited as British National Archives).

[34] For more detailed discussion of the Soviet Union's opposition to nuclear waste disposal at sea, see Jacob Darwin Hamblin, "Environmental Diplomacy in the Cold War: The Disposal of Radioactive Waste at Sea during the 1960s," *International History Review* 24 (2002): 348–75.

[35] A. H. K. Slater, Note for Record, June 5, 1959, AB 16/3000, British National Archives.

[36] H. J. Dunster to J. M. Hartog, "International Agency Conference on Disposal of Nuclear Waste," April 20, 1959, AB 16/3000, British National Archives.

[37] W. G. Marley to J. F. Jackson, "International Atomic Energy Conference on Disposal of Radioactive Wastes," April 27, 1959, AB 16/3000, British National Archives.

[38] Dunster to Hartog, "International Agency Conference on Disposal of Nuclear Waste" (cit. n. 36).

Sterling Cole, had written that the conference was to be a discussion between atomic energy officials and oceanographers and geologists. In his letter to the British Foreign Office (and other nations' diplomatic offices), Cole explicitly suggested that the oceanographers and geologists might contribute to the solution of the waste disposal problem.[39] When they tried to do so, atomic energy officials resisted, exacerbating divisions that were more institutional and disciplinary than national. As one British official later chastised his own delegation's attitude, it was likely that most countries saw the Monaco conference as a step toward consensus about standards and procedures, and that the representatives of the AEA should not have been so indifferent to what others had to say.[40]

The Monaco conference turned out to be as much a failure for atomic energy officials as it was a successful meeting of minds for oceanographers. Sir John Cockcroft, director of the AEA's research establishment at Harwell, openly declared it a victory for the exchange of ideas but privately lamented that the result was much sharper distinctions between positions.[41] Two separate camps had emerged, the oceanographers and the atomic energy officials, and each had arrived at its own consensus. For example, French and Italian oceanographers believed that there was not sufficient knowledge of oceanographic conditions to justify dumping, especially not in nearby shallow seas such as the Mediterranean. By contrast, the representatives of the atomic energy establishments of three major dumping powers (the United States, Britain, and France) found common ground in their view that they already knew enough to make conservative estimates of what could be dumped and could make such estimates about any part of the sea. The leader of the British delegation, Dunster, dismissed the oceanographers' view as being "based more on prejudice than knowledge."[42] As Cockcroft acknowledged, few skeptics from other countries were convinced by the atomic energy officials' position, largely because of the influence of the oceanographers.[43]

The British saw Monaco as a failure because it publicly raised more questions about the validity of radioactive waste than it resolved, and it drew attention to the oceanographers' view that more study was needed. The *Sunday Times* summed it up nicely with a headline: "All at Sea on Atomic Waste."[44] The conference not only opened British practices to scrutiny but also left wide avenues for scientists to ask for more money and to assert a role in decision-making. British officials felt they had been outmaneuvered by oceanographers on the diplomatic front. In the aftermath of the meeting, one disappointed Ministry of Science official predicted that the discharge of waste into the sea was destined to become a "hobby-horse for the mischievous, the ignorant and the timid alike."[45] Another wrote that in the future, conferences on technical matters should be recognized as having wide political repercussions. As such, the government ought to ensure that the delegates "had a sufficiently high level of political competence."[46] And it ought to ensure that an oceanographer of international standing was

[39] Sterling Cole to Secretary of State for Foreign Affairs, Foreign Office, March 24, 1959, AB 16/3000, British National Archives.
[40] P. W. Ridley to F. J. Ward, Aug. 28, 1959, AB 16/3000, British National Archives.
[41] "Note by Sir John Cockcroft," Dec. 7, 1959, AB 16/3001, British National Archives.
[42] H. J. Dunster to John Cockcroft, Dec. 8, 1959, AB 16/3001, British National Archives.
[43] "Note by Sir John Cockcroft" (cit. n. 41).
[44] Stephen Coulter, "All at Sea on Atomic Waste," *Sunday Times,* Nov. 22, 1959, press clipping, AB 16/3001, British National Archives.
[45] P. W. Ridley to H. J. Dunster, Dec. 9, 1959, AB 16/3001, British National Archives.
[46] A. H. K. Slater to A. S. McLean, Jan. 15, 1960, AB 16/3001, British National Archives.

included, if only for show, to beef up the scientific credibility of the delegation—it was probably worth the cost of an airline ticket and hotel.[47] The message was clear: the AEA alone did not carry sufficient political or scientific authority, certainly not beyond Britain's shores.

MER FERMÉE! THE MEDITERRANEAN DEBACLE

By this time, France also had devoted major resources and political commitment to becoming a nuclear nation.[48] Like its American and British counterparts, the French Commissariat à l'Énergie Atomique (CEA) knew of the difficulties associated with nuclear safety and saw radioactive waste as a potential political problem.[49] The commissariat envisioned two possible solutions: land burial in "radioactive cemeteries" in the environs of its reprocessing plant in Marcoule, or sea disposal. The former's main drawback would be the disquietude of neighboring populations. CEA officials hoped that sea disposal would help them avoid public outcry. Thus in May 1960, the CEA decided to plan an experimental dump of about 2,000 tons of liquid and solid waste—contaminated work clothes, with assortments of plastic, wood, metal, glass, and other materials. These would be packed into 200-liter drums and dumped into the Mediterranean, at a site between the towns of Antibes (near Nice) and Calvi (in Corsica), in water about 2,500 meters deep, fifty miles from the coast of France and sixty miles from the coast of Italy.[50]

The CEA's official description of the plan observed that there were no currents at the sea's surface in the specified region. "One could hope," it added, "that the currents would be equally nonexistent at the bottom." Some measurements had been taken in 1959 that indicated that such might be the case. Thus the site seemed ideal: it was in relatively deep water, away from the coast, with no discernable current, and not near known fishing waters. There would be no notable risks, the CEA stated; even if all of the drums burst, which was unlikely, the surrounding water would dilute the material so much that the danger to human health would be "completely negligible."[51]

French newspapers picked up the story in October 1960, and the negative response from oceanographers was immediate. Jean Furnestin, director of the Institut Scientifique et Technique des Pêches Maritimes, pointed out that all of the physical oceanographers and biologists at the recent IAEA meeting in Monaco, without exception, had underlined the formidable dangers that confronted humanity from ocean dumping of radioactive waste. No one had been able to demonstrate that there were in fact "dead zones" in the ocean. The Soviet work aboard the *Vityaz* during the IGY,

[47] H. J. Dunster to A. H. K. Slater, Dec. 4, 1959, AB 16/3001, British National Archives.

[48] On the importance of nuclear power in French politics and culture, see Gabrielle Hecht, *The Radiance of France: Nuclear Power and National Identity after World War II* (Cambridge, Mass., 1988). For a French insider's view on the political and diplomatic issues related to nuclear programs, see Bertrand Goldschmidt, *Le Complexe atomique* (Paris, 1980). The origins of the CEA are discussed in Spencer R. Weart, *Scientists in Power* (Cambridge, Mass., 1979).

[49] On the history of nuclear safety in France, see M. Cyrille Foasso, "Histoire de la Sûreté de l'Énergie Nucléaire Civile en France, 1945–2000" (Ph.D. diss., Univ. Lumière, Lyon II, 2003).

[50] Most of the elements would have been Ce-144 and Pr-144 (about 350 curies), Sr-90 (about 40 curies), and Pu (about 14 curies). "Projet de Rejet en Mer en Méditerranée de Déchets Radioactifs," May12, 1960, Folder "Effluents/Rejets en Méditerranée," Box F2/23-18 80-1217 (HC), Archives, Commissariat à l'Énergie Atomique, Fontenoy aux Roses, France (hereafter cited as CEA Archives).

[51] Ibid.

Furnestin pointed out, had proven that even the deepest Pacific waters moved to the surface much faster than previously believed. In the Mediterranean site, there were no instruments sufficiently sensitive to tell whether there were currents at that depth or not. Moreover, there was plenty of evidence to suggest that such currents might exist on a seasonal basis, and that the region was unstable. Besides, biologists were unanimous in pointing out the perils of radioactive concentration in deep flora and fauna that could be passed to other creatures at shallower depths. Furnestin argued that approval should not be given without first consulting oceanographers and fisheries specialists.[52]

Back in Paris, oceanographers at the Centre de Recherches et d'Études Océanographiques also criticized the CEA's action because the commissariat had planned it without any consultation. Vsevolod Romanovsky, the director, learned of the plans from the newspapers and was appalled to discover that the CEA had suggested to the press that oceanographers—including Romanovsky specifically—approved of the idea. Romanovsky had conducted the 1959 studies cited by the CEA, but he believed they had been inconclusive, and some were still ongoing. True, he had recommended an experimental dumping, but he had in mind something on the order of 10 drums. The CEA was planning to dump 6,500.[53]

Closer to the (proposed) action, scientists at the Institut Océanographique in Monaco seethed with anger at the CEA. In a widely circulated letter, the institute's director, Louis Fage, wrote that he had been stupefied to read the news that the CEA was planning to dump 2,000 tons of waste into the Mediterranean. He described the Mediterranean in two words, exclaimed on the page: *mer fermée* (closed sea)! Aside from directing the institute, Fage also was the president of the Committee for the Exploitation of the Sea, a quasi-international body consisting of scientists from France and Monaco. In that capacity, he registered strong protest against the CEA's decision, giving several rebuttals to its scientific assumptions.[54]

In reading the protests of Furnestin, Romanovsky, Fage, and others, one is left with the indelible impression that if their rage came partly from the scientific foolishness of the dumping experiment, it came also from the fact that they had been left out of the decision-making process. They were reading about the experiments from the newspapers like everyone else. Fage wrote that he could not vouch for the veracity of the newspapers' claims, since no specialists in *marine biology* (Fage underlined this also in his letter) seemed to have been consulted. These scientists, not atomic energy officials, had already established the crucial questions on the issue, namely the ecological connections between marine life and human beings. Rather than look purely at circulation, Fage insisted, one should look at plankton, "for they are at the base of the chain in which we occupy the summit." He quoted the findings of American oceanographer Bostwick Ketchum, who had shown that the concentration of radioactive substances in plankton could be up to 500 times that of sea water. This was a living environment (*milieu vivant*) for which the introduction of radioactive waste

[52] Jean Furnestin to Secrétaire Général de la Marine Marchande, Oct. 6, 1960, Folder "Effluents/Rejets en Méditerranée," Box F2/23-18 80-1217 (HC), CEA Archives.

[53] V. Romanovsky to Haut Commissaire, CEA, Oct. 8, 1960, Folder "Effluents/Rejets en Méditerranée," Box F2/23-18 80-1217 (HC), CEA Archives.

[54] Louis Fage to Le Délégué Général à la Recherche Scientifique et Technique, Oct. 13, 1960, Folder "Effluents/Rejets en Méditerranée," Box F2/23-18 80-1217 (HC), CEA Archives.

could be destructive. Fage insisted that it would have been better to hear the marine biologists prior to the decision.[55]

The most formidable opponent was Jacques-Yves Cousteau, a member of Fage's committee and director of the Musée Océanographique de Monaco. According to Fage and Cousteau, undersea photographs had revealed that the deep water in the dumping area did indeed move. With underwater breathing gear, and with manned and unmanned submersibles, Cousteau and his colleagues took photographs of the depths and published them internationally. In the years ahead, Cousteau's books and films about undersea life would make him an international household name. By 1960, he already had established his fame with his book *Le Monde du Silence,* translated into several languages in the 1950s, and a 1955 documentary film of the same name that won the Palme d'Or at the 1956 Cannes Film Festival.[56]

On the pages of French newspapers, Cousteau blasted the CEA. He proclaimed that scientists had not been consulted. He made it clear that neither the Musée Océanographique nor his ships were involved in any way. He accused the CEA of having acted behind the backs of scientists after the Monaco conference, when it had become clear that profound differences of judgment existed between atomic scientists (*atomistes*) and an international group of oceanographers. The latter, he claimed, had categorically condemned sea disposal, on the grounds that sufficient studies had not yet been made. He then included a list of the important scientific bodies that had not been consulted, such as the International Commission for the Scientific Exploration of the Mediterranean, the (French) Academy of Sciences, and the Centre National de la Recherche Scientifique (CNRS). It was, he claimed, a bit like announcing that tomorrow morning there would be an experiment to dispose of allegedly inoffensive nuclear waste at the Place de l'Opéra, without first consulting with the mayor of Paris.[57] With such complaints, Cousteau appealed to local officials (who also had not been consulted) and implied that oceanographers, not atomic energy officials, were the true custodians of the seas and the protectors of local interest.

Meanwhile, mayors and city councils of towns all along the French Riviera sent protests to the CEA's high commissioner, Francis Perrin. Of course, the CEA sent reassuring replies to them, pointing out that the experiment rested on the firmest scientific grounds. However, the mayor of Nice, Jean Médecin, sent back a telegram that cut right to the heart of the issue, underscoring the power of international scientific consensus. He baldly stated that whatever the CEA's scientific competence might be, it would certainly not prevail over the numerous French and foreign scientists of contrary opinion. Whoever marked this telegram (in the CEA's archives), Perrin or a subordinate, underlined that statement in red and penciled two exclamation marks in the margin. The marks must have signified an array of frustrations. If Nice was any indication, clearly the battle for scientific authority was being lost to the oceanographers. The people of Nice, as Médecin said, stood ready to oppose the CEA and to stop the dumping "by all possible means." Other nearby towns voiced similar sentiments. One

[55] Ibid.

[56] A brief biography of Cousteau is Axel Madsen, *Cousteau: An Unauthorized Biography* (New York, 1986).

[57] The CEA's records include various newspaper clippings reprinting Cousteau's remarks. Quoted here is an unpaginated excerpt from *Le Patriote,* a newspaper in Nice, dated Oct. 11, 1960. Folder "Effluents/Rejets en Méditerranée," Box F2/23-18 80-1217 (HC), CEA Archives.

anonymous letter writer suggested to Perrin that if the waste was so inoffensive, perhaps he should put it in his own breakfast.[58]

The CEA's plans became worldwide news. Prince Rainier of Monaco urged the government of Charles de Gaulle to put off the experiment until scientists knew more about the dangers. The leading French newspaper, *Le Monde,* quoted Cousteau's statements that the CEA did not understand anything about the problems of the sea.[59] Cousteau tried to make it an international issue, saying that it involved all of the countries bordering the Mediterranean, not just France. The *New York Times* called Cousteau the "unofficial leader of the anti-dumping campaign." In the face of the publicity assault, less than a week after making the announcement of the experiment, the CEA backed down and decided to put the project off for a while.[60] In the weeks that followed, Cousteau gave more interviews, stressing that the issue was really an international one, one that could be resolved only by oceanographers. It might end up as a choice for all humanity, he said, between using the sea as a waste dump or preserving the riches within it.[61]

For its problems regarding ocean dumping, the CEA did not blame the press, or the general public's irrationality, or local mayors. It laid the blame entirely at the feet of oceanographers, particularly Cousteau. In an internal note, CEA officials dismissed the idea that the press could have mounted such an offensive or that the population could have spontaneously reacted so negatively. Instead, the escalation of the issue's importance "is a direct function of the declarations, acts, and positions taken by M. J.-Y. Cousteau, director of the Musée Océanographique du Monaco." Quoting American newspapers, they lamented the fact that Cousteau suddenly seemed to be internationally recognized as the leading figure against ocean dumping. The department outlined specifically the steps Cousteau had taken to undermine them. He began to critique the experiment "violently"; he then sent telegrams to all the mayors in the area; he attended all the important local meetings to discuss the issue; he attacked Perrin in the press; he acted as a "scientific expert" at a major regional meeting; he then acted indirectly at other local political meetings to oppose the project. In addition, he had given "innumerable interviews" to reporters for newspapers, radio, and television. For the CEA, the Cousteau nightmare brought into question their ability to speak for the sea. Cousteau focused on the CEA's incompetence, calling it childish; its scientists were incapable of understanding the sea, he claimed, making mathematical calculations that would not even measure up to the standards of first-year oceanography students.[62]

From the CEA's point of view, the task ahead was not to change plans to dump radioactive waste but to repair relationships with oceanographers by ceding to them

[58] The letters of protest from towns on the coast are all collected in the same folder, near Jean Médecin to Francis Perrin, Oct. 12, 1960, Folder: Effluents/Protestations, Box F2/23-18 80-1217 (HC), CEA Archives.

[59] The statements about Rainier and Cousteau in *Le Monde* are from clippings saved by the CEA dated Oct. 11, 1960. Folder "Effluents/Rejets en Méditerranée," Box F2/23-18 80-1217 (HC), CEA Archives.

[60] "France to Delay Atomic Disposal," *New York Times,* Oct. 13, 1960, 20.

[61] Extract from interview of Cousteau by Pierre Ichaac, Oct. 26, 1960, Fonds M. A. Gauvenet, Folder "Déchets dans la Mer," Box 156/256, CEA Archives.

[62] "Éléments d'Appréciation de l'Importance de l'Action de M. J.-Y. Cousteau dans la Campagne d'Opposition au Projet de Rejet Expérimental en Méditerranée," Department of External Relations, CEA, Nov. 1960, Folder "Effluents/Rejets en Méditerranée," Box F2/23-18 80-1217 (HC), CEA Archives.

some scientific authority and possible financial support. Perrin left for a three-week visit to the United States and promised Fage that they would meet upon his return.[63] In the meantime, Henri Baïssas (of the Department of External Affairs) asked leading French geophysicist Jean Coulomb to help facilitate a rapprochement with oceanographers. Coulomb felt that it would be easy to bring physical oceanographers into the CEA's camp but more difficult with biological oceanographers. He tried to arrange a meeting with Cousteau, but was ignored. A leading CEA physicist, Bertrand Goldschmidt, did finally meet with Cousteau and informally promised to support more scientific work under the IAEA. Baïssas and a colleague met with Furnestin and had an informal conversation about the importance of supporting research. Baïssas followed up with a formal letter stating that the CEA would not proceed with a dump without proper studies by, guidance from, and agreement with, oceanographers.[64]

Perrin met with Fage upon his return from the United States, taking other CEA officials with him. According to a CEA internal memorandum, the meeting was very pleasant. In fact, Fage declared himself in support of the commissariat's activities, being convinced that they were harmless. In return, the CEA promised to lower the number of drums dramatically to keep it in line with biologists' views.[65] Baïssas went with a colleague to Monaco to reenlist the scientists there and soon reported his "mission to Monaco" as a success. His strategy was to admit candidly to the oceanographers that the CEA, despite being convinced that its plans were harmless, had committed the error of not sufficiently consulting the scientific community. He promised much closer collaboration in the future, declaring the Mediterranean as a place for experiments (on the order of ten to twenty tons), not for massive dumping. As a result, Furnestin "incontestably" wanted to help them, as did other scientists present. Even Cousteau, cornered by Baïssas during a prelunch cocktail, privately assented to the CEA's plans.[66] In the coming months, Baïssas worked hard to cajole Cousteau and others, careful not to ruffle any scientific feathers, to create an experiment that helped the CEA but also drew on outside scientists' expertise. Such conciliatory maneuvers were necessary, Baïssas wrote, to rupture "the mystical charm that paralyzes us."[67]

ATOMIC ESTABLISHMENTS FIND COMMON GROUND

One of the suggestions at the Monaco conference was the creation of a permanent international laboratory to study ocean disposal of radioactive waste. Publicly, Sir John Cockcroft grandly stated to the IAEA that Britain supported future scientific work at Monaco. But in reality the British did so only grudgingly. Leading health physicist H. J. Dunster acknowledged that more research could be done, "if only to demonstrate that current practice by Great Britain in this connection is safe." The value of international research, he wrote to a colleague, was principally to allay the fears of people swayed more by political arguments than scientific ones. At the national level, they

[63] Francis Perrin to Louis Fage, Oct. 14, 1960, Folder "Effluents/Rejets en Méditerranée," Box F2/23-18 80-1217 (HC), CEA Archives.
[64] "Déchets Radioactifs," no author, Nov. 15, 1960; Henri Baïssas to Jean Furnestin, n.d., Folder "Effluents/Rejets en Méditerranée," Box F2/23-18 80-1217 (HC), CEA Archives.
[65] "Déchets Radioactifs" (cit. n. 64).
[66] Henri Baïssas, "Compte Rendu de Ma Mission à Monaco," Dec. 15, 1960, Folder "Effluents/Protestations," Box F2/23-18 80-1217 (HC), CEA Archives.
[67] Henri Baïssas, note, Feb. 6, 1961, Folder "Effluents/Rejets en Méditerranée," Box F2/23-18 80-1217 (HC), CEA Archives.

should start to monitor their principal dumping grounds, such as the Hurd Deep in the English Channel, not because they expected to find significant levels of radioactivity but because they did not wish to face the criticism that they had no data. From the AEA's point of view, such environmental surveys were superfluous, costly, and purely political.[68]

After the Monaco conference, American, British, and French atomic energy establishments tried to find common ground. To ensure that they could do so, the AEA and the AEC helped each other by sharing copies of their national delegations' secret instruction briefs for meetings at the IAEA. The Americans discovered that the British bristled at the thought of more mushrooming scientific projects, but for political reasons they did not want to be seen as the only nation to oppose the new laboratory. Cockcroft's instructions were to "do what he can to curb the [International Atomic Energy] Agency's natural proclivities in this matter" without blatantly taking too strong a stand against international scientific studies. One official urged the British delegation to show a "conspicuous lack of enthusiasm," but admitted that it had become politically impossible not to support such studies.[69]

The British learned through this cooperation that the United States, by contrast, strongly supported the international laboratory. According to the U.S. delegation briefing (shared with the British by physicist Isidor Rabi), its primary reason for doing so was to counter Soviet propaganda. The Soviet Union opposed dumping in principle, and the United States feared that the environmental effects of waste disposal would form the basis of a major propaganda campaign against the West. The Americans certainly were correct on this score, and the Soviets spent much of the 1960s reproaching the United States and the United Kingdom for poisoning the seas.[70] In 1960, however, the Americans hoped that their support for international scientific work could give more credibility to their policies, particularly if scientists could identify problems and be seen to be researching them. The draft trading between the AEC and the AEA allowed officials to see that their apparent disagreement was more of a difference in tactics and immediate diplomatic necessity, rather than a genuine divergence of views. Neither side believed that a new international laboratory was necessary to solve real problems of waste disposal.[71]

In fact, the AEA and AEC had done more than trade briefings. For the AEA, physicist Sir William Penney and two colleagues went to the United States in October 1960 and met with General Alvin Luedecke, the general manager of the AEC, along with several of his staff. They appeared to have agreed that an IAEA laboratory in Monaco, with a mandate to study radioactive waste, might become a source of scientific criticisms, not solutions. But Luedecke pointed out that the problem was manageable, because the laboratory would be staffed with plenty of American scientists whose paychecks would come from the AEC. They also agreed that it was a very good idea for

[68] H. J. Dunster to F. A. Vick, "Radioactive Waste Disposal at Sea," July 27, 1960, AB 6/2087, British National Archives.

[69] D. E. H. Peirson to John Cockcroft, Nov. 4, 1960, AB 16/3002, British National Archives. Quotation taken from I. G. K. Williams to J. C. Walker, "Collaboration between the IAEA and the Government of Monaco," Nov. 7, 1960, AB 16/3585, British National Archives.

[70] For more on the diplomatic aspects of radioactive waste in the 1960s, see Hamblin, "Environmental Diplomacy in the Cold War" (cit. n. 34).

[71] The brief for the United States delegation is titled "Scientific Collaboration with the Government of Monaco on Research on Disposal of Radioactive Waste into the Sea," n.d., AB 16/3002, British National Archives.

the two establishments to continue collaborating in this way, prior to meeting with scientists and certainly prior to any international meeting. They felt that they needed to see if there was any way to do the same with the French, to understand fully their position before coming to the IAEA. Although the French had not been close partners on atomic energy matters in the past, the Mediterranean fiasco gave them, the Americans, and the British reason to believe that the CEA was confronting similar issues and might want to collaborate.[72]

The French were more than willing to compare notes on how to deal with troublesome oceanographers. Francis Perrin met with American atomic energy colleagues in November, after being forced by public opinion to shelve the Mediterranean experiment. By the next month, the French were collaborating directly with the British. The French and British atomic commissions met to deal specifically with their mutual public relations problems. It turned out to be a productive meeting of minds. They felt they had to do something to establish common practices that they could defend against critics, even scientists. The British gave talks on subjects ranging from the technical details of disposal to relations with local authorities. In fact, each side specified two people to liaise directly with their foreign opposite on public relations and technical issues in the future, to avoid any semblance of disagreement that could be exploited in international meetings.[73]

The existing notes of these meetings between the CEA and the AEA, held in the British National Archives, reveal an atmosphere of mutual understanding, not of diplomatic negotiation. Their common problem was public relations, and their common headache was the oceanographic community. Geopolitical difficulties seemed nearly trivial by comparison. In the course of one meeting, for example, the British found that their French counterparts felt the same as they did about the Soviets—that they discharged quite a bit into rivers and oceans, despite their public pronouncements, and were not to be taken seriously. More important were the oceanographers who (they believed) made mountains out of molehills to serve themselves. Cousteau, French public relations official Jean Renou told his British counterparts, might be well known for his literary work, but he was not much of an oceanographer. What was needed, he and others agreed, was a push for education about atomic energy so that officials and the general public would not be so easily swayed by prejudiced scientists such as Cousteau.[74]

The solidarity among atomic energy establishments would continue into the 1960s, particularly between the British and the French, who soon began to coordinate their dumping operations with each other and with other European countries.[75] But the antagonism with oceanographers at the international level also would continue. Oceanographers' scientific claims had little direct effect on dumping practices, but their claims about the dearth of knowledge on the effects of radioactivity in the oceans exerted enormous political pressure, contributing to a series of international agreements

[72] "Record of Discussion of Anglo-US Collaboration on Waste Disposal Problems Prior to November Meeting of IAEA Committee," Oct. 14, 1960, AB 16/3002, British National Archives.

[73] Public relations liaison was to be handled by Jean Renou (France) and Eric Underwood (the United Kingdom), while technical aspects would be handled by André Gauvenet (France) and Ian Willams (the United Kingdom). See Bertrand Goldschmidt to D. E. H. Peirson, Dec. 20, 1960, AB 16/3588, British National Archives.

[74] "Notes of Talks with CEA 14th–15th December: Public Relations Aspects of Waste Disposal," n.d., AB 16/3588, British National Archives.

[75] See Hamblin, "Environmental Diplomacy in the Cold War" (cit. n. 34).

beginning in the 1970s that limited the disposal of radioactive waste.[76] For the time being, the victory went to the oceanographers, who won both scientific authority and the promise of patronage. All three establishments (U.S., British, French) publicly claimed that they needed to support more research in oceanography, to ensure that their practices were indeed safe. Moreover, they agreed to the creation of a permanent laboratory financed by national governments through the International Atomic Energy Agency. When it was founded in 1961, the Monaco laboratory looked exactly as oceanographers had hoped and as the atomic energy officials cynically had expected. Its first director, oceanographer Ilmo Hela, was the former director of Finland's Institute of Marine Research and definitely an academic scientist, and there were close institutional ties to Cousteau's Musée Océanographique. According to an IAEA press release, the laboratory's first goal was to understand the movement of water and marine organisms and the deposition of organic and inorganic matter, a pretty broad agenda that made no specific reference to radioactive waste. Second was the study of the distribution of radioactive materials in organisms, and last of all was the study of the effects of radioactive materials on marine ecology.[77] If the AEC, AEA, and CEA saw oceanographers as interested in power and patronage rather than solving the waste disposal problem, the establishment of the Monaco laboratory only strengthened this view.

CONCLUSION

Environmental controversy is familiar territory when examining cases of contested authority. Given the breadth of uncertainty, even the most incorruptible and irreproachable of "nature's experts," as Stephen Bocking calls them, would walk on tenuous political ground because scientific authority is constantly questioned.[78] It may be that contested scientific authority is inextricably tied to democratic values. Sociologist Dorothy Nelkin wrote on such themes during much of her academic career; when she published *Technological Decisions and Democracy* in 1978, claiming that political struggles are inherent to technological decisions,[79] longtime British science policy advisor Sir Solly Zuckerman agreed, recalling the story of British nuclear power. One consequence of democracy, he wrote, was that anything claiming to be scientific would appeal to the segment of the population wanting to believe it, resulting in political pressure.[80] On the national level, individuals or institutions had to play the role of experts in mediating such political questions; it was not an easy job, nor was it an easy role to assert, because scientific authority was fiercely contested terrain.[81]

[76] These agreements are part of the Convention on the Prevention of Marine Pollution by Dumping of Wastes and Other Matter, the initial version of which was signed in 1972. Details are available at http://www.londonconvention.org.

[77] International Atomic Energy Agency, press release, "Finnish Scientist to Direct Major Research Program in Monaco," June 8, 1961, AB 16/3585, British National Archives.

[78] Bocking, *Nature's Experts* (cit. n. 10).

[79] Dorothy Nelkin, *Technological Decisions and Democracy* (London, 1978).

[80] Lord Zuckerman, "Science Advisers and Scientific Advisers," *Proceedings of the American Philosophical Society* 124 (1980): 241–55.

[81] On science advising, see ibid.; Gregg Herken, *Cardinal Choices: Presidential Science Advising from the Atomic Bomb to SDI* (Stanford, 2000); Sheila Jasanoff, *The Fifth Branch: Science Advisers as Policymakers* (Cambridge, Mass., 1990); Zuoyue Wang, "Responding to *Silent Spring:* Scientists, Popular Science Communication, and Environmental Policy in the Kennedy Years," *Science Communication* 19 (1997): 141–63.

Furthermore, the same battles for authority could replay on the international scale with entirely different results, especially if one side could lay claim to international consensus.

Were oceanographers the robber barons of the sea that many atomic energy officials thought they were, capitalizing on international opinion to assert their authority and sap money from reluctant patrons? Or had they become the new "hallowed lords," taking on responsibility for high-stakes policy decisions of international importance that previously had been the province of atomic energy establishments? We cannot ignore the fact that oceanographers did succeed in gaining money from a variety of patrons during the cold war, from military, atomic energy, foreign policy, and fisheries sources, to name a few, and that international consensus among scientists strengthened their ability to assert power at home.[82] As this essay has shown, oceanographers' scientific claims often were interpreted as opportunistic appeals for political influence and research money. Such efforts to secure patronage were not always welcome, particularly when they implied neglect or ignorance on the part of those scientists and officials already dealing with the related problems.

When taking part in international affairs, oceanographers did not necessarily serve the interests of the state. Quite to the contrary, by contesting scientific authority about the oceans, oceanographers became national liabilities for the state interests of the United States, Britain, and France. Their activities provoked no small amount of antipathy from government scientists and officials, who saw those activities as veiled grasps for money. The undeniable existence of such attitudes turns the conventional question of patronage upside down, provoking questions about the direction of pressure. Instead of generous patrons hampering science through undue influence, scientists in the pursuit of funding wrought havoc on existing policies by insisting, publicly, upon the need for more research. With international agreements resting on their advice, oceanographers wielded considerable power. As a remarkable response to the apparent international consensus among oceanographers, atomic energy establishments forged their own bonds of collaboration across borders, to help each other deal with the potential diplomatic and public relations problems brought on—in their view—by oceanographers' opportunism. Such perspectives reinforce the need to look carefully at the importance of international affairs in understanding some of the major questions in the historiography of science. Certainly they highlight the reciprocal effects of patronage, at national and international levels, during the postwar era. The siege mentality within atomic energy establishments is telling evidence that the pressures sometimes went in the opposite direction than we might expect and that the historiographic question that often consumes us—the effects of patrons' priorities upon the practice of science—can rob scientists of much of the power and influence they undoubtedly possessed.

[82] These efforts are discussed in greater detail in Hamblin, *Oceanographers and the Cold War* (cit. n. 8).

Meteorology as Infrastructural Globalism

*by Paul N. Edwards**

ABSTRACT

This chapter explores the history of a global governance institution, the World Meteorological Organization (WMO), from its nineteenth-century origins through the beginnings of a planetary meteorological observing network, the WMO's World Weather Watch (WWW), in the 1960s. This history illustrates a profoundly important transition from voluntarist internationalism, based on shared interests, to quasi-obligatory globalism, based on a more permanent shared infrastructure. The WMO and the WWW thus represent infrastructural globalism, by which "the world" as a whole is produced and maintained (as both object of knowledge and unified arena of human action) through global infrastructures.

INTRODUCTION

Intense debates about the nature and trajectory of globalization have consumed historiography and international relations theory in recent years. Is globalization really global? Is it new or old? What are its causes and consequences? No one who has followed these debates can fail to notice the prominence of information and communication technologies (ICTs) in virtually all accounts. For example, Manuel Castells defines the global economy as one "whose core components have the institutional, organizational, and technological capacity to work *as a unit in real time, or chosen time, on a planetary scale*" via ICT infrastructures,[1] and every chapter of *Global Transformations: Politics, Economics and Culture,* a major survey of globalization, discusses the role of communication infrastructures.[2]

In an important variation on this theme, Martin Hewson offered a three-phase notion of "informational globalism": systems and institutions dedicated to the production and transmission of information on the planetary scale. In the first, nineteenth-century phase, national informational infrastructures (NIIs), such as telegraph systems, postal services, and journalism, were linked into interregional and intercontinental (if not yet fully global) networks. Between 1914 and 1960 (Hewson's second phase), the pace of infrastructural linking diminished, and some delinking occurred. Yet simultaneously, world organizations such as the League of Nations and the International

* School of Information, 3078 West Hall, University of Michigan, 550 East University Ave., Ann Arbor, MI 48109-1107; pne@umich.edu.

I thank Kai-Henrik Barth, John Krige, and Gabrielle Hecht for extremely useful comments.

[1] Manuel Castells, *The Rise of the Network Society* (Cambridge, Mass., 1996) (emphasis in original).

[2] David Held, Anthony McGrew, David Goldblatt, and Jonathan Perraton, *Global Transformations: Politics, Economics, and Culture* (Stanford, 1999).

Monetary Fund "established the legitimacy of producing globalist information"—that is, information about the whole world—in such areas as health, armaments, and public finances (although they did not in fact attain that goal). Hewson's third phase brought general achievement of the two previous eras' goals, beginning with the establishment of worldwide civil communication networks (from the 1967 inauguration of the Intelsat system) and global environmental monitoring (from the UN Conference on the Human Environment, 1972). Throughout, Hewson sees global governance institutions such as the United Nations and the International Telecommunications Union, rather than an autonomous technological juggernaut, as chiefly responsible for informational globalism.[3]

In this chapter, I explore the history of one such global governance institution, the World Meteorological Organization (WMO). The WMO's story confirms the pattern Hewson discerned, but it also has special characteristics. Arguably, the weather data network and its cousins in the other geophysical sciences, especially seismology and oceanography, are the oldest of all systems for producing globalist information in Hewson's sense. When the young John Ruskin wrote, in 1839, that meteorology "desires to have at its command, at stated periods, perfect systems of methodical and simultaneous observations . . . to know, at any given instant, the state of the atmosphere on every point on its surface,"[4] he was only giving voice to his contemporaries' grandest vision. By 1853, the Brussels Convention on naval meteorology had created a widely used standard meteorological logbook for ships at sea; these logs constitute the oldest continuous quasi-global meteorological record.

By 1950, the informational-globalist imperative of planetary monitoring in meteorology was already far stronger than those of many other putatively global systems emerging around that time. When computerized weather forecasting arrived later in the decade, ambitions quickly grew for *real-time* planetary data to feed the forecast models. Achieving these became the early WMO's fundamental goal. To meet them, in the 1960s it extended and linked existing data networks to form a global information infrastructure (GII). Decades before the World Wide Web, this became the first WWW: the World Weather Watch, a global network for the automatic collection, processing, and distribution of weather and climate information for the entire planet.

I contend that the WMO and the WWW illustrate a profoundly important, though messy and incomplete, transition: from voluntarist internationalism, based on an often temporary confluence of shared *interests,* to quasi-obligatory globalism based on a more permanent shared *infrastructure.* Therefore I will speak not only of informational globalism but also of *infrastructural globalism.* By this I mean the more general phenomenon by which "the world" as a whole is produced and maintained—as both object of knowledge and unified arena of human action—through global infrastructures.

THE INTERNATIONAL METEOROLOGICAL ORGANIZATION AND THE *RÉSEAU MONDIAL*

In the 1850s, telegraphy permitted meteorologists for the first time to create synoptic weather maps, that is, "snapshots" of observations taken simultaneously over very

[3] Martin Hewson, "Did Global Governance Create Informational Globalism?" in *Approaches to Global Governance Theory,* ed. Martin Hewson and Timothy J. Sinclair (Albany, N.Y., 1999), 97–113.

[4] Quoted in Napier Shaw, *The Drama of Weather* (Cambridge, 1939).

large areas. Quite suddenly, it became possible to "watch" storms and other weather phenomena develop and move, as well as to warn those downwind. Empirically based synoptic forecasting did not achieve great accuracy, but its dramatic God's-eye views brought new visibility to meteorology. States—especially their military and agricultural services—began to take a strong interest in weather science. By the end of the nineteenth century, most nations with telegraph networks had established national weather services responsible for charting and predicting the weather.

As Frederik Nebeker has shown, theoretical meteorology diverged from the more practice-oriented national weather services, developing separately in mostly academic institutions. Until the 1920s, theory provided little guidance to forecasters. Well into the twentieth century, the major predictive technique was a form of pattern matching: forecasters hunted through huge libraries of past weather maps, seeking similar situations and making their predictions based on how those patterns had evolved. Really it was not until the 1940s that meteorological theory and forecast practice began to converge.[5]

For over a century, then, national weather services focused principally on collecting and charting data. Beginning with existing networks of military, astronomical, and amateur observers, they added professional observers of their own and built operational data networks. Issues of calibration and standardization immediately became salient. To be useful in synoptic forecasting, data must be collected by instruments calibrated to a single standard, and recorded in similar units (of temperature, velocity, pressure, and so forth). Therefore, national weather services established standards. However, in a story endlessly repeated throughout the history of infrastructure, agreement on and enforcement of standards proved remarkably difficult, even within a country.[6] Implementing international standards was even more problematic.[7]

The relatively small nations of Europe soon understood that because weather moves quickly, data from within their own borders would never be enough for really useful prediction. With international telegraphy, national data could be easily shared. By the 1860s, the first of Hewson's phases of informational globalism was already well under way. Naval weather logs were being collected and shared, while Paris served as a hub for Pan-European telegraphic data exchange. Wherever national standards differed, however, these data exchange systems posed new problems.[8] To address them, in 1873 national weather services throughout Europe and the United States founded an International Meteorological Organization (IMO). The new organization's chief agenda was to coordinate international standards for meteorological measurement and data exchange. Despite hiatuses during the world wars, when virtually all international data exchange ceased, the IMO persisted until 1949.

[5] The Bergen School developed the crucial theory of polar fronts around 1920, but the same school's actual forecast techniques remained chiefly empirical. See Robert Marc Friedman, *Appropriating the Weather: Vilhelm Bjerknes and the Construction of a Modern Meteorology* (Ithaca, 1989); and Frederik Nebeker, *Calculating the Weather: Meteorology in the 20th Century* (New York, 1995).

[6] On implementing a uniform time standard for meteorological observations in the United States, see Ian R. Bartky, "The Adoption of Standard Time," *Technology and Culture* 30 (1989): 25–56.

[7] On problems with the Brussels Convention standard naval logbook, see James R. Fleming, *Historical Perspectives on Climate Change* (New York, 1998).

[8] For example, some nations used British units, while others used the metric system. Standard observing hours, methods of sea surface temperature measurement, and a myriad of other seemingly small details produced tremendous headaches for those trying to integrate data sets across national borders.

The IMO case was typical of pre–World War II scientific internationalism. For seventy-five years, the organization remained a cooperative nongovernmental association of national weather services. The principle of interaction was explicitly voluntary. As a result, IMO standards and policies functioned only as recommendations, which nations were at liberty to refuse or simply ignore. In practice, national identity and independence often mattered more than international standards, though the polite language of scientific exchange muted national rivalries. Each national weather service chose its own balance between IMO standards and its own, sometimes diverging techniques. Ambivalence about intergovernmental status among national weather service directors, who feared bureaucratic meddling, kept the organization frozen in this state until just before World War II.

The tension between national technical systems and internationalist aspirations frustrated early efforts to build a global meteorological data network. At the IMO's founding in 1873, U.S. and Swiss delegates pressed for, and received, general acceptance that existing data networks should be extended into a complete global observing network—in other words, they endorsed the principle of informational globalism. Christophorus Buys Ballot advocated "an International Fund for the establishment of meteorological observatories on islands and at distant points of the Earth's surface."[9] Buys Ballot's proposal failed, but in 1875, under the IMO's aegis, the U.S. Army Signal Office began publishing a regular *Bulletin of International Meteorological Observations Taken Simultaneously,* containing worldwide synoptic charts based on sporadic national contributions. However, coverage beyond the United States and Europe was poor, especially in the Southern Hemisphere.[10]

The first, rudimentary, but partially successful, *global* effort began in 1905, with French meteorologist Léon Teisserenc de Bort's proposal for a telegraph-based global weather data system rather grandly named the *Réseau Mondial* (worldwide network).[11] Simplifying Teisserenc de Bort's ambitious vision, the IMO decided that the network should collect, calculate, and distribute monthly and annual averages for pressure, temperature, and precipitation from a well-distributed sample of meteorological stations on land—creating, in effect, a global climatological database. The distribution standard was two stations within each ten-degree latitude/longitude quadrangle, an area about twice the size of France. Ultimately, the network comprised about 500 land stations between 80°N and 61°S. Seemingly modest in concept, in practice this early project proved ferociously difficult.

Why was it so challenging to gather data from 500 stations and calculate a few averages? The explanation lies in the lack of settled standards, the limited reach of communications infrastructure, and the voluntary, essentially private mode of scientific

[9] Howard Daniel, "One Hundred Years of International Co-Operation in Meteorology (1873–1973): A Historical Review," *WMO Bulletin* 22 (1973): 164.

[10] U.S. Army Signal Office, *Bulletin of International Meteorological Observations Taken Simultaneously* (Washington, D.C., 1875–1884).

[11] Similar efforts to build very large-scale data networks began around the same time in other geophysical sciences, including seismology and oceanography. All of them suffered from standardization problems and long publication delays. See R. D. Adams, "The Development of Global Earthquake Recording," in *Observatory Seismology,* ed. J. J. Litehiser (Berkeley, 1989), 3–24; Elisabeth T. Crawford, Terry Shinn, and Sverker Sœrlin, *Denationalizing Science: The Contexts of International Scientific Practice* (Boston, 1993); Robert Stoneley, "The History of the International Seismological Summary," *Geophysical Journal of the Royal Astronomical Society* 20 (1970): 343–9. I am grateful to Kai-Henrik Barth for drawing this literature to my attention.

internationalism prior to World War II. Although telegraph services relayed weather data free of charge, the network's technical capabilities were not yet robust enough to support Teisserenc de Bort's vision of global, real-time data collection. Instead, the *Réseau* collected most data by mail. The problem of nonstandard observing and recording techniques remained considerable. Retrieving and confirming information from remote observers required great effort. As a result, the *Réseau Mondial's* first annual data set, for 1911, did not appear until 1917. Delays of up to thirteen years marked the publication of subsequent volumes, which ceased with the data for 1932. Without governmental commitments, IMO backing provided little institutional leverage.[12]

Prior to World War II, governmental powers *were* invoked to promote international standards in one particular area: aeronautical meteorology. The rise of international air travel after World War I led to the Paris Convention Relating to the Regulation of Aerial Navigation, which laid out the legal basis for international air traffic and effectively codified the vertical extent of the nation-state. Under the 1919 convention,[13] each nation retained sovereign rights over its own airspace. This would later become a crucial object of cold war maneuver and diplomacy regarding overflight by earth-orbiting satellites.[14] Among other things, the convention specified guidelines for international meteorological data exchange, to be carried out several times daily by radiotelegraph.

In the same year, the IMO established a Technical Commission for the Application of Meteorology to Aerial Navigation. But the Paris convention ignored it, establishing a separate, *intergovernmental* International Commission for Air Navigation (ICAN), charged in part with implementing the convention's meteorological standards. Participating governments officially recognized only ICAN. By 1935, this led the IMO to transform its technical commission into an International Commission for Aeronautical Meteorology (CIMAé) *with members appointed by governments.* CIMAé thus became the first, and until after WWII the only, IMO entity to acquire official intergovernmental status. In the event, most CIMAé members also sat on ICAN; the former functioned more as an IMO liaison than as an independent organization.[15]

This episode reflects both the IMO's endemic institutional weakness and the relative infancy of professional meteorology. Despite the vast scientific, technological, and political changes sweeping around it, the IMO administrative structure constructed in 1889 remained largely unchanged until after WWII. Throughout, the Conference of Directors of national weather services did most of the detail work, supplemented by IMO technical commissions covering specific areas. The larger, broadly inclusive International Meteorological Committee met infrequently to discuss general policy and directions. Both of these bodies met as scientists and forecasters, rather than as government representatives. The IMO had no policy-making powers, serving only as an advisory and consensus body. Between its infrequent meetings, the organization itself did little. The IMO did not acquire a permanent Secretariat until 1926, and the latter's annual budget never exceeded $20,000.[16]

[12] Great Britain Meteorological Office, *Réseau Mondial, 1910: Monthly and Annual Summaries of Pressure, Temperature, and Precipitation at Land Stations* (London, 1920), iv–v.

[13] Part of the Treaty of Versailles, the convention entered into force in 1922.

[14] Walter A. McDougall, *The Heavens and the Earth: A Political History of the Space Age* (New York, 1985).

[15] Daniel, "One Hundred Years" (cit. n. 9).

[16] Ibid.

In this era—before the advent of heavy state investment in scientific research—meteorologists themselves remained divided over the desirability of government involvement. In part this reflected the conflicting loyalties of national weather service directors. Though serving national governments, they saw their primary identity as scientists, and IMO meetings as apolitical spaces for scientific discussion. Intergovernmental status, they feared, might change this, turning them into representatives of their governments, reducing their independence and prerogatives, and perhaps subverting IMO proceedings toward the fulfillment of political agendas. For this group, in other words, scientific internationalism served as a way to *bypass* the nation-state and keep science separate from politics. Another faction, however, saw governmental commitment as the only road to the permanent, fully integrated international data exchange that would aid forecasters and climatologists, especially in Europe. As long as the IMO lacked official status, its decisions could not bind government weather services. As a result, many standardization problems remained unresolved or progressed only slowly toward solutions. For this second group, the road to better science lay *through* political commitment.

There were at least four reasons for the increasing dominance of this view within the IMO. First, as we have just seen, ICAN had challenged the organization's control of meteorological standard setting and threatened its status. Second, by the 1930s a breakthrough—the Bergen School's polar front theory—had focused meteorologists' attention on the hemispheric dynamics of weather.[17] As a result, global information became more than a far-off, abstract goal; reports from the whole Northern Hemisphere above the tropics could be used directly in national weather forecasts.

Third, rapid technological change in the interwar period vastly expanded the possibilities for, and geographic reach of, real-time data exchange. Throughout Europe and the United States, weather services traded data via Teletype, a kind of automated telegraph widely adopted in the 1920s. In the 1930s, data from more remote locations began to arrive via shortwave radio broadcasts, from ships at sea, remote island and land stations, and other locations beyond the reach of the telegraph network. "Bounced" off the ionosphere, shortwave allowed instantaneous data transmission over thousands of kilometers—even across the oceans, under some conditions—though noise in the broadcasts frequently caused errors and incomplete transmission. Telegraph and radio authorities prioritized weather data. By the late 1930s, a rudimentary real-time weather data network covered the Northern Hemisphere between the Arctic Circle and the tropics.

Finally, technoscientific changes had begun to overwhelm the institutional and organizational context. Data traveled widely, even globally—but in a bewildering variety of forms. As late as 1945, the *Handbook of Meteorology* declined to attempt a worldwide survey of meteorological data transmission and coding because

> the currently used codes are far too numerous. There are many reasons for the complexity in weather codes and reports: the diversity of elements to be observed, the various techniques of observation, variation in the information desired by analysts in different

[17] Tor Bergeron, "Methods in Scientific Weather Analysis and Forecasting," in *The Atmosphere and the Sea in Motion,* ed. Bert Bolin (New York, 1959), 440–74; Friedman, *Appropriating the Weather* (cit. n. 5); Nebeker, *Calculating the Weather* (cit. n. 5). The polar fronts, a marked feature of the global atmospheric circulation, are the boundaries between polar and mid-latitude air in each hemisphere. Their locations vary seasonally between about 30° and 60°.

> parts of the world, and lack of uniformity in the codes adopted by separate political units are some of the reasons . . . [M]any political units use International [IMO] codes, [but] others use portions of these codes or have devised forms of their own.[18]

Multiple data transmission techniques exacerbated the problem. Fast, reliable Teletype was the gold standard, but standard telegraph, shortwave broadcast, point-to-point microwave links, and other technologies were also in widespread use. In the precomputer age, collecting and integrating data from all these channels and media remained slow, labor intensive, and error-prone.

In hopes of conquering this meteorological babel and regaining central control of standard-setting processes, IMO leaders came to agree on the need for powers like those of ICAN. In 1929, the organization posted a letter to governments seeking intergovernmental status. Arriving on the eve of the Great Depression, this proposal was generally ignored. The IMO revisited the issue with renewed vigor at its 1935 meeting in Warsaw. This time, in an attempt to acquire government endorsement by stealth, the group decided to submit future meeting invitations directly to governments, asking them to appoint each weather service director as an official government representative. At the same time, led by France and Norway, the organization began drafting a World Meteorological Convention that would secure intergovernmental status.

A preliminary version of this convention was presented to the 1939 meeting of the IMO before World War II—held, ironically, in Berlin. IMO president Theodor Hesselberg's comments reflected the general frustrations with the IMO's unofficial status:

> In view of the steadily increasing practical importance of meteorology, it is desirable that governments . . . should have a greater influence on the work of the Organization. The resolutions of the Organization should be binding on the countries to a greater extent. The Organization must be able to rely on adequate resources so that efficient cooperation should not be hampered by financial difficulties. It is abnormal for one of the Organization's commissions [the intergovernmental CIMAé] to have a more official status than the Organization itself. Similar organizations [such as CIMAé rival ICAN] have a more official status than IMO, a circumstance which has its drawbacks.[19]

The conferees forwarded the draft World Meteorological Convention to a committee for refinement. Plans called for final approval at a 1941 Conference of Directors, and the stage seemed set for meteorology's transformation. War, of course, intervened.

FROM INTERNATIONAL TO GLOBAL: THE WORLD METEOROLOGICAL ORGANIZATION

The IMO could not meet again until 1946. Already primed for a major change by its prewar activism, the organization worked at a furious pace, building on the draft convention written seven years earlier. Agreement was by no means unanimous; many participants remained skeptical of the value of intergovernmental status. The key, perennial issues were whether the change might lead to control of meteorology by politicians rather than professional meteorologists and whether the new organization would reduce the prerogatives of national weather services to function as they saw fit.

[18] G. R. Jenkins, "Transmission and Plotting of Meteorological Data," in *Handbook of Meteorology,* ed. F. A. Berry Jr., Eugene Bollay, and Norman R. Beers (New York, 1945), 574.

[19] Quoted in Daniel, "One Hundred Years" (cit. n. 9), 174.

Nonetheless, in the postwar atmosphere of optimism, conferees resolved the major outstanding questions in just over a year. Reassured by negotiators that professional status would remain primary, that nations would retain equal rights as members, and that governments would not control its deliberations, the final drafting conference in Washington, D.C., drew to a close in October 1947.

The new organization would be one among many "specialized agencies" of the United Nations, so it would have to conform to UN rules of membership. As a result, during the final proceedings an important change occurred. American legal experts advised the conferees that membership in the new organization should be accorded only to "sovereign states," as recognized by the UN. Defined in Article 3(c) of the World Meteorological Convention as a nation's "being fully responsible for the conduct of its international relations," this criterion excluded from full membership not only divided nations such as Germany—the major issue immediately after the war—but also the People's Republic of China (PRC), colonial territories, and individual Soviet republics. As Clark Miller has observed, for meteorologists this "new vocabulary of 'States' instead of 'countries' superimposed a *geopolitical* imagination of the world over the *geographical* imagination that had previously organized meteorological activities."[20] These debates mirrored those occurring simultaneously in the UN itself. In the end, representatives of thirty-one governments signed the World Meteorological Convention in October 1947. The convention entered into force in early 1950, and in 1951 the International Meteorological Organization officially became the World Meteorological Organization (WMO).

The explicit and fundamental purpose of the new organization was informational globalism. As outlined in the convention's opening paragraphs, WMO goals were:

> (a) To facilitate world-wide co-operation in the establishment of networks of stations for the making of meteorological observations or other geophysical observations related to meteorology . . . ;
> (b) To promote . . . systems for the rapid exchange of weather information;
> (c) To promote standardization of meteorological observations and to ensure the uniform publication of observations and statistics;
> (d) To further the application of meteorology to aviation, shipping, agriculture, and other human activities; and
> (e) To encourage research and training in meteorology and to assist in co-ordinating the international aspects of such research and training.[21]

Although these goals differed little from those of the IMO, now meteorologists could call upon the power of government, via the authority (and the finances) of the UN, to implement them.

Committed as they already were to informational globalism, those drafting the convention must have been struck by the irony of a "world" organization that excluded some nations. The convention did specify a mechanism by which states not belonging to the UN could apply to join the WMO; approval required a two-thirds majority

[20] Clark A. Miller, "Scientific Internationalism in American Foreign Policy: The Case of Meteorology, 1947–1958," in *Changing the Atmosphere: Expert Knowledge and Environmental Governance*, ed. Clark A. Miller and Paul N. Edwards (Cambridge,, Mass., 2001), 167–218.

[21] World Meteorological Organization, *Basic Documents (Excluding the Technical Regulations)* (Geneva, 1971), 9.

vote. Territories (i.e., colonies and protectorates) could also join, under the sponsorship of their governing states. Membership grew quickly. By the mid-1960s, most nations were represented.[22]

The exceptions to this rule remained, however, extremely significant, and the issue of sovereign statehood as a requirement for membership would dog the new organization for decades. The First World Meteorological Congress, in 1951, immediately moved to soften the rebuff of the PRC's exclusion by inviting that nation to participate as an "observer." This decision became a general policy: any nonmember nation could send official observers to World Meteorological Congresses. Further, the director of the nation's meteorological service could attend or be represented at technical commission meetings.

This uneasy compromise avoided overt conflict with UN policy and the United States, but it did not satisfy the desire of many states for full recognition. Only five nonmember nations sent observers to the Second World Meteorological Congress, in 1955.[23] For many years the second-class "observer" status and the exclusion of divided nations provoked anger. For example, during the 1971 sixth congress, held at the height of the Vietnam War, Cuban delegate Rodriguez Ramirez insisted on reading into the minutes a statement denouncing the exclusion of "the socialist countries" from full membership. Ramirez accused the WMO of hypocrisy:

> The World Weather Watch would have more amply fulfilled its objectives had the WMO opened its doors to all countries. . . . The WMO . . . is rejecting UN agreements on the peaceful uses of the World Weather Watch. Viet-Nam, in particular, has suffered the destruction of nearly half of its meteorological stations, loss of the lives of more than 100 scientists and meteorological workers, terrible destruction of its forest wealth by the use of chemical products which have altered its ecology and biology . . . , at the hands of the armed invasion forces of the United States and its allies. This declaration, Mr. Chairman, has been supported by the socialist countries of Byelorussia, Bulgaria, Czechoslovakia, Hungary, Mongolia, Poland, Romania, Ukraine and the Soviet Union.[24]

U.S. representative George Cressman responded heatedly that such statements "served no purpose other than to interrupt the proceedings with political propaganda." Still, he could not resist venting some propaganda of his own, justifying the U.S. intervention in Vietnam as an invited response to "coercion, organized terror and subversion directed by North Viet-Nam."[25] Though such confrontations remained rare at WMO meetings, they marked the subterranean antagonism between the informational globalism inherent in the organization's scientific and operational goals and the conflicted, voluntarist internationalism inherited from the IMO.

Under the Westphalian internationalist model prevailing at the time of the IMO's founding, states retained absolute control over affairs within their territories and had

[22] For membership figures, see Daniel, "One Hundred Years" (cit. n. 9), 187; and Sir Arthur Davies, *Forty Years of Progress and Achievement,* WMO-721 (Geneva, 1990), 151–2.

[23] "Second World Meteorological Congress," *WMO Bulletin* 4(3) (1955), 94.

[24] For Ramirez, as for other members of the Communist bloc, this category included individual republics of the Soviet Union, which were still arguing (unsuccessfully) for separate representation at the United Nations. It also incorporated the Communist governments of divided nations such as Germany (not admitted to the UN until 1973), Vietnam (1977), and Korea (1991).

[25] The Ramirez-Cressman exchange is recorded in World Meteorological Organization, "Sixth World Meteorological Congress, Geneva, April 5–30, 1971: Proceedings," in *Congress Proceedings,* WMO CP-6 (Geneva, 1972), 162–3 (Ramirez), 164 (Cressman).

none whatsoever over the affairs of other states.[26] No state recognized any authority higher than its own. International associations existed simply to promote mutual (and shifting) interests. The paradigm cases were military alliances. Accordingly, the IMO had sought to secure common standards through persuasion via an ethic of shared, universal scientific interests. As in other organizations following the internationalist model, the IMO's constituents had cooperated when it served their mutual interests but readily ignored IMO directives when their goals diverged.

The UN system simultaneously perpetuated and eroded this voluntaristic internationalism. On the one hand, the UN strengthened the nation-state framework by codifying the rights of states against each other and creating explicit criteria for legitimate sovereignty. On the other hand, these very acts also limited sovereign power, implicitly asserting the UN's authority to challenge the legitimacy of governments. Its status as a world organization made withdrawal from the UN system difficult and costly. Contemporaries clearly experienced these contradictions as acute challenges, frequently hedging their commitments to avoid even the appearance of surrendering sovereign powers. Therefore, like most accords in the early years of the post-WWII international order, the WMO convention carefully avoided any claim to absolute authority. Rather than dictate to its member states, the WMO would "promote," "encourage," "facilitate," and so on. Under Article 8 of the convention, members were required to "do their utmost" to implement WMO decisions.

However, members could still *refuse* to adopt any WMO recommendation simply by notifying the WMO secretary-general and stating their reasons. Such deviations instantly became a ubiquitous issue at WMO technical meetings. For example, the Soviet Union and some other countries, "for practical reasons," elected to continue their standard two-hourly observing times of 02, 04, 06 GMT, and so forth, despite a majority view that a three-hourly system at 03, 06, 09 GMT, and so forth, would be sufficient. A compromise "placed emphasis on" the three-hourly times. At its first meeting in 1953, the Commission on Synoptic Meteorology expressed confusion about the contradiction between Article 8 of the WMO Convention and Resolution 15(I) of the First World Meteorological Congress (1951), which spoke of "*obligations* to be respected by meteorological administrations."[27] Debate ensued over whether to frame regulations in terms of "shall" or "should." Ultimately, the commission put off any decision. Nor did the WMO Executive Committee feel ready to impose stronger language. Both bodies deferred to the full World Meteorological Congress.

The Second World Meteorological Congress, in 1955, spent considerable time confronting this problem. Finally, the congress decided to issue two separate sets of WMO regulations. All WMO members were expected to conform, "within the sense of Article 8," to the "standard" regulations, while a second set of "recommended" regulations could be implemented at members' discretion. The criterion dividing these two sets was whether a given practice was considered "necessary" to the collection of a minimal global data set or merely "desirable."[28] Nonetheless, deviations even from "standard" practices remained common for many years.

[26] For a review see Held et al., *Global Transformations* (cit. n. 2), chap. 1.

[27] WMO Commission for Synoptic Meteorology, *Abridged Final Report of the First Session, Washington, 2nd–29th April, 1953,* CSM-1/WMO-16 (Geneva, 1953), 41 (my italics).

[28] "Second World Meteorological Congress" (cit. n. 23), 95.

INFRASTRUCTURAL GLOBALISM

Very slowly, the new WMO chipped away at the Herculean task of integrating the unruly complexity of national weather observing and communication systems into a functional planetary infrastructure. It accomplished this by embedding social and scientific norms in worldwide infrastructures, in two complementary ways. First, as the process of decolonization unfolded, the WMO sought to *align individuals and institutions with world standards* by training meteorologists and building national weather services in emerging nations. Second, the WMO worked to *link national weather data reporting systems into a single, increasingly automated global data collection and processing system.* In the early 1960s, as we will see below, the WMO began explicitly planning a global information infrastructure, the World Weather Watch.

As mentioned above, most theorists of globalization discuss information and communication technologies. Few, however, distinguish between ICTs as nonspecific channels and ICT infrastructures dedicated to specific forms of globalist information. Even Hewson's insightful discussion confounds these. Although the two are clearly related, and both are important, I argue that the latter have special significance. International communication channels, such as post, telegraph, and telephone, facilitate global flows of information, but they neither produce information nor seek to control its quality. Specifying world standards for linking communication systems facilitates globalization, but specifying uniform standards for globalist information *actively produces a shared understanding of the world as a whole.* This is why I believe we should see the meteorological project not only as informational but also as *infrastructural globalism.*

This concept refers to efforts to achieve globalist goals *by building permanent, unified world-scale institutional-technological complexes.* If Hewson's notion of informational globalism captures the emergent idea that knowledge about the whole world has practical value and sociopolitical legitimacy, then infrastructural globalism describes the material dimension of this imperative. The value of the term "infrastructure" here is manifold. Even in everyday usage the word comprehends both institutions and technological systems.[29] It expresses the invisibility that systems acquire as they become embedded in ordinary life and work as well as the reliance placed on them by whole societies. It also captures the endurance of some sociotechnical systems and institutions, whose momentum and long lifespans limit and shape human agency even as they are shaped by it.[30] (We might call this mutual shaping "infrastructuration," suggesting its substantial resonance with Anthony Giddens's structuration theory.)[31] Enduring, reliable global information infrastructures build both

[29] For example, the *American Heritage Dictionary* (New York, 2000) defines infrastructure as "the basic facilities, services, and installations needed for the functioning of a community or society, such as transportation and communications systems, water and power lines, and public institutions including schools, post offices, and prisons."

[30] On these concepts, see Paul N. Edwards, "Infrastructure and Modernity: Scales of Force, Time, and Social Organization in the History of Sociotechnical Systems," in *Modernity and Technology,* ed. Thomas J. Misa, Philip Brey, and Andrew Feenberg (Cambridge, Mass., 2002), 185–22; Geoffrey C. Bowker and Susan Leigh Star, *Sorting Things Out: Classification and Its Consequences* (Cambridge, Mass., 1999); Thomas P. Hughes, "The Evolution of Large Technological Systems," in *The Social Construction of Technological Systems,* ed. Wiebe Bijker, Thomas P. Hughes, and Trevor Pinch (Cambridge, Mass., 1987), 51–82.

[31] Anthony Giddens, "Agency, Institution, and Time-Space Analysis," in *Advances in Social Theory and Methodology: Toward an Integration of Micro- and Macro-Sociologies,* ed. Karin Knorr-Cetina and Aaron V. Cicourel (Boston, 1981), 161–74; idem, *The Constitution of Society* (Berkeley, 1984).

scientific and political legitimacy for the knowledge they produce. Similarly, long-term dependency on global information infrastructures can subtly erode expectations of state sovereignty, as many have noted in connection with more recent GIIs such as the Internet and the World Wide Web. Thus infrastructural globalism (to the extent that it succeeds) is a particularly effective agent of globalization.

The WMO began its project in infrastructural globalism by exerting three kinds of institutional power. First, like the IMO before it, the organization served as a central site for negotiating technical standards. WMO technical commissions worked more vigorously than their predecessors, in part because constant effort was required simply to keep abreast of the many new instruments and techniques arriving in the 1950s. The technical commissions and quadrennial World Meteorological Congresses provided the necessary opportunities to resolve differences over standards. Over time, these institutional decisions became embedded in the emerging infrastructure, built into instruments and technological systems—a trend that continues into the present. Weather balloons and automated weather stations, for example, take readings and broadcast them for processing by meteorological centers. WMO standards govern how these instruments are constructed, used, and calibrated, as well as how their data are interpreted.

The second institutional power exerted by the WMO was simple peer pressure. Its founding members clearly hoped that the organization's new status would produce conformity to standards almost by force. Instead, as we have already seen, the process took considerable time. Lacking any kind of police power, the WMO exerted peer pressure chiefly through meetings and official publications. At first, like its predecessor, the central organization in Geneva maintained only a skeleton staff. Except for the congresses, most efforts coordinated by the WMO took place elsewhere. The Secretariat conducted no research and played no part in managing data networks; all of that was still done by national weather services. Its only activities were to facilitate meetings and to print and distribute WMO publications.

However, the organization's budget grew rapidly in its first two decades. Annual spending, only about $300,000 in the early 1950s, had quadrupled to about $1.3 million twelve years later, and by 1968 the annual budget was nearly $4 million. The WMO Secretariat acquired permanent offices in Geneva in 1955, moving into its own building in 1960. On a symbolic level, the increasingly substantial presence of a central organization mattered enormously. The series of WMO-coordinated international ventures beginning with the International Geophysical Year (IGY), 1957–1958, and culminating in the 1970s with the World Weather Watch and the Global Atmospheric Research Program placed increasingly stringent requirements for standardized observations on participants.

The third and most direct of the WMO's institutional powers was its technical assistance program. At the time of the First WMO Congress, in 1951, the impending independence of Libya, formerly an Italian colony, created the possibility of a break in meteorological services there as the existing weather service was staffed mainly by non-Libyan personnel. The congress directed the WMO Executive Committee to propose a plan for continuing services and "to express the willingness of the WMO to provide all possible technical assistance within its available resources."[32] From mod-

[32] World Meteorological Organization, *Final Report: First Congress of the World Meteorological Organization, Paris, 19 March–28 April 1951* (Geneva, 1951), 10.

est beginnings—$23,000 contributed to four countries in 1952—the Voluntary Assistance Program (VAP) soon became one of the WMO's most significant activities.

Decolonization, accelerating after 1955, created some forty new nations, multiplying the problem posed by Libya manyfold. Newly independent, poor countries, with inexperienced leaders and shaky governments, typically had few resources and less attention for meteorology. Throughout the decolonization period, the United Nations Expanded Program of Technical Assistance for the Economic Development of Under-Developed Countries (EPTA) invested in a variety of meteorological assistance projects under WMO guidance. (EPTA, established in 1950, was absorbed into the larger United Nations Development Programme [UNDP] in 1966.)[33] Though hardly the most substantial of EPTA/UNDP expenditures, neither were these projects negligible, typically comprising 1–3 percent of EPTA/UNDP's overall budget.[34]

Initially, the WMO had hoped to rely entirely on EPTA for funds, but the latter's small budget and shifting priorities made it an unreliable ally. Therefore the WMO established its own Voluntary Assistance Program in 1956. Although the majority of funding continued to flow from EPTA, between 1956 and 1959 the WMO's own VAP contributed some $430,000 in aid to thirty-four countries, mostly in the form of on-site expert assistance and fellowships for meteorological training. In the next WMO financial period, 1960–1963, this figure reached $890,000; in 1964–1967, it rose to $1.5 million, with EPTA and its successor, the UNDP, contributing another $6.5 million. By 1972, the WMO and ETPA/UNDP together had spent a total of about $55 million on meteorological assistance to developing nations, including some 700 expert missions, 1,500 fellowships, and numerous seminars and training courses in some 100 nations.[35] Wealthier WMO members also donated large amounts of equipment to less developed nations.

Who paid for all this? Contributions to the VAP varied from year to year, but as a rule the large majority of the WMO portion came from the United States. The Soviet Union typically provided roughly half the U.S. contribution (almost all of it in kind rather than in direct financial aid). The United Kingdom and France were the third and fourth largest contributors, each donating amounts roughly one-tenth of the U.S. amount. Sweden led the list of other European countries that provided most of the rest. Altogether, some fifty nations—including some of those that also received aid—made monetary or in-kind contributions to the fund during the 1960s.[36]

The WMO perceived these activities as purely technical. As Miller has argued, however, at a larger level they formed part of a new politics of expertise. Recipients, particularly those engaged in nation-building, often understood them as part of a political program. For example, by helping the new Israeli state to provide expert advice to its (mostly immigrant) citizens, WMO assistance to the Israeli weather service simultaneously promoted the legitimacy of the new state.[37]

It would be absurd to claim any major role for the WMO in nation-building. Yet the organization certainly helped to construct an international community of civil servants,

[33] Ruben P. Mendez, *United Nations Development Programme,* http://www.yale.edu/unsy/UNDPhist.htm.

[34] Miller, "Scientific Internationalism" (cit. n. 20).

[35] Daniel, "One Hundred Years" (cit. n. 9).

[36] See, e.g., *Consolidated Report on the Voluntary Assistance Programme Including Projects Approved for Circulation in 1971,* WMO-323 (Geneva, 1972).

[37] Miller, "Scientific Internationalism" (cit. n. 20).

science and technology administrators, scientists, and engineers who carried the banner of their native countries. The Voluntary Assistance Program furthered the representation of weather expertise as a basic *and apolitical* element of the infrastructures furnished by modern sovereign states to modern citizens. Multiplied across many forms of scientific and technical expertise, this representation promoted the integration of expert institutions into emerging liberal states. Additionally, by creating channels and even requirements for the two-way flow of scientific information, the practice helped reduce the chance—much feared by early cold warriors—of being "scooped" in critical areas of science or technology by insular or secret state-sponsored Communist institutions. Ultimately, these and the myriad of similar intergovernmental scientific and technical bodies that arose after World War II heralded "a significant shift of foreign policy responsibilities from Departments of State to other government agencies as the participation of experts in international institutions has become central to international affairs."[38]

The technical assistance program also served as a key conduit for the WMO's standardization efforts. The training and expert advisory programs accomplished this not only through their educational content but also by building human relationships and participatory norms. WMO documents on training frequently stressed the importance of communicating to newly trained meteorologists the value of their contribution to the global effort. Efforts were made (and also resisted) to standardize syllabi for WMO-sponsored training courses.[39] Equipment donated through the VAP functioned to carry WMO standards, embodied in the machines, from donors to recipients.

In summary, the early WMO did not immediately fulfill the expectations of its founders. Instead, joining the UN system actually inhibited the WMO's informational globalism by preventing all nations from joining on equal terms, and it involved the organization in cold war politics in ways its leaders probably did not anticipate. However, cold war geopolitics *also* worked in the organization's favor in several important ways. First, the superpowers themselves began to seek global information in arenas that included weather. Second, two of the cold war's most central technologies—computers and satellites—would become the most important tools of meteorology as well. Finally, international scientific cooperation would become part of the cold war's ideological dimension.

METEOROLOGY, COLD WAR POLITICS, AND NEW TECHNOLOGY

The geopolitical context of post-WWII geophysics was the desire of both superpowers for a multidimensional form of global reach. This necessarily involved the collection of certain kinds of globalist information. As I showed in *The Closed World,* the United States' foreign policy of "containment" conceptualized the cold war as a global struggle, reading all conflicts everywhere in the world as part of the contest for military and ideological advantage. Containment strategy materialized in specific technological forms. High-technology weapons, in the form of thermonuclear bombs, long-range bombers, nuclear submarines, and missiles, would project U.S. power across the globe, while computers, radar, and satellites would enable centralized, real-

[38] Ibid.

[39] J. Van Mieghem, *Problem of the Professional Training of Meteorological Personnel of All Grades in the Less-Developed Countries,* WMO TN-50 (Geneva, 1963).

time surveillance and control. Heavy investment in military equipment would reduce reliance on men under arms, something the American public was never keen to support. The extremely rapid improvement of computers between 1945 and 1960 owed much to cold war goals. By enabling centralized command and control on a vast scale, computers and ICT infrastructure also helped to shape American global ambition.[40]

Mathematician John von Neumann, one of the major figures in both computer and nuclear weapons development, promoted the new machines for "numerical weather prediction" (NWP), or weather forecasting by means of mathematical models. Von Neumann helped found the Joint Numerical Weather Prediction (JNWP) Unit, which brought operational computerized forecasting to the United States starting in 1955. "Joint" here meant combined support from the U.S. Weather Bureau, the U.S. Air Force, and the U.S. Navy, with the military backers providing the lion's share. The idea of weather *control*—techniques such as cloud seeding and hurricane steering—was frequently deployed (not least by von Neumann himself) to justify NWP, receiving about half the total U.S. government budget for weather research throughout the 1950s.[41] The prospect of using weather as a weapon remained very much on the military agendas of both superpowers well into the 1970s, when it was finally abandoned.[42]

Military technological change also increased the superpowers' appetites for global weather data and forecasts. High-flying jet aircraft needed information on the jet streams and other high-altitude weather phenomena, which could also affect ballistic missiles. Tactical nuclear strategy depended on knowing the likely path of fallout clouds and the distances they might travel on the wind. In the 1950s, the U.S. Air Force Air Weather Service (AWS) grew to be the world's largest weather agency, employing an average of 11,500 staff. During this period, approximately 2,000 of these AWS personnel were officers with some degree of formal training in meteorology. By the end of the decade, military officers accounted for over half of the total enrollment in meteorology programs at American universities.[43]

Geostrategy and technological change—mutually reinforcing—thus aligned military interests with the informational globalism of scientists involved in NWP research. Global data procurement grew into a joint, unified effort of the Weather Bureau and the navy and air force weather services and (later) of NASA as well.[44] American military weather observations, especially from radiosondes and reconnaissance aircraft

[40] Paul N. Edwards, *The Closed World: Computers and the Politics of Discourse in Cold War America* (Cambridge, Mass., 1996).

[41] In practice, NWP and weather control remained separate research tracks. See Paul N. Edwards, "The World in a Machine: Origins and Impacts of Early Computerized Global Systems Models," in *Systems, Experts, and Computers,* ed. Thomas P. Hughes and Agatha C. Hughes (Cambridge, Mass., 2000), 221–54; James R. Fleming, "Fixing the Weather and Climate: Military and Civilian Schemes for Cloud Seeding and Climate Engineering," in *The Technological Fix,* ed. Lisa Rosner (New York, 2004), 175–200; Kristine C. Harper, "Research from the Boundary Layer: Civilian Leadership, Military Funding, and the Development of Numerical Weather Prediction (1946–55)," *Social Studies of Science* 33 (2003): 667–96; Chunglin Kwa, "The Rise and Fall of Weather Modification: Changes in American Attitudes towards Technology, Nature, and Society," in Miller and Edwards, *Changing the Atmosphere* (cit. n. 20), 135–66; John von Neumann, "Can We Survive Technology?" *Fortune* (June 1955): 106–8, 151–2.

[42] Kwa, "Weather Modification" (cit. n. 41).

[43] John F. Fuller, *Thor's Legions: Weather Support to the U.S. Air Force and Army, 1937–1987* (Boston, 1990); Committee on Atmospheric Sciences, U.S. National Research Council, "The Status of Research and Manpower in Meteorology," *Bulletin of the American Meteorological Society* 41 (1960): 554–62.

[44] Gerald L. Barger, *Climatology at Work* (Washington, D.C., 1960).

flying from remote Arctic and Pacific island airbases such as Alaska, Greenland, Hawaii, the Philippines, Midway, and Guam, became very important sources of upper-air data in sparsely covered regions. Bases in France, Germany, Japan, and Korea provided coverage of the surrounding regions independent of national weather services. The increasingly worldwide forays of American military vessels supplemented coverage of the oceans. A key purpose of this military network in the 1950s was to monitor atmospheric tests of nuclear weapons and the spread of fallout.[45]

Military observing networks freely shared most of their synoptic data, but they also produced their own separate, secret forecasts and data.[46] Still, there is little evidence that these were any better than those produced by their civilian counterparts. Indeed, military weather services experienced ongoing threats to their survival from commanders who found them redundant or sought to cut costs by relying on civilian forecasts instead. Around 1960, both the U.S. Air Weather Service and the Royal Swedish Air Force in fact discontinued some internally produced forecasts in favor of publicly available results.[47] But the separate military weather networks had another purpose: to provide an independent basis for forecasting in case war stopped the flow of data through the civilian network. In the end, international cooperation in data sharing continued with few interruptions throughout most of the cold war. Even at the cold war's most dangerous moment—the Cuban missile crisis of 1962—the weather services of the opposing nations (i.e., Cuba and the United States) continued their routine exchanges of data.[48]

Cold war politics did sometimes impede the free exchange of weather data from civilian weather services. Before 1956, the People's Republic of China—excluded from full WMO membership—shared no weather data at all. The Soviet Union did provide most information but withheld the locations of some weather stations near its northern borders, presumably for military reasons. The U.S. Air Weather Service was able to determine the probable location of these stations by modeling their fit to several months' worth of weather analyses.[49]

By the end of the 1950s, the most pressing weather-related military issues involved satellites. Intelligence has always been crucial to military strategy. During the cold war, it reached new levels of urgency as well as technological sophistication. Uncertainty about the other side's real capabilities drove an accelerating race to build ever more, faster, and longer-range bombers and missiles. The United States, in particular, found it difficult to penetrate the closed Soviet society with human agents, while the Soviet propaganda machine produced convincing images of that nation's rapidly advancing high-tech armed forces. From 1956 on, American knowledge of Soviet military capabilities came largely from secret reconnaissance flights by high-altitude U-2 spy planes. These flights were illegal under the 1922 Convention Relating to the Reg-

[45] Fuller, *Thor's Legions* (cit. n. 43); David M. Hart and David G. Victor, "Scientific Elites and the Making of U.S. Policy for Climate Change Research," *Soc. Stud. Sci.* 23 (1993): 643–80.

[46] For example, the Defense Meteorological Satellite Program (DMSP) operated independent military weather satellites—nearly identical to their civilian counterparts—from 1962 until 1994, when the civil and military meteorological satellite programs were combined. R. Cargill Hall, "A History of the Military Polar Orbiting Meteorological Satellite Program," *Quest* 9(2) (2002): 4–25.

[47] Fuller, *Thor's Legions* (cit. n. 43).

[48] Gerald S. Schatz, *The Global Weather Experiment: An Informal History* (Washington, D.C., 1978).

[49] Richard Davis, U.S. National Climatic Data Center, interview with author, 1998, Asheville, N.C.

ulation of Aerial Navigation, which reserved sovereign rights to national airspace. When a Soviet antiaircraft missile shot down a U-2 over Svedlovsk in May 1960, finding another way to spy from above became an urgent United States priority. Satellites offered a bulletproof alternative, more difficult to detect and virtually invulnerable to attack.

The first proposals for intelligence satellites, in RAND Corporation studies of the late 1940s and early 1950s, almost immediately noted the tight links between military satellite reconnaissance and weather forecasts. To have intelligence value, photographs from space would have to be taken on clear days, with little or no cloud cover. How better to improve forecast quality than with satellites? Explorer VII, launched in 1959, carried both the first meteorological instruments for measuring radiation balance and the first camera from which film was successfully recovered for the secret CORONA spy program.[50] As on many other occasions, public meteorological purposes thus provided cover for secret military ones. The famed U.S. Television and InfraRed Observation Satellite (TIROS) weather series began life in 1956 under an army "weather reconnaissance" program, but it became a "civilian" project after being transferred to NASA.[51] In the end, the U.S. Weather Bureau received responsibility for the TIROS operation, with NASA retaining responsibility for engineering and launch.[52] By 1966, thirteen TIROS satellites had been placed into orbit, and other meteorological satellite programs were underway.

Would spy satellites violate the sovereignty of nations they overflew? International law did not specify an upper limit to national airspace. Perhaps there *was* no limit, although no nation then possessed the ability to intercept a foreign satellite. Walter McDougall has shown that a covert purpose of John F. Kennedy's proposals for "peaceful uses of outer space" was to preempt legal challenge on this issue. A world reaping the daily benefits of weather and telecommunications satellites would be much less receptive to objections against satellite photography of airbases and missile silos. In the event, the Soviets never raised the issue, probably because they had the same objective. Both American and Soviet satellite programs continued on their dual course, with large, secret military programs looming quietly behind the huge fanfare accorded the public, peaceful space race, in which weather and telecommunications satellites led the way.

Infrastructural globalism in meteorology also benefited from the IGY. Scientists from some 67 nations conducted global cooperative experiments to learn about world-scale physical systems, including Earth's oceans, ionosphere, magnetic field, and geologic structure. The IGY's atmospheric component claimed even broader participation, from some 100 nations. The IGY served significant ideological purposes for both superpowers, which used their scientific collaboration to promote their technological prowess and their commitment to peaceful coexistence.[53] Internationalism may have

[50] John Cloud, personal email, July 7, 1997.

[51] Abraham Schnapf, "The Tiros Meteorological Satellites—Twenty-Five Years: 1960–1985," in *Monitoring Earth's Ocean, Land, and Atmosphere from Space: Sensors, Systems, and Applications*, ed. Abraham Schnapf (New York, 1985), 51–70.

[52] Margaret E. Courain, *Technology Reconciliation in the Remote-Sensing Era of United States Civilian Weather Forecasting: 1957–1987* (Ph.D. diss., Rutgers Univ., 1991).

[53] Miller, "Scientific Internationalism" (cit. n. 20); Clark A. Miller and Paul N. Edwards, "Introduction: The Globalization of Climate Science and Climate Politics," in Miller and Edwards, *Changing the Atmosphere* (cit. n. 20), 1–30.

been indigenous to science, but in the IGY it would be used, in a marriage of convenience, to help guarantee American (and more broadly Western) political interests.[54] As Ron Doel has observed, this argument implies a kind of co-optation.[55] Instead, the situation was one of what I have called "mutual orientation."[56] Science would be used to promote a particular vision of world order, but in exchange scientists could better promote their own. Their involvement in government and governance would, in the long run, produce pressures to which governments would be forced to respond. Ozone depletion and climate change are major cases in point.

The WMO became heavily involved in early IGY planning. A general consensus emerged across the many sciences represented that the IGY's overarching purpose should be to study the Earth as a "single physical system." This fit well with meteorology's increasing orientation toward large-scale, hemispheric or global atmospheric motion. Hence the IGY's meteorological component focused most of its attention on observations of the global general circulation. Three pole-to-pole chains of observing stations were established along the meridians 10°E (Europe/Africa), 70°–80°W (the Americas), and 140°W (Japan/Australia). Dividing the globe roughly into thirds, these stations coordinated their observations to collect data simultaneously on specially designated "regular world days" and "world meteorological intervals." In addition to balloons and radiosondes, an atmospheric rocketry program, initially proposed by the Soviet Union, retrieved information from very high altitudes. In addition to Sputnik—nominally an IGY experiment—six other satellites were successfully launched, though the meteorological data they returned had little value. Here as elsewhere, the IGY marked a transition that would be fully achieved only later. Extensive efforts were made to gather information about the Southern Hemisphere from commercial ships, as well as (for the first time) from the Antarctic continent. In total, the IGY meteorological network claimed some 2,100 synoptic surface stations and 650 upper-air stations—far more than the networks for ionospheric and magnetic studies, which counted only about 250 stations each.[57] The program proved so popular that many of its operations were extended through 1959 under the rubric of an International Geophysical Cooperation (IGC) year.

A look at the WMO's data strategy during the IGY reveals the cusp of change from voluntarist internationalism to infrastructural globalism. Plans called for depositing complete collections of IGY/IGC meteorological data at three world data centers (WDCs): one in the United States (WDC-A), the second in the Soviet Union (WDC-B), and a third at WMO headquarters in Geneva (WDC-C).[58] Each WDC would be funded by the host nation. The data centers did not undertake to process the data but merely

[54] Ronald E. Doel and Allan A. Needell, "Science, Scientists, and the CIA: Balancing International Ideals, National Needs, and Professional Opportunities," in *Eternal Vigilance? Fifty Years of the CIA*, ed. Rhodri Jeffreys-Jones and Christopher Andrew (London, 1997), 59–81; Allan A. Needell, *Science, Cold War, and the American State: Lloyd V. Berkner and the Balance of Professional Ideals* (Amsterdam, 2000).

[55] Ronald E. Doel, "Constituting the Postwar Earth Sciences: The Military's Influence on the Environmental Sciences in the USA after 1945," *Soc. Stud. Sci.* 33 (2003): 635–66.

[56] Edwards, *The Closed World* (cit. n. 40), chap. 3.

[57] Aleksandr K. Khrgian, *Meteorology: A Historical Survey,* 2nd ed., ed. Kh. P. Pogosyan and trans. Ron Hardin (Jerusalem, 1970).

[58] Comité Spécial de l'Année Géophysique Internationale, "The Fourth Meeting of the CSAGI," in *Annals of the International Geophysical Year,* ed. M. Nicolet (New York, 1958), 297–395; Sir Harold Spencer Jones, "The Inception and Development of the International Geophysical Year," in *Annals of the International Geophysical Year,* ed. Sydney Chapman (New York, 1959), 383–414.

to compile and distribute it. Each national weather service and research group was responsible for reporting its data to the WDCs.

In an echo of the prewar *Réseau Mondial,* merely collecting and compiling these data took years. The full set was not completed until 1961. Though planners knew that electronic methods of data processing would soon become the norm, standardization in computer storage techniques remained in the future. Therefore the IGY data sets were produced and distributed on microcards, a miniature photographic reproduction method similar to microfiche. This choice, based chiefly on economy and convenience, reflected the WMO commitment to informational globalism as it allowed more members affordable access to the IGY data.[59] At the same time, the lack of a standard format for electronic data processing reflected the continuing technical challenges to infrastructural globalism.

This lack, and the enduring internationalist ethic of voluntarism, profoundly affected the fate of the IGY data infrastructure. Instead of continuing with centralized global data collection, the WMO urged each national weather service to publish its own data. Centralized data collection, opponents feared, might reduce the resources available to national services and duplicate effort. As an official WMO history rather delicately put it, at the Third World Meteorological Congress in 1959,

> some delegates expressed the view that the WMO Secretariat should continue to discharge the functions of the IGY [Data] Centre on a permanent basis. The prevailing view was however that the responsibility for the regular publishing of meteorological observations, and thus for making them readily available for research workers, should be in the hands of national Meteorological Services.[60]

Instead, the organization undertook to catalog all the data residing in national repositories around the world. These immense volumes took years to compile. The data catalog for the IGY appeared in 1962. Two fat volumes listing various additional data sets from around the world arrived in 1965, while a third—"meteorological data recorded on media usable by automatic data-processing machines"—did not see print until 1972.[61]

CONCLUSION: THE FIRST WWW

During the 1950s, dramatic new possibilities for informational globalism emerged as a result of both scientific and technological change. The advent of digital computers meant that physics-based simulations of weather could be performed fast enough to be useful in forecasting. By the end of the decade, computerized forecasting was planned or operational in the United States, the United Kingdom, the Soviet Union, Japan, Sweden, Israel, West Germany, Belgium, Canada, and Australia. Computer-forecast models required data from regular three-dimensional grids over very large

[59] Davies, *Forty Years* (cit. n. 22), 74–5. A full set of 16,500 IGY microcards cost about $6,000. The same data set required about 10 million punch cards, weighing approximately thirty tons. See Barger, *Climatology at Work* (cit. n. 44).

[60] Davies, *Forty Years* (cit. n. 22), 75.

[61] *Catalogue of IGY/IGC Meteorological Data,* IGY-4//WMO-135 (Geneva, 1962); *Catalogue of Meteorological Data for Research,* parts 1 and 2, WMO-174 (Geneva, 1965); *Catalogue of Meteorological Data for Research,* part 3, *Meteorological Data Recorded on Media Usable by Automatic Data-Processing Machines,* WMO-174 (Geneva, 1972).

areas. Initially regional or continental, by the end of the 1950s forecast models were already moving to the hemispheric scale.

For technical reasons, data voids mattered much more to these models than they had to the human synoptic forecasters who preceded them.[62] This made improvement in data networks imperative. Real-time data exchange on a planetary scale would require better infrastructure—not just better technology, but more uniform conformance to WMO standards, better coordination, and more widely distributed knowledge and skills in weather services across the globe. The IGY, conceived as a temporary collaboration, came up short of building a permanent infrastructure, though it represented an important "proof of concept."

By the end of the decade, however, meteorologists were explicitly conceiving an infrastructure to support their informational globalism. Satellites were the spur. In 1959, the WMO convened the Panel of Experts on Artificial Satellites, consisting of U.S. and Soviet representatives. In 1961, as the panel was completing its first report, U.S. president Kennedy delivered an unrelated address to the United Nations General Assembly. He outlined an arms control agenda that included the demilitarization of outer space and promised soon to bring forward new proposals for "further cooperative efforts between all nations in weather prediction and eventually in weather control." These words were soon backed by action. Kennedy's national security adviser, McGeorge Bundy, directed the secretary of state and numerous relevant U.S. government agencies to pursue this objective actively.[63]

In December 1961, the UN General Assembly approved Resolution 1721 encouraging all nations to participate, via the WMO, in efforts "to advance the state of atmospheric science and technology so as to provide greater knowledge of basic physical forces affecting climate and the possibility of large-scale weather modification," as well as to improve weather forecasting.[64] The WMO satellite panel's report, along with a U.S. National Research Council proposal, were rapidly integrated into the World Weather Watch. The project envisaged automatic, global data collection; a global telecommunication system; and computerized data processing for forecasts and climate studies. Implementation began in 1967, following an intensive planning process. Planning focused on making the system as automatic and as global as possible.[65] From that point on, the World Weather Watch has been the WMO's central *raison d'être* and its primary activity.

[62] Jule G. Charney, R. Fjørtoft, and John von Neumann, "Numerical Integration of the Barotropic Vorticity Equation," *Tellus* 2 (1950): 237–54; P. D. Thompson, "A History of Numerical Weather Prediction in the United States," *Bull. Amer. Meteor. Soc.* 64 (1983): 755–69.

[63] McGeorge Bundy, "National Security Action Memorandum No. 101: Follow-up on the President's Speech to the United Nations General Assembly on September 26, 1961," available in the Federation of American Scientists' Intelligence Resource Program online archive of official documents (Washington, D. C.), http://www.fas.org/irp/offdocs/nsam-jfk/index.html.

[64] Available online via the "Index of Online General Assembly Resolutions Relating to Outer Space," United Nations Office for Outer Space Affairs, Vienna, http://www.oosa.unvienna.org/SpaceLaw/gares/.

[65] N. G. Leonov, H. P. Marx, and WMO, *Requirements and Specifications for Data-Processing System,* WWW Planning Report No. 8 (Geneva, 1966); T. Thompson, *Telecommunications Problems in Computer-to-Computer Data Transfer,* WWW Planning Report WWW-PR-3 (Geneva, 1966); U.S. Weather Bureau, *The World Weather Watch: An International System to Serve All Nations* (Washington, D.C., 1965); WMO, *Planning of the Global Telecommunication System,* WWW Planning Report No. 16 (Geneva, 1966); idem, *The Role of Meteorological Satellites in the World Weather Watch,* WWW Planning Report No. 18 (Geneva, 1967); idem, *World Weather Watch: Status Report on Implementation* (Geneva, 1968).

The World Weather Watch *could* be understood simply as a series of incremental improvements in an existing global network, driven by an inexorable process of technological change (and some of its planners have described it this way). But this is to misconstrue its significance as a technopolitical achievement. It marked the successful transfer of key standard-setting and coordinating powers from national weather services to a permanent, globalist intergovernmental organization. Unlike its many predecessors, this global data network has persisted now for four decades, gathering momentum as it grows. It is a genuinely global infrastructure that produces genuinely global information. Virtually all nations contribute data and receive, in turn, WWW data products.

Has infrastructural globalism in meteorology limited the power of national governments? Generally, the answer must be yes. The globalization of data networks makes it almost unthinkable, not to mention unaffordable, for most nations to develop separate, independent networks or standards. Even military meteorology now relies heavily, though not exclusively, on data provided by public, civilian networks. As an example, in the 1990s some European governments began to contemplate recovering costs by selling meteorological data (in contravention of centuries-old traditions), draining data from the WWW. In response, the WMO—prodded by the United States—defined "basic" data that member states are required to share freely.[66] Despite some ongoing resistance, most governments are complying.

A different question is whether this particular project for infrastructural globalism actually matters outside the meteorological community. Here the politics of global warming provide a unique metric. For three decades, scientific predictions of anthropogenic climate change were resisted by political leaders from many countries, who argued that such predictions were based on theories and computer simulations—not on observations, which still failed to show a convincing signal against the noise of natural climatic variation. From the late 1980s on, the Intergovernmental Panel on Climate Change (IPCC), established jointly by the WMO and the UN Environment Programme, began providing state-of-the-art reports on climate change. By the time of its *Second Assessment Report,* released in 1995, the organization could state that "the balance of evidence" supported the theory of anthropogenic climate change.[67] The 2001 IPCC report went further, claiming "new and stronger evidence that most of the warming observed over the last 50 years is attributable to human activities."[68] These conclusions were endorsed not only by the IPCC's scientists but also by most IPCC member governments. The 2005 ratification of the Kyoto Protocol solidifies this acceptance.

I would argue that this consensus has come about precisely because the weather data network has grown into a well-developed, highly standardized GII. As it improved and endured, measurements accumulated. Uncertainty was reduced. Seemingly contradictory sets of observations from different instrument sets were reconciled. At the

[66] WMO Resolution 40, adopted at the Twelfth World Meteorological Congress (1995), reafffirms the principle of free exchange of basic data and defines a set of shared "supplemental" data that cannot be used commercially in the country of origin. For example, French supplemental data cannot be used to make French forecasts that are sold instead of freely distributed.

[67] *Intergovernmental Panel on Climate Change, IPCC Second Assessment—Climate Change 1995: A Report of the Intergovernmental Panel on Climate Change* (Geneva, 1995), 5, http://www.ipcc.ch/pub/sa(E).pdf.

[68] *Climate Change 2001: Synthesis Report,* ed. Robert T. Watson and the Core Writing Team (Geneva, 2001), 5, http://www.ipcc.ch/pub/un/syreng/spm.pdf.

same time, virtually all nations were enrolled, first at the agency level, through weather services, and then, via the IPCC, at the executive and legislative levels as well. Linking *governments* to environmental *governance* by means of a global data-producing infrastructure has made it increasingly difficult for the former to ignore the latter. The IPCC's conclusions are not ones that any nation or its political leaders would ever seek to reach. Barring heroic and extremely costly changes to the world's energy economy, anthropogenic global warming cannot be easily controlled. Yet even the ideologically driven George W. Bush administration finally abandoned its wait-and-see position, acknowledging that global warming is real and is caused in significant part by human activity even while declining to ratify the Kyoto Protocol.[69]

Of course, I am not arguing that scientific evidence by itself determined this change in a key political position. Global scientific organizations cannot force political action on the issue. Yet their extraordinary success in promoting highly unwelcome conclusions shows how infrastructural globalism has helped transfer power from states to global science-based organizations.

No one who has studied the global warming debate can ignore the long-term convergence of the observational evidence. This convergence has everything to do with the increasing precision, power, and scope of the global weather data network. No infrastructure this complex or this large will ever be perfectly integrated or seamlessly smooth.[70] Debate will continue over the quality of the data it generates. Nonetheless, it has established global warming as an accepted, highly consequential fact. Without the infrastructural globalism that produced it, this knowledge would remain far more heavily contested.

[69] *U.S. Climate Action Report* (Washington, D.C., May 2002).

[70] Thomas C. Peterson, David R. Easterling, Thomas R. Karl et al., "Homogeneity Adjustments of *In Situ* Atmospheric Climate Data: A Review," *International Journal of Climatology* 18 (1998): 1493–517; Thomas C. Peterson and Russell S. Vose, "An Overview of the Global Historical Climatology Network Temperature Database," *Bull. Amer. Meteor. Soc.* 78 (1997): 2837–49.

Globalization and Regulation in the Biotech World:
The Transatlantic Debates over Cancer Genes and Genetically Modified Crops

By Jean-Paul Gaudillière[*]

ABSTRACT

Biotechnology has become a major issue in international affairs. This paper follows the scientific and public debates on genetic testing for breast cancer predispositions and genetically modified organisms, which surfaced in France and in the United States during the 1990s. It focuses on the contrasting ways in which intellectual property rights and issues of risks were addressed in both countries. The two case studies suggest that transatlantic tensions about biotechnology are rooted in different forms of regulation. While civic regulation has emerged on both sides of the Atlantic, France remains a privileged terrain for professional regulation under the umbrella of the state, whereas the United States has increasingly relied on industrial regulation.

INTRODUCTION

Toward the end of the year 2000, as America was preparing for the presidential election, President Bill Clinton's health secretary received a special report on the situation of genetic tests in the United States. This document, written by the recently nominated Secretary's Advisory Committee on Genetic Testing, surprised many experts who regarded the United States as something of a wonderland for the development of biotechnology. The committee actually recommended strong regulation for newly invented genetic tests, with the creation of an FDA-based review procedure. At the same time, a group of French oncologists working in the area of genes and cancer was considering this very issue of testing practices. The clinicians working at the Institut Curie in Paris who had called for this meeting were not, however, so much concerned about the clinical utility of diagnosing genetic predispositions to cancer as angered by the control over the uses of the breast cancer genes exercised by the U.S. biotechnology

[*] Centre de Recherche Médecine, Sciences, Santé et Société (INSERM), Paris Site CNRS, 7 rue Guy Moquet, F-94801 Villejuif Cedex, France; gaudilli@vjf.cnrs.fr.

Earlier versions of this chapter were discussed with Pierre-Benoît Joly and with my colleagues at Cermes, Ilana Löwy and Maurice Cassier, with whom the research on breast cancer genetic testing was conducted. I would like to express my appreciation for their help, insights, and many comments. The chapter also owes much to the stimulus of John Krige.

start-up firm Myriad Genetics. The local practitioners wanted to launch a lawsuit against the European approval of Myriad's patents.

The events described above are typical of the new form of biotechnology that emerged in the last quarter of the twentieth century, as the life sciences became critical assets in industrial innovation, market competition, and international affairs. In the past thirty years, the fate of biology has been deeply affected by "globalization," that is, the increased circulation of people, goods, and capital as well as its political ramifications. In the mid-1980s, approval of the Human Genome Project was, for instance, associated with tense debates about the putative productivity gap between Japan and the U.S., and about the need for America to take the lead in biotechnological research to restore its economic power. The multiple initiatives taken during the 1980s and 1990s to enlarge the patentability of living entities and reinforce intellectual property rights have been elements in this strategy. In parallel, the rise of intermediate structures supplementing states and markets, such as the World Trade Organization (WTO), the world economic summits, and the United Nations conferences, has led to new forms of governance that affect both the ways in which the knowledge of the living is constituted and the ways in which this knowledge is used in medicine and agriculture, a trend illustrated by the various conventions on biodiversity signed since the Earth Summit in 1992.

Although globalization undoubtedly undermines the twentieth-century basis for the development of the life sciences, namely, the nation-state and its activities as scientific entrepreneur, the practice of biological research remains embedded in highly specific contexts. It remains shaped by locally defined styles of thought, professional interests, and political cultures. As a consequence of these local roots, in the past ten years debates have emerged over dozens of biotechnological issues, which include the property of rain forest biological resources, the validity of not only gene patents but also growth hormones, genetically modified organisms (GMOs), and the use of human embryos for stem cell research and production. These tensions have revealed the mounting importance of nongovernmental actors, transnational associations as well as local communities, in international affairs. Nation-states, however, have not disappeared. As the case of the breast cancer genes suggests, most attempts at regulating the invention and uses of biotechnological goods have taken place within national borders. As a consequence, they have resulted in very different outcomes. In the case of the breast cancer genes, the proposal of the committee was adopted by the U.S. Public Health Service but did not survive the arrival of the Bush administration. By contrast, the French oncologists have been highly successful with their opposition. This contrast in outcomes is a striking illustration of the different ways in which the new biomedicine is perceived, valued, and practiced within national borders.

To analyze the new biotech order, and more specifically the regulatory configurations associated with globalization and its consequences, this chapter follows two controversies that emerged in the 1990s in Europe and in the United States and led to significant transatlantic tensions. The first is the case of genetic testing for cancer predisposition; the second is the case of genetically modified crops and their uses in food production. These innovations gave rise to controversies over the norms of intellectual property rights, the role of the state in supporting science and regulating its uses, the risks such biotechnologies create, and the nature of their evaluation, including the parts users and laypeople should play in this process. In Europe and in the United States, these debates were fueled by different alliances and have resulted in contrasting attitudes and choices. In order to discuss these different modes of regulation, the last part of this chapter compares the two trajectories in a more systematic way. Before turning to the history of these entities, however, it is necessary to present some essential background information concerning the old world of biology.

MOLECULAR BIOLOGY AND THE *ANCIEN RÉGIME* OF BIOMEDICAL RESEARCH

In the early 1960s, the search for the mechanism by which DNA sequences could be turned into proteins was at the very center of a new burgeoning domain whose practitioners called themselves "molecular biologists." Research activity focusing on the role of genes and the properties and function of biological macromolecules, which had previously been associated with biochemistry, microbiology, or virology, was increasingly perceived as a new form of knowledge.[1] Thus, the 1960s witnessed a rapid institutionalization of the field, with the formation of the discipline being manifested in the creation of institutes, departments, programs, committees, and journals. The specificity of molecular biology within the biological sciences was then perceived as being a question not only of specific objects, that is, DNA and proteins, but also of style. Molecular biology was considered more fundamental, more theoretical, and more closely associated with the physical and chemical sciences than other forms of work in the life sciences.

Molecular biology was seen as an intellectual achievement based on disciplinary hybridization. It was also seen as an exemplary form of science practiced at the international level, emerging as it did out of centers distributed throughout the Western world. Neither French, British, American, nor Swiss, it was considered truly transnational.[2] Of course, biological "internationalism," like that of its disciplinary cousin high-energy physics, was a combination of competition and collaboration. Competition was impossible to deny when representatives of "molecular biology" at the national level, such as John Kendrew in the United Kingdom and Jacques Monod in France, made strong pleas for the local development of the discipline to limit the *brain drain* (the perceived threat in the United Kingdom) or to avoid *American dominance* of the field (the perceived threat in France).[3]

These two characteristics of the early history of molecular biology—its basic and international characters—should not, however, be overemphasized. They are part of the way molecular biologists perceived themselves and contributed to the image they

[1] For recent work on molecular biology in the postwar era see: Pnina Abir-Am, "The Politics of Macromolecules: Molecular Biologists, Biochemists, and Rhetoric," *Osiris* 7 (1992): 164–91; Soraya de Chadarevian, *Designs for Life: Molecular Biology after World War II* (Cambridge, 2002); Angela Creager, *The Life of a Virus* (Chicago, 2002); Lily E. Kay, *The Molecular Vision of Life: Caltech, the Rockefeller Foundation, and the Rise of the New Biology* (Oxford, 1993); idem, *Who Wrote the Book of Life? A History of the Genetic Code* (Stanford, 2000); Michel Morange, *A History of Molecular Biology,* trans. Matthew Cobb (Cambridge, Mass., 1998); Hans-Jörg Rheinberger, *Toward a History of Epistemic Things: Synthesizing Proteins in the Test Tube* (Stanford, 1997).

[2] Pnina Abir-Am, "From Multidisciplinary Collaboration to Transnational Objectivity: International Spaces as Constitutive of Molecular Biology, 1930–1970," in *Denationalizing Science: The Context of International Scientific Practice,* ed. Elisabeth Crawford, Terry Shinn, and Sverker Sörlin, Sociology of Science Yearbook 16 (Dordrecht, 1993), 153–86.

[3] De Chadarevian, *Designs for Life* (cit. n. 1); Jean-Paul Gaudillière, *Inventer la biomédecine: La France, l'Amérique et la production des savoirs du vivant (1945–1965)* (Paris, 2002) (English translation forthcoming at Yale University Press).

wanted to display to the "outside" world. So while this image does reflect actual research practices to some extent, the complete picture is much more complex. The reasonably well-studied historiography of medically oriented biological research bears witness to this complexity. The biochemists, virologists, and biophysicists working on macromolecules after 1945 were operating in scientific arenas dominated by the state in its role of scientific entrepreneur and by the development of the biomedical research complex.[4]

Changes were not purely institutional; they had cognitive dimensions, which are easy to see when one traces the postwar "molecularization" of biology and medicine. This molecularization had at least two aspects: the first associated with the work on macromolecules, and the second, more practical, aspect epitomized by the discovery of penicillin and the 1950s race to develop new "magic bullets." Medical progress after 1945 became increasingly synonymous with the use of molecules with therapeutic properties. This included the use of chemicals originating in the chemist's laboratory as well as the preparation of biological drugs, such as killed viruses employed as vaccines and derivatives of hormones such as the sex steroids and cortisone. These two aspects of molecularization were, however, closely related. The manipulation of macromolecules, in the first place viruses (although discourses about the uses of DNA boomed in the 1960s), was perceived as critical to the quest for new therapeutic agents. The self-identified "basic" molecular biologists thus participated in complex networks linking biological laboratories, research hospitals, and pharmaceutical companies.[5]

Similarly, one may wonder how influential international affairs were for this postwar growth of biomedical research. There is little doubt that general trends in the history of the cold war played a part. The post-Sputnik era was, for instance, a time of increasing support for European biologists from both American institutions and NATO.[6] In the same vein, during the 1960s, Charles de Gaulle's commitment to a nationalistic foreign policy that would balance France's participation in the Atlantic community with enlarged autonomy and occasional cooperation with the Soviet Union had scientific consequences. National *grandeur* and independence were translated into large technoscientific programs. These not only focused on space, electronics, and nuclear research but also included genetics and cancer research.[7] However, in Europe, while space and atomic research combined national policy and highly visible transnational initiatives, such as the creation of CERN (Conseil Européen pour la Recherche Nucléaire) and the European Space Agency, biomedicine remained rooted in the interventions of the nation-state. It is therefore not unrealistic to suppose that

[4] Paul Starr, *The Social Transformation of American Medicine* (New York, 1982); Victoria Harden, *Inventing the NIH: Federal Biomedical Research Policy, 1887–1937* (Baltimore, 1986); Nicolas Rasmussen, "The Midcentury Biophysics Bubble: Hiroshima and the Biological Revolution in America Revisited," *History of Science* 35 (1997): 245–93.; Gaudillière, *Inventer la biomédecine* (cit. n. 3); Samuel Paul Strickland, *Politics, Science, and the Dread Disease* (Cambridge, Mass., 1972); James Patterson, *The Dread Disease: Cancer and Modern American Culture* (Cambridge, Mass., 1987).

[5] Gaudillière, *Inventer la biomédecine* (cit. n. 3); Creager, *The Life of a Virus* (cit. n. 1); Ilana Löwy, *Between Bench and Bedside* (Cambridge, Mass., 1996); Jordan Goodman and Vivian Walsh, *The Story of Taxol* (Cambridge, 2001).

[6] John Krige, "La science et la sécurité civile de l'Occident," in *Les sciences pour la guerre, 1940–1960,* ed. Amy Dahan-Damedico and Dominique Pestre (Paris, 2004), 369–97.

[7] François Jacq, *Pratiques scientifiques, formes d'organisation, et représentations politiques de la science dans la France d'après-guerre* (Ph.D. diss., Ecole des Mines, Paris, 1996).

up until the early 1970s, the "Atlantic" system of science had not established a true transatlantic biological community but remained what it was during the time of the Marshall Plan, namely, the juxtaposition of national spaces between which scientific resources could circulate. Men, instruments, and data actually moved between the United States and Europe. The traffic was rather uneven, however. Most European biologists were principally importing (and benefiting from) tools, instruments, protocols, and ways of experimenting that originated in the practice of U.S. biomedical scientists.

This overall picture of the biological research system remained applicable until the mid-1970s; molecular biology operated at the national level, was state-based, and was legitimized through a rhetoric of basic science. It was also technology-driven but remained loosely connected to the market. It was linked to medical applications but barely associated with industrial research. Subsequent to its early appearance in a context typical of the postwar period, the development of applied molecular biology, and DNA-oriented biotechnology, has taken place in a rapidly changing scientific world. Analysts such as Michael Gibbons and Helga Nowotny, for instance, argue that a "new mode of knowledge production" has emerged, pointing to the decreased importance of state and disciplinary organizations, the integration of pure and applied research, the new roles played by global markets and organized users, and the public debates on risks and unintended effects, which have brought a more reflexive tone in science and technology.[8] In order to discuss this putatively new regime and its roots and effects on biology, the rest of this paper compares two recent controversies about the regulation of gene technologies.

PRACTICING RESEARCH IN A GLOBAL SPACE: THE QUEST FOR THE BRCA GENES

In 1996, the media reported the cloning of a gene linked to a hereditary form of breast cancer.[9] Called BRCA2 (for breast cancer), this gene was the second of its sort. Two years earlier, a group of researchers at Myriad Genetics, a start-up affiliated with the University of Utah, had announced the cloning and sequencing of a first BRCA gene. Until the late 1980s, the hereditary transmission of breast cancer had been considered of little importance and was investigated in only a limited numbers of laboratories and medical services.[10] By the end of the 1990s, however, the situation had changed radically. Thanks to the development of techniques in molecular biology, the cloning of genes responsible for human diseases had become highly competitive. In 1990, a group of researchers based in San Francisco had been the first to identify a gene involved in

[8] Michael Gibbons, Helga Nowotny, Camille Limoges et al., *The New Mode of Knowledge Production: The Dynamics of Science and Research in Contemporary Societies* (London, 1994).

[9] The case study presented here originates in a research project supported by the Mission Interministérielle à la Recherche and the Fondation de France. It has been conducted with Ilana Löwy and Maurice Cassier. The material discussed below was made available in French in a preliminary report: Jean-Paul Gaudillière and Maurice Cassier, *Production, valorisation et usage des savoirs: La génétique du cancer du sein* (Paris, 2002). The detailed analysis of testing practices has been presented in Jean-Paul Gaudillière and Ilana Löwy, "Science, Markets, and Public Health: Contemporary Testing for Breast Cancer Predisposition," in *Medicine, Markets, and Mass Media: Producing Health in the Twentieth Century,* ed. Virginia Berridge and Kelly Loughlin (London, 2004).

[10] Henry T. Lynch, *Dynamic Genetic Counseling for Clinicians* (Springfield, Ill., 1969).

hereditary forms of breast cancer.[11] This discovery launched a race for the identification and sequencing of these genetic factors.

In the 1970s, human genetic research had been free of any marketing considerations. This was clearly no longer the case in the 1990s.[12] Property rights and the "privatization" of biological knowledge had come to play a central role in the life of the international consortia that organized contemporary genetic research. Based on the history of the BRCA genes, one may oppose two models of research organization. One model can be termed the clinical model and is well illustrated by cancer research in France. The second model can be labeled the biotech model and is, in turn, exemplified by the practices of the private start-up company Myriad Genetics.

The first breast cancer gene (BRCA1) was not cloned and sequenced by the public laboratories in the international breast cancer consortium but by this private company. Myriad's success was based on its privileged access to the pedigrees and genetic material from Mormon families that had been gathered by the University of Utah. It was also based on aggressive appropriation of genetic knowledge. Like most start-ups, Myriad was not financially viable at the beginning; the creation and management of a significant research infrastructure was only made possible by the injection of venture capital. Development depended as well on signing research contracts with large pharmaceutical companies. Having no actual products, Myriad could only sell its preliminary results and promises of patents. Thus, work on BRCA expanded after a large contract had been signed with Eli Lilly. The division of labor and proprietary rights between the pharmaceutical company and the start-up company was such that Eli Lilly would receive exclusive rights on all therapeutic developments, while Myriad would retain the rights to the easier and more short-term diagnostic developments. The two parties based their agreement on the assumption that Myriad would be the first one to sequence the breast cancer genes and that the company could secure large "umbrella" patents on the use of such data.

These financial intricacies are of interest not only for the investors involved but also as an illustration of the research dynamics created by new patterns of appropriation in the biotech world, in particular the recent development of patents that protect gene sequences. Myriad's patent on BRCA1 was no ordinary patent and belongs to a type of claim that might be termed a "gene" patent as it covers a very broad definition of the invention. The basic information documenting the invention—and asserting its novelty—is the description of the molecular sequence of the gene, and it is often the case that more than half of the text of such patents consists of sequencing data. In the case of BRCA, the first claim is accordingly an umbrella demand covering all uses of both the nucleotide sequence of BRCA1 and the derived protein sequence. Specific claims then focus on all possible applications of the sequence: the design of mutation tests for cancer predisposition, the production of genetically modified animals, and the development of gene therapies. The mere existence of such patents is a vivid testimony of the very broad redefinition of patentability that has taken place since the 1980s.[13]

[11] Kay Davis and Michael White, *Breakthrough: The Quest to Isolate the Gene for Hereditary Breast Cancer* (London, 1995).

[12] Paul Dasgupta and Paul David, "Towards a New Economics of Science," *Research Policy* 23 (1994): 487–521; Richard Eisenberg, "The Move towards the Privatization of Biomedical Research," in *The Future of Biomedical Research,* ed. C. E. Barfield and L. R. Smith (New York, 1997).

[13] Arnold Thackray, ed., *Private Science: Biotechnology and the Rise of the Molecular Sciences* (Philadelphia, 1998); Daniel J. Kevles, *A History of Patenting Life in the United States with Compar-*

The understanding of two criteria—novelty and industrial utility—has accordingly been revised in turn. Changes in the structure of intellectual property have been considerable, at least when one considers the case of living entities in general, genetic material in particular.[14]

Gene patents would not have been granted fifteen years ago as the entities they protect, that is, the DNA sequences, would have been considered "natural" entities and their identification a process of discovery rather than one of invention. The argument, which nowadays justifies the appropriation of such sequences, is that the work involved in identifying DNA and finding its chemical structure produces laboratory artifacts that are substantially different from the *functional* gene in the organism with its noncoding, regulatory, and dispersed structural sequences. This perception is a direct consequence of the molecularization of the gene. The transformation of molecular biology in the 1970s and 1980s resulted in a critical extension of the biologists' ability to manipulate, that is, isolate and synthesize, DNA molecules. One consequence was to displace the notion of the gene toward a technical, operational definition: rather than a functional unit, the gene became a sequence, an assemblage of bases, the product of procedures that take place in the organism as well as at the laboratory bench. The legal corollary of this transformation was the acceptance of an equation linking operationally defined DNA sequences with established patterns of patents applied to chemicals.

The second root of the "gene" patents was creation of a real market for biotechnological knowledge, a global, rather than a national, market. This development is part of the commercial globalization of the 1980s and 1990s. It is rooted in the close articulation of international agreements and legal changes in the United States that linked the patentability of biological entities and the alignment of intellectual property regimes with the U.S. situation. One early step in the direction of the valorization of research results in the United States was creation of the Nasdaq, in 1971, and other changes in the regulation of the stock market to increase the presence of firms in the "innovation" sectors.[15] The passage of the Bayh-Dole Act in 1980 played a much more decisive role as it enabled universities to protect the results of investigations conducted with public funds. The new patterns of appropriation that originated in the United States have assumed international dimensions through a series of conventions and institutional innovations, first at the national, and later at the international, level. In 1984, the U.S. Congress reacted to the widespread concerns about Japan's technological advances by modifying the Trade Act to enforce the intellectual property rights owned by American firms outside the country. A special section of the act stated that the U.S. government could take special economic and foreign policy measures against countries in which the rights associated with U.S.-based patents would not be recognized, even if such nonrecognition would not infringe on international trade agreements. The key step was the WTO negotiations of the Uruguay Round, which led to

ative Attention to Europe and Canada: Report to the European Group on Ethics (Brussels, 2002); Benjamin Coriat, ed., *Les droits de propriété intellectuelle: Nouveaux domaines, nouveaux enjeux,* special issue, *La Revue d'Economie Industrielle,* no. 99 (2002).

[14] Kevles, *A History of Patenting Life in the United States* (cit. n. 13); Maurice Cassier, "Private Property, Collective Property, and Public Property in the Age of Genomics," *International Social Science Journal* 171 (2002): 83–98.

[15] Benjamin Coriat, Fabienne Orsland, and Oliver Weinstein, "Does Biotech Reflect a Science-Based Innovation Regime?" *Industry and Innovation* 10 (Sept. 2003): 231–53.

the 1994 agreement on intellectual property. Although some provisions were preserved (e.g., for public health motives), the Marrakech convention basically transformed the U.S. patent system into a global one: all countries that had no other forms of intellectual property laws agreed to reform their legislation accordingly.[16] Nongovernmental organizations (NGOs), governments from developing countries, and researchers' organizations often describe these agreements as a *coup de force* giving the United States patenting practices a global validity in order to maintain technological and commercial hegemony. Whether or not this is the case, the internationalization of enlarged patent protection has generated considerable tensions between the United States, Europe, and countries such as Brazil and India, with emerging capabilities in agricultural and pharmaceutical research. On the transatlantic scene, the BRCA story has accordingly been transformed in a test case.

REGULATING USES OR REGULATING PROPERTY? THE TRANSATLANTIC CONFLICT

In 1998, after the legal situation regarding the ownership of the second BRCA gene had been settled, Myriad was free to exploit its monopoly over testing for breast cancer genetic predispositions. It considered developing the diagnostic market as a way of getting quick revenues from knowledge about predisposing genes. Usually, holders of patents for diagnostic procedures either commercialize diagnostic kits or sell licenses. Myriad decided instead to centralize BRCA testing by creating a sister firm and constructing a local "test factory." This decision was partly based on technical reasons: as the BRCA gene is very large, it is difficult to manufacture diagnostic kits. Moreover, having successfully established a large sequencing platform for research purposes, Myriad could readily build on this experience. In addition, selling a service, rather than a kit, was a cheap way to avoid regulation. While diagnostic kits need to be evaluated by the U.S. Food and Drug Administration (FDA), medical services in the United States are neither reviewed nor subjected to marketing authorizations.[17]

The development of Myriad's testing service reflects another important dimension of the start-up model of genetic innovation, namely, the consumer-based organization of access. In 1997, Myriad started an aggressive promotion of the tests for BRCA mutations in the United States. The campaign included direct marketing to potential users. Under the slogan "understanding your risk can save your life," the company laid great emphasis on the fact that women have the right to have access to their own genetic information, that is, they have a right to know whether they are mutation carriers. Myriad's diagnostic service is even advertised on the Web.[18] Women can, therefore, assess their family status to determine whether they are part of the group that "should" consider testing. If this is the case, the person can find the address of cancer centers or physicians who might write a prescription for such a test. Such broad access to tests is justified not only by women's "right to know" but also by the possible benefits of early detection of malignancies in high-risk women and better targeting of preventive measures for mutation carriers.

The service model has been contested by both medical practitioners and breast can-

[16] Jerome H. Reichman and David L. Lange, "Bargaining around the TRIPS Agreement," *Duke Journal of Comparative and International Law* 9 (1998): 11–23.

[17] Neil Holtzman, "Are Genetic Tests Adequately Regulated?" *Science* 286 (1999): 409–11.

[18] See http://www.myriad.com.

cer activists. It is nonetheless based on current practice in the field of medical biology. Laboratory procedures are conducted in specialized units that have no direct connection with clinical services. Testing for blood glucose and cholesterol levels is done in financially autonomous units that operate in a discrete segment of the medical market. The extension of this configuration to BRCA testing in the United States is well illustrated by the following two features of Myriad. First, company officials do not consider that the writing of prescriptions should be limited to specialists in genetics or cancer. They argue that genetic testing is just like any other biological examination, and so any medical doctor should be considered competent to prescribe it. Second, and more important, it is not the role of a service center such as Myriad to organize follow-up and care. The service laboratory provides controlled, accurate results concerning the mutations; what happens subsequently is up to the person tested and her physician.

This close articulation between cancer research and the biotechnology markets is not, however, a universal feature of cancer-related genetic research. Other research models have operated in the field and have played an important role as illustrated by developments in France. Most cancer research in this country is conducted either in public hospitals or in the Centres de Lutte Contre le Cancer (CLCC). These centers are hybrid organizations operating at the interface between the public and private sectors, that is, they are charity-based centers, although most of the money they receive comes from the state.[19] They were created on a regional basis and associate laboratory research facilities with hospital beds for patients who participate in clinical trials or receive more routine treatments. In the 1990s, as the search for BRCA genes became more visible, some centers (Paris, Lyon, and Marseille) established specialized oncogenetic clinics.[20]

One critical aspect of this model of oncogenetics is the lack of any entrepreneurial quest to make money out of this research. French physicians working in cancer genetic units simply did not care about commercializing their tests, which were developed using research funds allocated on the basis of grants and umbrella appropriations for the cancer centers. Like most clinicians working in elite research hospitals in France, cancer specialists do not need to take into account strong cost-benefit constraints or to handle the administrative and commercial development of their inventions. The prevailing feeling is that everything medically justified should somehow be made available, meaning that neither technological nor proprietary issues have attracted much attention. Thus, for instance, while molecular biologists working in local state research agencies (Institut National de la Santé et de la Recherche Médicale [INSERM] and Centre National de la Recherche Scientifique [CNRS]) have been increasingly influenced by new patterns in biotechnology, gene patents have remained outside the scope of French physicians working in the field of cancer genetics. No patent applications for the "invention" of BRCA sequences or the development of testing techniques were filed in France.

In France, testing practices did not expand through the efforts of a new diagnostic market but were increasingly used in the CLCC and their specialized oncogenetic clinics. The initial aim of using these tests in such settings was to enhance research. People with significant family histories of cancer would come to these services, where

[19] Patrice Pinell, *Naissance d'un fléau: La lutte contre le cancer en France* (Paris, 1994).
[20] Gaudillière and Cassier, *Production, valorisation et usages des savoirs* (cit. n. 9).

they would give samples, help track down other family members, and in return, receive information concerning their risk status. As these clinics were founded in treatment centers, the majority of the women followed for inherited forms of breast cancer had already been diagnosed with the disease. Thus, French laboratories combined routine tests for mutations in BRCA genes, the search for new predisposing genes, genetic counseling for individuals perceived to be at risk, the establishment of pedigrees, the collection of DNA samples, and the centralization of data.[21]

The notion that molecular genetics should be regulated can be traced back to debates about the risks associated with the early transfer of DNA between species, that is, the genetic engineering controversy.[22] Critics in the 1970s considered laboratory work as a possible source of danger (e.g., a source of new pathogens). They questioned the ability of professional scientists to recognize and control the risks they created, especially when pursuing technological goals or when collaborating with industry. In the 1990s, critics of biomedicine were more concerned about the failure of systems, bureaucratic entrenchment, identities, and domination. This displacement led to new alliances and practices, best illustrated by the mobilization of AIDS advocacy groups, especially in their attempts to control the aims and methods of clinical research.[23] These displacements are also illustrated by the contemporary discussions over the regulation of genetic testing.

Questions regarding the legitimate uses of genetic knowledge have become ubiquitous since the launching of the Human Genome Project.[24] Breast cancer predisposition testing has introduced a different type of problem since the practice seeks to justify investigating healthy individuals in terms of reducing individual risk. The discussion over predisposition testing developed in two phases in the United States. The first was a professional phase that focused on issues of quality control. The second phase, initiated by public health authorities, targeted problems of marketing and clinical utility and efficacy. The first phase began in the mid-1990s with an accumulation of guidelines produced by various medical groups. During the second phase, which began in the late 1990s, regulations developed along very different lines as a consequence of the mobilization of patient advocacy groups, with the National Breast Cancer Coalition (NBCC) assuming a leading role.

The critical attitude of the NBCC has to be understood within the framework of a more general development of the practice of obtaining a second opinion, which has affected the cancer scene in the United States. This move has complex roots, including the rise of environmental activism, and its associated interest in carcinogenic chemicals, and the feminist movement. The women's health movement developed ways of conducting autonomous evaluations of medical claims and fostered a form of woman-centered medical culture. Beginning in the 1990s, a new wave of mobiliza-

[21] Maurice Cassier and Jean-Paul Gaudillière, "Recherche, médecine, et marché: La génétique du cancer du sein," *Sciences Sociales et Santé* 18 (2000): 29–49.

[22] Susan Wright, *Molecular Politics: Developing American and British Regulatory Policy for Genetic Engineering* (Chicago, 1994).

[23] Samuel Epstein, *Impure Science: AIDS, Activism, and the Politics of Knowledge* (Berkeley, 1996); Nicolas Dodier, *Leçons politiques de l'épidémie de Sida* (Paris, 2003); Sebastien Dalgalarrondo, *Sida: La course aux molecules* (Paris, 2004).

[24] Daniel Kevles and Leroy Hood, eds., *The Code of Codes: Scientific and Social Issues in the Human Genome Project* (Cambridge, Mass., 1992); Robert Cook-Degan, *The Gene Wars: Science, Politics, and the Human Genome Project* (New York, 1994).

tion in favor of women's health issues emerged in the form of local breast cancer organizations. Many of these organizations originated in self-help or support groups, some linked with specialized cancer clinics. Following the example of AIDS activists, they criticized existing cancer policies and cancer treatment practices.[25] They articulated new themes, ones that diverged from the classical attitude of supporting biomedical research that had always been adopted by organizations such as the American Cancer Society. In 1990, these local groups established an umbrella organization to help coordinate their action at the federal level. This umbrella structure became the National Breast Cancer Coalition. One of the targets of their cooperative action was to lobby the Congress and the federal administration for "more money" to fight breast cancer. This proved highly successful, with the coalition managing, in 1993 (in the early days of the Clinton era), to reroute $300 million from the defense budget into cancer research. Following this achievement, the NBCC initiated a discussion about the type of research that should be supported. This led the NBCC not only to collaborate with specific groups of researchers and clinicians but also to augment their advisory function, organize controversial debates, acculturate members to the biomedical discourse, and develop their ability to participate in various evaluation committees. The discussion about genetic testing emerged out of these practices.

As a consequence of this form of scientific empowerment, the culture of advocacy and counter-expertise shared by the NBCC members was echoed in the governmental debates about the regulation of genetic practices. In 1994, the National Institutes of Health (NIH) established the Task Force on Genetic Testing, as part of the Ethical, Legal and Social Implications (ELSI) Research Program of the NIH-based human genome project.[26] Although no agreement could be found concerning appropriate surveillance, the Task Force on Genetic Testing nonetheless framed the U.S. debate in one essential way: it specified why genetic tests should receive special attention. The argument was that these tests provided a special form of information with wide-ranging effects in terms of personal identity as well as medical and social status. This view implied that some tests (predisposition testing, in particular) would require "stringent scrutiny," one of the aims being the evaluation of the clinical utility of the test under consideration. In 1998, the health secretary announced the establishment of the Secretary's Advisory Committee on Genetic Testing (SACGT), with most of its members being users and producers of tests, rather than governmental and public health officials.[27] The SACGT adopted a set of proposals, including the creation of a system of marketing permits that would resemble the mechanism being employed for drugs.[28] This white paper, echoing many of the concerns raised by the NBCC and public health specialists, clearly stated that "the FDA should be involved in the review of all genetic tests" but that "the review should be appropriate to the level of complexity of the information generated by the test." BRCA-like tests were perceived as the most problematic, given the possibly adverse consequences of testing, including feelings

[25] Maureen Hogan Casamayou, *The Politics of Breast Cancer* (Washington, D.C., 2001); Sandra Morgen, *In Our Own Hands* (New Brunswick, N.J., 2002).

[26] Neil Holtzman and Michael S. Watson, *Promoting Safe and Effective Genetic Testing in the United States: Final Report of the Task Force on Genetic Testing* (Baltimore, 1998).

[27] See http://www4.od.nih.gov./oba/sacgt.htm.

[28] SACGT, *Enhancing the Oversight of Genetic Tests: Recommendations of the SACGT* (Washington, D.C., 2000).

of anxiety for the persons labeled at high risk, the multiplication of useless operations, the existence of irreversible and controversial means of prevention, and the social stigma associated with being identified as a person at risk. The recommendation for marketing permits was endorsed by the administration shortly before the 2000 presidential election. With the return of a Republican administration hostile to the development of state "control," there is very little chance that SACGT's proposals will be implemented. It is likely, therefore, that testing for BRCA mutations will remain in the domain of market regulation.

When compared with the U.S. situation, the French regulatory debates offer an interesting combination of similarities and differences. To begin with, a group of geneticists and oncologists assembled by INSERM and the national federation of cancer centers came up with equivalent professional rules of good practice.[29] The regulatory alliance took the classical form of a convergence of elite clinicians and state administrators. This alliance pushed for the official oversight of the internal organization of the profession, with two fears justifying the idea of a state-based surveillance system. First, molecular geneticists feared that anyone could begin testing without having the necessary (analytical) knowledge. Second, this classic issue of expert knowledge was linked to another organizational issue, namely, the fear that after the first wave of creation of medical genetics laboratories within research hospitals, the increase in the number of small centers would result in technical inefficiency and inaccurate results. Ironically, the intervention of the state proved indispensable for establishing a form of technical control by the profession.

When it comes to the question of public debates, the French scene presents an even more dramatic contrast to the scene in the United States. Discussions about the clinical utility of DNA analysis and the management of "high-risk" persons were held, and have remained, behind the closed doors of the consultation room in France. The one issue discussed in public spaces was the question of DNA patents. In 1999, roughly 2,000 clinicians in France and Germany signed a petition protesting the acceptance of sequence patents at the European level. The petition asked for a change in the patent directive adopted by the EU Commission in 1998. As the European directive on biotechnological patents contains several provisions for taking into account ethical and public health concerns, one may wonder why these problematic patent applications were accepted. Nevertheless, the legal framework supplied a clear approval of the patenting of genetic sequences. The clinicians asked the French and German governments not to translate this text into national laws. Although they were not mentioned in the text, the cancer genes were the principal consideration for many of the practitioners who signed the petition, since there was a mounting conflict between the French oncologists and Myriad over the use of BRCA sequences.

The existence of an alternative clinical basis for the organization of a testing service had led Myriad into attempts to control the diagnostic market and enforce its monopoly, a campaign that started in the United States immediately after the creation of Myriad's testing company. Myriad proposed a licensing contract for the laboratories

[29] INSERM, Expertise collective, *Risques héréditaires des cancers du sein et de l'ovaire: Quelle prise en charge?* (Paris, 1998); François Eisinger, Nicole Alby, Alain Bremond et al., "Recommendations for Medical Management of Hereditary Breast and Ovarian Cancer: The French National Ad Hoc Committee," *Annals of Oncology* 9 (1998): 939–50; Association des Praticiens de Génétique Moléculaire, *Livre blanc* (Paris, 1998).

located in medical schools that were conducting investigations into BRCA mutations at the frontier between research and routine analysis. This policy triggered the transatlantic conflict over the use of BRCA genes. In October 1998, Myriad invited a series of European oncologists to visit its services in Salt Lake City. The official aim of the conference was to "explore ways to expand the availability of breast and ovarian cancer genetic testing in Europe."[30] Myriad offered to do all the sequencing for a fee of $2,400 per test, leaving the European cancer centers to search for identified mutations. For the latter, Myriad would only charge a flat fee of $45 per test. Within the French cancer centers, the firm's strategy was resented at two levels. First, they saw it as an abuse of intellectual property that threatened local practices and curtailed potential innovation in this area. Second, they perceived it as a danger for the model of clinically integrated biotechnology since any agreement with Myriad would disrupt the connection between the genetic platform, the cancer consultation, and the clinical management of risk. Rather than arguing at the level of quality of care or attempting to promote the local regulation of clinical practice (that could, for instance, have made integration mandatory), the oncologists at the Institut Curie chose to push the patenting issue. Following the acceptance of Myriad's applications by the European Patent Office (EPO), several third-party objections were initiated by a coalition under the leadership of the Institut Curie.[31]

Several arguments were used against Myriad's patents. One was the absence of priority, as Myriad used information drawn from public databases to win the sequencing race. Accordingly, Myriad had no strong claims on the "invention" of the BRCA1 sequence since most of the important information was collected within the framework of the International Breast Cancer Research Consortium.The final sequencing was a technical operation any scientist trained in the art could perform. Broad control of the sequence, opponents argued, was therefore illegitimate. A second argument was the claim that Myriad's patents did not describe the industrial applications of the invention with sufficient precision. Thus, the claim to "all therapeutic applications of BRCA1 and BRCA2" could be interpreted as covering some independently developed, novel innovations that the inventor would then be unable to use. Granting protection in such general terms would create a monopoly that would threaten to stifle public research. Finally, it was alleged that the method used by Myriad could not detect large deletions, and according to French experts, 10 to 20 percent of BRCA mutations were just those kinds of large deletions. European scientists explained that their methods provided comparable results at a lower cost. Ruling on this opposition recently, the EPO dismissed Myriad's claims. The office, however, took good care not to establish any legal precedent that could be used to challenge the European directive on biotechnological patents. It judged that patents on the BRCA1 sequences would have been acceptable if Myriad had provided more convincing evidence of its inventive activity. Thus, no monopoly of intellectual property concerning the BRCA genes has been established in Europe. This ruling has probably saved the clinical genetic testing centers from Myriad's monopolistic competition, without displacing the transatlantic power balance in biotechnology.

[30] Myriad Genetics to the European oncologists, letter, Oct. 26, 1998. Author's research data.
[31] Information on the judicial process may be found on the Institut Curie's Web site: http://www.curie.net/actualities/myriad/.

BEYOND THE *EXCEPTION SANITAIRE*: THE DEBATE OVER GENETICALLY MODIFIED CROPS

The trajectory of the breast cancer genes on the two sides of the Atlantic reveals a striking contrast of attitudes, values, and choices regarding the status of biotechnology and its industrial, medical, and social uses. To borrow from recent debates in the cultural arena, one might want to speak of an *exception sanitaire* to describe the attitude of French cancer specialists toward technology- and commerce-driven globalization. Although they are often presented as another form of resistance against "Americanization," demands for the regulation of property rights are better understood as a result of a renewal of the tensions between the medical professions and the health industry, which were particularly acute in the late nineteenth century and early twentieth century in France.[32] Beyond the specific case of biomedicine, it is therefore tempting to link the *exception sanitaire* with the long-term alliance between the "professions" and a French bureaucratic state that plays a prominent role in defining public good and exercises considerable power in the administration and regulation of "civil society." One should nonetheless be cautious about reading transatlantic tensions as the result of different and conflicting national cultures. Heterogeneous forms of regulation exist within national borders, as exemplified by the second topic of this chapter, namely the debates around GMOs.

During the 1980s, the application of recombinant DNA technologies to plants seemed to epitomize the new globalization. On the one hand, the creation of the first genetically modified plants was achieved at the same time on both sides of the Atlantic. On the other hand, mergers and alliances within the food and seed industry concentrated research and development within a half-dozen international groups that dominated the markets. The assessment of the risk associated with the dissemination of such "genetically modified" living entities was initially conducted behind closed doors by biotechnology experts rather than being the subject of public debates. Principles were discussed within the Organization for Economic Co-Operation and Development (OECD) without much conflict. As a result of those discussions, the OECD's 1986 guidelines stated that gene transfers did not introduce new risks when compared with classical genetic selection.[33]

It is true that the field use of modified bacteria had been a public issue in the United States, following the mobilization of environmental activists and, particularly, Jeremy Rifkin's Foundation of Economic Trends. Although the debate was tense, the issue was soon "black-boxed" with a White House decision that the products of genetic engineering did not deserve any specific oversight beyond the review processes already established for other biological material and administered by the appropriate federal agencies (the Environmental Protection Agency and NIH).[34] In France, field release of GMOs was comparatively restricted and remained experimental, with its oversight being assured by an expert committee created in the aftermath of the first recombinant

[32] Christian Bonah and Anne Rasmussen, eds., *Histoire et médicament* (Paris, 2005).

[33] National Research Council, *Genetically Modified Pest-Protected Plants: Science and Regulation* (Washington, D.C., 2000).

[34] Camille Limoges, Alberto Cambrosio, Frances Anderson et al., "Les risques associés au largage dans l'environnement d'organismes génétiquement modifiés: Analyse d'une controverse," *Cahiers de Recherche Sociologique* 21 (1993): 17–52.

DNA controversy. In 1986, the Ministry of Agriculture established the Commission du Génie Biologique (CGB), a gathering of molecular biologists who would discuss the release of genetically modified plants. The CGB focused on the molecular soundness and safety of the genetic constructs proposed by the applicants.[35] Professional regulation was dominating the scene in both countries.

In France, another arena for discussion emerged in the mid-1990s with the politicization of the GMO issue. The change was triggered by a conjunction that illustrates the two sides of globalization. In 1996, shipments of genetically modified soy started to arrive in European ports. Transnational environmental organizations such as Greenpeace International used this opportunity to frame their opposition in terms of a transatlantic conflict.[36] According to this point of view, the U.S.-based seed and food giants were imposing untested innovations on European consumers. Activists from Greenpeace tried to block the arrival of soy and targeted agribusiness firms such as Monsanto. Food safety and, more generally, the relationship between agriculture and industry were very sensitive issues at that time as the United Kingdom, France, and the EU were already entangled in the management of Bovine Spongiform Encephalopathy (BSE), or mad cow disease, and the resulting beef crisis.[37] Fearing another scandal that would put enormous pressure on the state administration and could result in a financial and judicial disaster (as had been the case with AIDS), the French government decided not to authorize the culture of genetically modified corn.

This choice had two major consequences. First, it reinforced the crisis of expertise. One effect of the mad cow affair was to reorganize the expertise about food safety around increased surveillance and tracing of animal tissues as a means of taking into account the precautionary principle.[38] Public concerns facilitated the importation of this framework within the biotech domain. The experts gathered in the CGB had previously stated that genetically modified corn was safe. Facing the government's decision, the commission's president resigned in protest. At the same time, a small group of biologists publicly protested the insufficiency of the risk assessment conducted by the CGB. They signed a petition against the commercialization of agricultural GMOs, arguing that, if authorized, commercial release would create an irreversible situation whose consequences had not been properly evaluated, thereby providing environmental groups with more legitimate scientific arguments. The second consequence of the position adopted by the French government was to push the European administration in the direction of exercising tighter control over GMOs. In 1998, governments with "green" ministers—Belgium, France, and Germany—were strong enough to obtain a European ban on the marketing of new varieties of genetically modified plants, at least until they could be proven safe and their fate brought under control, thus establishing a moratorium on the uses of GMOs.

The development of the French GMO controversy may be compared with the environmental debates that took place in the United States in the 1970s.[39] Two dimensions are worth considering: on the one hand, the mobilization of "civil society" organizations;

[35] Alexis Roy, *Les experts face au risque: Le cas des plantes transgéniques* (Paris, 2001).
[36] Hervé Kempf, *La guerre secrète des OGM* (Paris, 2003).
[37] Paul Benkimoun, *Démocratie et sécurité alimentaire: La peur au ventre* (Paris, 2000).
[38] Francis Chateauraynaud and Didier Torny, *Les sombres précurseurs* (Paris, 1999).
[39] Daniel Vogel, *Ships Passing in Night: GMOs and the Politics of Risk Regulation in France and in the United States* (Fontainebleau, 2001).

on the other, the reorganization of the system of expertise toward more "pluralism" and "openness." Observers have stressed the fact that the most important opponents to agricultural GMOs in France are not only the nongovernmental environmental organizations such as Greenpeace and Friends of the Earth but also the farmers' unions such as the leftist Confédération Paysanne and the very powerful and conservative Fédération Nationale des Syndicats d'Exploitants Agricoles.[40] This is new and contrasts sharply with the U.S. configuration, in which all the comparable professional organizations representing farmers are strong supporters of the use of GMOs. In France, the farmers' engagement played a critical role in the rapid politicization of the debates. GMOs have been targeted as a symbol of economic globalization and the negative effects it has had on French agriculture. GMOs are perceived not only as American imports but also as tools in the battle for economic domination. The use of GMOs, produced by U.S.-based seed companies, would increase the financial pressure on local producers as well as those producers' technological dependency. As articulated by the Confédération Paysanne, the battle against GMOs has come to mean both a battle against "bad food" and a battle to save small farmers from agricultural capitalism.[41]

The opening up of expertise in the GMO debate has taken two different forms. The first one was the integration of scientists from domains other than molecular biology, that is, ecologists and agricultural scientists. Their participation in the CGB and other evaluation committees somehow made it official that the GMO debate was laden with uncertainties, that different scientific viewpoints were necessary, and that controversies were inevitable, if not useful. This change was triggered by the public and political nature of the debate, but it was also rooted in the existence of two cultures of risks.[42] The classical framing of the problem was that of the molecular biologists who focus on "molecular" assumptions. Accordingly, DNA structure determines the properties and fate of a gene, and a genetic construction is better known and somehow cleaner than the old varieties patiently selected by local breeders and farmers. Since gene transfers are extremely rare events, almost impossible in nonhybridizing plants, the risk of genetic pollution, that is, of a transferred gene's dissemination, is minimal. The new risk framework originated in ecology and population genetics. It focuses on the long-term circulation of GMOs, on the global genetic structure of cultivated plants, and more broadly, on landscapes and ecosystems. From a population perspective, gene transfers are certain events, creating irreversible gene flow and changes in biodiversity, which need to be modeled and monitored to manage the complexity of agricultural practices.

The second opening of the expert system has been a cautious inclusion of interested parties and "lay" citizens. Having representatives of farmers' organizations participate in expert committees was new for molecular biologists, even if it was not unusual in agricultural research. What was, however, unprecedented was having laypeople without any role in science or in agriculture involved in the assessment of GMOs. Rather than an indication of an emerging trend, these changes were contextual, rooted

[40] Pierre-Benoît Joly, Gérard Assouline, Dominique Kréziak et al., *L'innovation controversée: Le débat public sur les OGM en France* (Grenoble, 2000).
[41] Kempf, *La guerre secrète des OGM* (cit. n. 36)
[42] Christophe Bonneuil, "Les recherches sur les impacts agro-écologiques des OGM dans les années 1990," in *Maîtrise des risques, prévention et principe de précaution* (Paris, 2002), 81–94.

in the series of crises that had marked France's recent past, that is, the aforementioned AIDS movement, the Chernobyl affair, and the beef crisis.[43] The most visible innovation was a consensus conference (called a citizen conference) formally organized by the French parliament in 1998.[44] The conference brought together a panel of so-called ordinary citizens (randomly selected according to criteria that ensured social diversity), who first received a crash-course in genetic engineering given by experts representing the range of specialties involved in the GMO arena before conducting hearings of experts from the sciences, the agricultural sphere, industry, environmental groups, and NGOs. According to both participants and outside observers, the experience demonstrated the possibility of integrating lay viewpoints and perspectives into a highly technical debate. The hearings led by the citizens' panel not only shed light on the nature of risks—for instance, the extent of gene flow in corn or rape—but also highlighted the economic and social dynamics of the agricultural uses of GMOs. The political outcome of the citizen conference was less convincing than the content of its final recommendations, however. The recommendations ranged from the need for more research (including fieldwork), to the necessity to reorganize the process of expertise (for instance, opening the CGB to a wider range of representatives of "society") and the need for strong regulation of GMO marketing.[45] These conclusions were presented in the form of a report to the government, which never responded to it. Regulation was subsequently enacted, but as a result of the discussions conducted through more traditional channels—namely, the negotiations of interested parties with the Ministry of Agriculture and the Ministry of Environment, on the one hand, and the various European committees and roundtables, on the other.

In the United States, the GMO debate was resurrected in 1998, partly as a result of the European opposition, which environmental groups rapidly used in their U.S. efforts, a move facilitated by the transnational nature of organizations such as Greenpeace. Greenpeace launched a campaign for the labeling of GM food, arguing that the U.S. government should give American consumers the same information the European consumers received. The first "affair" was, however, the so-called Terminator debate, which saw the environmental activists targeting Monsanto's "technology protection system." This innovation had been introduced to block further agricultural uses of the seeds produced by the plants grown from Monsanto's seeds. Echoing ancient debates about hybrid corn and the increased technical dependency of farmers, this technology was seen as a direct illustration of the monopolistic practices of global agribusiness.[46] This industry was not only pushing the "appropriation of life" through gene patents but also inventing "technologies to sterilize life" in order to exercise tighter control over the users of their systems. NGOs linked this issue as well to the relationship between the West and "third world" countries and, more generally, problems associated with the North-South technological gap. Terminator, as the technology was renamed by activists, was all the more visible as Monsanto started to publicly denounce farmers who used the seeds produced by their GM corn, launching lawsuits

[43] Chateauraynaud and Torny, *Les sombres précurseurs* (cit. n. 38); Janine Barbot, *Les malades en mouvement: La médecine et la science à l'épreuve du Sida* (Paris, 2002).
[44] Claire Marris and Pierre-Benoît Joly, "Between Consensus and Citizens: Public Participation in Technology Assessment in France," *Science Studies* 12 (1999): 3–32.
[45] Jean-Yves Le Deaut, *De la connaissance des gènes à leur utilisation* (Paris, 1998).
[46] Glen E. Bugose and Daniel Kevles, "Plant as Intellectual Property: American Practice, Law, and Policy in a World Context," *Osiris* 7 (1992): 75–104.

against them.[47] The criticisms of market practices, therefore, became as visible in the United States as they were in France, but they were based on different justifications, focusing on support for the autonomous and industrious small-scale farmer or entrepreneur rather than attacking the patent system or the industrial nature of agriculture.

The GMO debate in the United States never assumed the political dimensions it did in France. Nothing like a moratorium or attempts to introduce new regulations were seriously considered at the governmental level: surveillance was inscribed in the existing regulatory framework. Today GM seeds are widely used in U.S. agriculture without major concerns being voiced by either the institutions in charge of agriculture or the public at large. According to Pierre-Benoît Joly and Claire Marris, this cannot be explained by the lack of media attention paid to the issue or by the "acceptance" of genetically modified food among consumers (panel studies document similar levels of distrust on both sides of the Atlantic).[48] The differences between the American and French situations arose first from the political dynamics of the controversy, with one critical factor being the absence of opposition within the major organizations representing U.S. farmers. These bodies have, for instance, been quite supportive of the administration's position that food products originating in GM plants should carry special labels. A second factor, more significant when considering biotechnological innovation at large, is the nature of scientific expertise in the United States. In contrast to the post-AIDS reorganization of biomedical research, the notion that decisions have to be made by professionals on the basis of *sound science* has won many supporters within the system for food and agricultural expertise. This commitment is perceived as all the more necessary after the "excesses" of the 1970s. A major consequence of this vision was the choice (in contrast to the proposed regulation of genetic tests) not to place GMOs in any new special category of products that would require specific authorization. Products originating in GM biological material are considered equivalent to the same products obtained from different processes. A major turning point was a 1995 court decision that the meat and milk products from cows fed with a genetically recombined growth hormone did not need to carry a special label since the use of a natural form of the growth hormone in agriculture was not a matter of regulation but simply a question of good veterinary practice.[49] Following this logic, GMOs used to replace other equivalent substances do not need to be independently evaluated for either environmental or health-related risks.[50] In other words, GMOs do not require surveillance since they do not (administratively) exist.

CONCLUSION

Generalizing from the controversy over the BRCA genes, one can argue that in France a model of expertise dominated by the autonomy of the scientific profession—its emphasis on laboratory systems, its distrust of applied (commercial) developments—is still the rule. In contrast, the U.S. configuration is dominated by a form of expertise more open to the "civil society," which means at the same time more open to patient

[47] *The Ecologist,* Sept. 1998.

[48] Pierre-Benoît Joly and Claire Maris, "Les Américains ont-ils accepté les OGM? Analyse comparée de la construction des OGM comme problème public en France et aux Etats-Unis" (forthcoming).

[49] Sheldon Krimsky, *Biotechnics and Society* (New York, 1991).

[50] Limoges et al., "Les risques" (cit. n. 34).

organizations and to the health industry. In France, the influence of professional regulations resulted in leaving the question of the uses of the BRCA sequences in the physicians' hands, while the debate on state-based regulation focused on the appropriation of knowledge and its consequences on clinical practice. In contrast, in the United States patient organizations have pushed into the direction of strong surveillance, focusing on the utility of gene testing and building a regulatory alliance, which backed the failed proposal of an FDA-based system of premarketing permits. The comparison with the GMO controversy, however, shows that opposing France with its centralized state and elite scientists and the more democratic, but market-oriented, United States is too simple a picture. The dynamics of the debate over GMOs discredit any general conclusions about the weakness of "lay" activism or the lack of engagement of "civil society" in French technoscientific debates. In that case, regulation has been pushed forward by an alliance of consumers, ecological organizations, and farmers' unions. Their convergence originates in a highly politicized discussion triggered by both the conflicts about globalization in agriculture and previous technoscientific affairs (the BSE crisis), which weakened the traditional management of agricultural policy. Within this context, the assessment of risks became more diverse and controversial than in the United States, with statements opposing the molecular biological vision of biosafety and the implementation of a quasi-moratorium at the European level.

To advance the comparison, we can offer a summary of the ways in which regulatory issues were discussed and handled in the two controversies with Tables 1 and 2 for France and the United States.

In a recent book, John Pickstone has argued that the modern history of science reveals the emergence and the succession of various "ways of knowing."[51] For instance, he distinguishes an analytical way of knowing prevailing in most areas of the study of nature circa 1800, while an experimental way of knowing had superseded this earlier configuration by the end of the nineteenth century. Pickstone considers that these regimes of scientific practices are connected to peculiar "ways of doing," which define how we work, how we produce things, and how we exchange goods. Ways of knowing are useful in addressing the dynamics of particular scientific domains as well as their links to the different social worlds that mobilize scientific knowledge. Considering the close relations the sciences have established with the state in general, and the modern nation-state in particular, one can suppose that these various ways of knowing have not only been articulated with ways of doing but also been connected to "ways of regulating." In the context of this paper, regulation does not mean only a system of legal constraints; it includes all sorts of norms and guidelines regulating the uses of biotechnological goods in the laboratory, in the plant, in the hospital, and in the field. Regulation is, therefore, a system of practice based on peculiar forms of interventions, and peculiar forms of knowledge used either directly as a source of information and procedures or indirectly as a source of legitimacy.

On the basis of the two case studies discussed in this paper, one may suggest that the contemporary biotechnological world constitutes an arena in which four ways of regulating are alternatively competing, juxtaposed, or combined. These four types of regulation not only mobilize different groups of actors but also reflect different views about what is a public good and how the state intervenes in relation to markets, civil

[51] John Pickstone, *Ways of Knowing* (Manchester, 2000).

Table 1. Regulating Biotech in France: BRCA Testing and GM Crops

	BRCA	GMO
Issues	a. Patent monopoly b. Clinical autonomy c. Access	a. Industrial agriculture b. Ecological impact, precautionary principle c. Consumer's choice
Public debate	No (media reports)	Yes (committee work, citizens' conference)
Controversy among scientists	No	Yes (gene transfers)
Lay counter-expertise	No	Yes (NGOs)
Main actors	a. Professionals (molecular biologists, oncologists) b. State (Ministry of Health, agencies)	a. Associations (farmers' unions, ecological NGOs) b. State (Ministry of Agriculture)
Forms of intervention	a. Norms of good practice b. Judicial appeals (patents)	a. Lobbying b. Media campaigns c. Civic protests d. Court action
Regulatory aims	a. Quality control guidelines b. Open uses of knowledge	a. Premarketing tests b. Labeling of GM food c. Biosurveillance, biosafety
Form of regulation	Professional	State/Civic-consumer

society, and globalization.[52] The first and oldest type of regulation stipulating the appropriate uses of biotechnological products is professional regulation. Within this framework, the scientists, or some self-designated groups among them, benefit from an apparently complete autonomy in establishing norms. This model is well illustrated by the dominant role physicians and pharmacists played in drug evaluation until the 1960s. Professional competency and users' compliance are the most important issues. Regulatory targets consist in internal guidelines, recommendations of good practice written by expert committees, and disciplinary bodies. The second model of regulation is regulation by the industry. Within this perspective, innovations are deemed good when they satisfy a demand in an economical and productive way. Safety and quality are important issues since their absence may result in liability and consumer dissatisfaction. What they mean is, however, defined by the producers. Regulatory tools include in-house quality control protocols and, more importantly, the leaflets, brochures, packet inserts, Web sites, and other forms of scientific marketing the companies employ to shape markets and uses. This is the model powerfully illustrated by the activities of the large corporations with their research divisions or by the

[52] Pierre-Benoît Joly, "Les OGM entre la science et le public? Quatre modèles pour la gouvernance de l'innovation et des risques," *Economie Rurale,* no. 266 (2001): 11–29; Michel Callon, Pierre Lascoumes, and Yannick Barthe, *Agir dans un monde incertain* (Paris, 2001). For a more comprehensive discussion of these ways of regulating, see Jean-Paul Gaudillière, "Cancer, Hormones, and Risk: Four Ways of Regulating Drugs in the 20th Century," *Bulletin of the History of Medicine* (forthcoming).

Table 2. Regulating Biotech in the United States: BRCA Testing and GM Crops

	BRCA	GMO
Issues	a. Access to tests (discrimination) b. Clinical utility	a. Food safety, toxicity b. Productivity c. Control of seeds, farmer's autonomy
Public debate	Yes (committee work, hearings)	No (media reports, expert gatherings)
Controversy among scientists	Yes (risk management)	At the margins (ecological damage)
Lay counter-expertise	Yes (NBCC)	Yes (NGOs)
Main actors	a. Professionals (molecular biologists, physicians) b. Biotech industry c. Patient organizations	a. Biotech industry b. Ecological NGOs c. EPA
Forms of intervention	a. Marketing b. Lobbying c. Norms of good practice	a. Lobbying b. Media campaigns c. Product surveillance d. Court action
Regulatory aims	a. Premarketing evaluation b. User's informed choice	a. Labeling of GM food b. Biosurveillance
Form of regulation	Industrial/Civic-consumer	Industrial/Civic consumer

new start-ups. The third type of regulation is rooted in the activities of the state administrative bodies, which have historically been assigned the role of preserving public health, the quality of the environment, and social equity. Marketing authorizations are among their privileged regulatory tools, usually used in conjunction with public statements. The last, and most recent, form of regulation may be called civic-consumer regulation. Its emergence is rooted in the notion that the sciences should become more democratic and that research and expertise need to be opened up to a larger public. In this case, regulation is linked to the necessity of taking into account either the freedom of choice of individuals or the collective needs that are neither translated into market demands nor recognized as such by professionals. Civic-consumer regulation only surfaces when interested groups—end users, patients, laypeople—are given a say and practice some level of counter-expertise. Intervention takes place in public arenas, with media reports, court cases, consensus conferences, and political hearings.

Any particular biotech configuration of the sort analyzed in this paper may be associated with different ways of regulating, usually operating at different levels. Thus, for example, the fate of BRCA genes in France has been a typical case of professional regulation associated with state regulation, while the GMO debate in the same country combined professional regulation with elements of civic-consumer regulation. Our comparison suggests that the transatlantic tensions regarding the production and use of biotechnology are the product of alternative combinations of regulatory practices, which are themselves rooted in different sociopolitical cultures. Although the civic-consumer way of regulating has emerged on both sides of the Atlantic, France remains a privileged terrain for professional regulation under the umbrella of the state, while the United States has increasingly relied on industrial regulation. This contrast is less a testimony of the purely national basis of regulation in the biotech domain than a consequence of the contradictions associated with globalization. Globalization is not synonymous with worldwide, un-situated, homogeneous, and inevitable patterns of socioeconomic change. As illustrated by the fate of gene patents, globalization has local roots, in most instances the policies and ways of doing of the most powerful actors. Moreover, globalization is itself a battlefield. The biotechnological conflicts between Europe and the United States actually show that several globalizations are operating at the same time. The most obvious globalization is the one associated with the development of international research markets and the enlarged circulation of technoscientific goods. A less visible, but equally significant, trend is the attempt at "globalizing" local alternatives to this corporate globalization.

Biotechnology and Empire:
The Global Power of Seeds and Science

*By Sheila Jasanoff**

ABSTRACT

Following the cold war, interest has grown in the possible rise of new forms of imperial rule and in the likely role of science and technology in processes of global governance. In particular, just as the life sciences advanced the interests of bygone empires, so modern biotechnology is likely to support today's transboundary exercises of political, economic, and cultural power. Drawing on analyses of large-scale political and technological systems, this chapter suggests that contemporary biotechnology may be enrolled into empire-making by several different means, including bottom-up resistance, top-down ideological imposition, administrative standardization, and consensual constitutionalism. At present, biotechnology seems more likely to increase the power of metropolitan centers of science and technology than that of people at the periphery. Institutional innovations will be needed to bring global biosciences and biotechnologies under effective democratic control.

INTRODUCTION

Imperialism is back on the circuits of public debate, and it is back with a vengeance. Contributors to the twenty-first-century discourse of empire include historians and social theorists, political scientists and anthropologists, op-ed columnists and politicians in positions of power. Books about imperialism, many sporting the word "empire" in their titles, appeared by the dozen at the turn of the century.[1] Through them, and through endless journalistic commentaries,[2] the attention of much of the reading

* Harvard University, JFK School of Government, 79 J.F. Kennedy Street, Cambridge, MA 02138; sheila_jasanoff@harvard.edu.

I am grateful to the Universities of Wageningen, Netherlands, and Halle, Germany, for invitations to present earlier versions of this chapter.

[1] Influential contributions include Michael Hardt and Antonio Negri, *Empire* (Cambridge, Mass., 2001); idem, *Multitude* (New York, 2004); David Harvey, *The New Imperialism* (Oxford, 2003); Chalmers Johnson, *Blowback: The Costs and Consequences of American Empire* (New York, 2000); Niall Ferguson, *Empire: How Britain Made the Modern World* (London, 2003); and idem, *Colossus: The Price of America's Empire* (New York, 2004); David Cannadine, *Ornamentalism: How the British Saw Their Empire* (Oxford, 2001); Catherine Hall, *Civilising Subjects: Colony and Metropole in the English Imagination* (Chicago, 2002); Linda Colley, *Captives: Britain, Empire, and the World, 1600–1850* (New York, 2003); Rashid Khalidi, *Resurrecting Empire: Western Footprints and America's Perilous Path in the Middle East* (New York, 2004); and Anne-Marie Slaughter, *A New World Order* (Princeton, 2004).

[2] For one widely discussed example, see Michael Ignatieff, "The American Empire: The Burden," *New York Times Magazine,* Jan. 5, 2003, 22. See also Charles S. Maier, "Forum: An American Empire?" *Harvard Magazine* 104 (Nov./Dec. 2002): 28–31. This is the topic of Maier's forthcoming book, *Among Empires: American Ascendancy and Its Predecessors.*

 0369-7827/06/2006-0013$10.00

world has turned to a particular instance of imperial expansion: the post–cold war United States, driven by what many see as a runaway ambition to impose military dominance, ideological conformity, and cultural homogeneity on the rest of the world.[3] It is as if a potential left implicit by Ronald Reagan's famous appellation for the Soviet Union—"the evil empire"—has come to fruition in George W. Bush's Manichaean vision, which pits an actual, divinely blessed, "good" America against its "evil" enemies, the states that harbor terror. Global power struggles are recast as a fight to the finish between the imperial forces of light and of darkness. Presidential rhetoric reprises popular culture: George Lucas's hugely successful trilogy of intergalactic conflict, *Star Wars,*[4] provided not only the template for dividing the world into two vast opposing armed camps but also the visual and metaphorical resources for reducing warfare between them to the starkness of black and white.

Empires, however, are patchier constructs than the simple dualisms of presidential imaginations, shaped by Hollywood imagery, would have us believe.[5] Neither culturally nor normatively homogeneous, they invite analysis as spaces in which power is exercised through complex, often subterranean means. From the Roman imperium to the territories ruled by Britain at the height of its Victorian expansion, diversity rather than homogeneity has been the characteristic look of empire. Possibly the most successful empires have been those that allowed multiple divergences in language, religion, dress, diet, and customs to flourish, within an envelope held together by various consolidating moves that coordinated, but did not erase, difference. For insights into these processes, we may turn to scholars of colonialism and postcolonialism, who have pointed out the disparate moves made to differentiate, as well as integrate, the populations under the ruling regime's control. On the one hand were steps that clarified and firmed territorial boundaries, imposed common linguistic and educational standards, and produced shared categories to reason and rule with.[6] On the other hand were strategies for preserving hierarchies of power, including rules of cohabitation allowing or disallowing mixing between the rulers and the ruled.[7]

Empires then were places of hybrid identities, with all the tensions for regularity

[3] American progressives would like to detach what many see as the illegitimate path of unilateral militarism from the legitimate, indeed desirable, path of economic and social globalization driven by the "soft power" of culture and markets. See Joseph S. Nye, *Soft Power: The Means to Success in World Politics* (New York, 2004). Celebrations of America's role in leading the world to free-market democracy include Thomas L. Friedman, *The Lexus and the Olive Tree: Understanding Globalization* (New York, 1999).

[4] Directed by George Lucas, the trilogy opened in 1977 with *Star Wars,* the film that gave its title to the series. It was succeeded by *The Empire Strikes Back* (1980) and *Return of the Jedi* (1983). Appearing in the waning years of the cold war, the films exercised a particular influence on Ronald Reagan, America's first Hollywood president. The idea of a satellite-based missile defense shield was initially broached in the Reagan era, and the project, which remained mired in conflict during his presidency, was nicknamed Star Wars.

[5] On this theme, see Tony Judt, "Dreams of Empire," *New York Review of Books,* Nov. 4, 2004, 38–41.

[6] On these points, see Benedict Anderson, *Imagined Communities,* 2nd ed., rev. and exp. (London, 1991); Sarah Radcliffe, "Imaging the State as Space: Territoriality and the Formation of the State in Ecuador," in *States of Imagination: Ethnographic Explorations of the Postcolonial States,* ed. Thomas Blom Hansen and Finn Stepputat (Durham, N.C., 2001), 123–45.

[7] Ann L. Stoler, *Carnal Knowledge and Imperial Power: Race and the Intimate in Colonial Rule* (Berkeley, 2002); idem, "Making Empire Respectable: The Politics of Race and Sexual Morality in 20th-Century Colonial Cultures," *American Ethnologist* 16 (1989): 634–60.

and order that hybridity entails.[8] The wonder is that they nonetheless held and that similar formations may yet hold in other times and places. In this respect, empires can be seen as analogous to large technological systems, like electric power grids[9] or civil aviation: so complex, heterogeneous, loosely pinned together, even jerry-built on close inspection that their stability is the thing that needs explanation. By contrast, as illustrated by the terrorist attacks of September 11, 2001, in the United States, mundane technological systems such as high-rise buildings, regarded as not seriously vulnerable to external threats, can reveal deep structural faults under unexpected attack.[10]

Viewing empires as social technologies, that is, as human-made assemblages that enable power to extend beyond its original spatial and cultural locations,[11] raises for us a critically important set of questions. What is the role of conventional technological systems, those built around material components such as guns, butter, and newspapers, in the production and maintenance of new forms of transnational rule? How, in particular, might the human capacity to instrumentalize nature influence the possibilities for politics in a globalizing world? Will the major technological revolutions of our time—in the life sciences, information and communication technologies, computers and weaponry, and most recently nanotechnology—favor emancipation or recolonization? Will they make people around the world more or less connected, more or less free, more or less comfortable, and most important for our purposes, more or less democratic? Will the radically unequal distribution of wealth and privilege in the contemporary world reinscribe itself through technological means, continuing older forms of hegemony and dominance? If that danger exists even in principle, are there institutions or processes through which a global citizenry can assert the right to shape the technologies that may, if widely deployed, shore up global regimes of control?[12]

I approach these questions in this chapter through the lens of modern agricultural biotechnology. Still in its infancy more than three decades after its first experimental successes in western laboratories, so-called green biotechnology has rapidly become a global industry promising enormous benefits to the world's poor. Its proponents claim it has the capacity to overcome nature, making plants that can resist drought, ward off insects, and with the ability to produce micronutrients engineered into their

[8] See, e.g., the account of collectors and collecting in the eighteenth-century British and French proto-empires, Maya Jasanoff, *Edge of Empire: Lives, Culture, and Conquest in the East, 1750–1850* (New York, 2005).

[9] Thomas Hughes, *Networks of Power: Electrification in Western Society, 1880–1930* (Baltimore, 1983).

[10] 9/11 Commission, *Final Report of the Commission on Terrorist Attacks upon the United States* (New York, 2004).

[11] This way of thinking about empires is consistent with contemporary work in science and technology studies. See, in particular, Sheila Jasanoff, ed., *States of Knowledge: The Co-Production of Science and Social Order* (London, 2004); Bruno Latour, "Drawing Things Together," in *Representation in Scientific Practice,* ed. Michael Lynch and Steve Woolgar (Cambridge, Mass., 1990), 19–68. Richard Drayton adopts a similar perspective when he speaks of empire as "an ecological system," stressing the interconnections among politics, economy, and nature that define empires. See, particularly, Drayton, "Imperial Science and a Scientific Empire: Kew Gardens and the Uses of Nature, 1772–1903" (Ph.D. diss., Yale Univ., 1993).

[12] For an argument that such demands are already being expressed through a tacit and unwritten form of global constitution-making, see Sheila Jasanoff, "In a Constitutional Moment: Science and Social Order at the Millennium," in *Social Studies of Science and Technology: Looking Back, Ahead,* ed. Bernward Joerges and Helga Nowotny, Yearbook of the Sociology of the Sciences (Dordrecht, 2003), 155–80.

genes, even transcend the "normal" dividing line between food and pharmaceuticals. Biotechnology by some definitions is as old as "second nature," the first successful prehistoric attempts by human societies to harness nature's growth to serve their basic needs for food, fuel, clothing, and shelter. Under another definition, the one I use here, biotechnology is much newer. It is the name given to an array of manipulative techniques based on alterations of the cellular and subcellular structures of living things enabled by the 1953 discovery of the structure of DNA.[13] These techniques include, most notably, not only genetic engineering, gene splicing, but also operations such as cell fusion and cell culturing carried out at levels of structure significantly smaller than the whole organism. How will these technological developments, heralding what some have called a second Green Revolution,[14] affect flows of power and opportunities for self-determination around the world?

In looking for answers, I begin in effect with a typology of empire, based on the diverse ways in which the extension of imperial power has been conceptualized by analysts of large-scale political, as well as technological, systems. The life sciences, as much research has shown, have long been implicated in serving the designs of empire builders. Modern biotechnology, I suggest, can similarly be drawn into the service of possible imperial constructions, and I ask in what ways this particular global production system is likely to influence today's transboundary exercises of political, economic, and cultural power. This analysis suggests that, without institutional innovations, biotechnology as currently governed may increase the power of metropolitan centers of science and technology in relation to people at the periphery. In conclusion, I reflect on the prospects for democratic governance of technological systems such as agricultural biotechnology that are centrally involved in contemporary processes of globalization.

IMPERIAL CONSTRUCTIONS

How are empires held together? Not, as I have suggested, through homogenized identities and uniform allegiances that make the residents of imperial territories carbon copies of one another. Clues may be found in those areas of the social sciences that occupy themselves with the stability of heterogeneous constructs, in such fields as international relations and law, science and technology studies, colonial and postcolonial history, and cultural anthropology. Work in all these domains suggests that the fabrication of empire proceeds not through any single grand gesture of unification, nor by a revolutionary process of mass struggle as suggested by two theorists of the Left, Michael Hardt and Antonio Negri,[15] but through a series of contingent, overlapping, altogether human practices that build coherence and cohesion while staving off dispersal. As shown in Table 1, we can discern five distinct modes of imperial governance—that is, five mechanisms, not mutually exclusive, through which the unruly

[13] Robert Bud, *The Uses of Life: A History of Biotechnology* (Cambridge, 1993).

[14] The first Green Revolution was the introduction worldwide of high-yielding grain varieties pioneered by Nobel laureate Norman Borlaug and other plant biologists. Their work was sponsored in part by the Rockefeller Foundation. For accounts of the scientific and social dimensions of the Green Revolution, see Lily E. Kay, *The Molecular Vision of Life: Caltech, the Rockefeller Foundation, and the Rise of the New Biology* (New York, 1993); J. R. Anderson, R. W. Herdt, and G. M. Scobie, *Science and Food* (Washington, D.C., 1988); P. B. R. Hazell and C. Ramasamy, *The Green Revolution Reconsidered* (Baltimore, 1991).

[15] Hardt and Negri, *Empire* (cit. n. 1).

Table 1. Modes of Imperial Governance

Empires of resistance	Emergent, agentless form of rule, constituted in possibly violent opposition between global ruling institutions and resisting citizens ("the multitude")
Empires of ideology and force	Communal norms and beliefs imposed through force, persuasion, surveillance, and sanctions
Empires of legibility	Communal standards imposed through administrative simplification and efficiency (Weberian)
	Communal standards achieved through classification, normalization, and erasure (Foucauldian)
Empires of identity	Imagined communities built through mass media, official representations, political and cultural symbols
Empires of law and constitutions	Rule of law under constitutional principles, enabling liberal individualism and free movement of goods and people

heterogeneity of empires can be made more orderly and therefore more tractable to rule.

The vision of empire put forward by Hardt and Negri stands in a somewhat anomalous relation to the others in Table 1, partly because the empire they envision is a global formation lacking any particular sovereign at the head, and partly because of the authors' disregard for the micro-processes of agency and governance that have loomed large in the work of other theorists of national and imperial power.[16] The empire whose emergence Hardt and Negri ambitiously prophesy is a revolutionary construct, propelled in part by the consolidation of a global multitude whose demands nation-states are no longer able to satisfy. Bottom-up political action in an inchoate field, mediated through the Internet, is seldom strategic or coordinated, but, through repeated, decentralized gestures, it can achieve something of the character of continuous mass protest. Hardt and Negri's account has drawn vigorous criticism for its lack of clarity, inattention to specifics, denial of agency, and leftist nostalgia for violence as a means of radical social change. At the same time, it provides a vision of uncoordinated, multicentric, populist, political, and normative action—propelled by ideas and beliefs—that is, in some ways, more appealing than the tight, and equally faceless, administrative networking of the world contemplated by some analysts.[17] Something resembling the dynamics of the multitude, as we will see below, is not altogether absent in the contemporary global politics of biotechnology.

Turning to more conventional articulations of empire, those constituted by (or as) an identifiable sovereign state, we note that the processes and practices that sustain imperial rule do not have to be consensual or responsive to the popular will and that violence remains very much an instrument of top-down domination. This is clearest in the case of empires of ideology and force, such as the former Soviet Union and perhaps the American empire currently taking shape, in which adherence to a common

[16] Contrast in this respect Hardt and Negri, *Multitude,* with Slaughter, *A New World Order.* (Both cit. n. 1.) See also Thomas N. Hale and Anne-Marie Slaughter, "Hardt and Negri's 'Multitude': The Worst of Both Worlds," *Open Democracy,* May 26, 2005, http://www.opendemocracy.net/globalization-vision_reflections/marx_2549.jsp.

[17] Slaughter, *A New World Order* (cit. n. 1).

ideology (socialism and market capitalism, respectively) has been achieved through the forceful subordination of countervailing belief systems and forms of life. Technology, historically, played a central role in the effectuation of such extended ideological dominion: not only military technologies, though these were of course essential, but also technologies of surveillance, punishment, and mass communication. Built to control hybridity, such control technologies are themselves hybrid, marrying the hardware of computers or cameras, for instance, with social supports from law and administration, and increasingly the mass media.[18] In this way, technologies of force shade into technologies of legibility and standardization, which are tools of imperial construction in their own right.

When we speak of standardization as a form of political control, we begin inevitably with Max Weber. Empires were, in the first instance, vast bureaucracies. They were administered domains, and their management called forth the production and diffusion of professionals of every stripe: scientists, engineers, surveyors, physicians, lawyers, linguists, archaeologists, and archivists, among others. Their task was to make government more efficient and rational, enable communication and exchange, and—in the more beneficent of imperial imaginations—extend the virtues of knowledge, reason, and productivity equally throughout the empire. Less clear in Weber's time, however, was the extent to which the imposition of administrative rule altered, or even created, the identities of the subjects being governed.

Michel Foucault's work on "governmentality" fills that gap, and it provides another indispensable starting point for understanding imperial standardizations today.[19] Foucault represented governmentality as a specific form of rule that emerged with European modernity, coincident with the waning of absolute monarchical power and the rise of science. In this social order, the governors and the subjects to be governed became part of the same enterprise, linked through their allegiance to new truth regimes, grounded in technical disciplines (preeminently the human sciences) that provide the means for authoritatively characterizing both social bodies and social problems. Experts trained in professional discourses can identify populations and, through clinical work, their individual members as healthy or sick, sane or mad, normal or deviant, racially pure or impure, criminal or socially responsible. These definitions become essential, not only to those who exercise power to keep illness, insanity, deviance, racial commingling, and criminality at bay, but also to their subjects, who, as what the philosopher Ian Hacking has called "interactive kinds,"[20] come to see and recognize each other in terms of the dominant classification systems of their time and place. Bureaucracies fitted out with elaborate expert support systems develop norms and regulations based on the experts' classifying knowledge.[21] Government (the project of the rulers) and mentality (the state of mind of the ruled) then fuse, as both begin to perceive the world in identical conceptual terms and reinforce each other's perceptual frames.

Governmentality, despite its pretensions of neatness, seldom divides the world into cleanly defined categories. It takes work of a special sort—specifically, boundary

[18] Consider, for example, the U.S. military's practice of "embedding" journalists with ground forces during the conduct of the 2003 Iraq war.

[19] Michel Foucault, "Governmentality," *Ideology and Consciousness* 6 (Summer 1986): 5–21.

[20] Ian Hacking, *The Social Construction of What?* (Cambridge, Mass., 1999).

[21] Geoffrey C. Bowker and Susan Leigh Star, *Sorting Things Out: Classification and Its Consequences* (Cambridge, Mass., 1999); on the dynamics of bureaucratic expertise, see also Sheila Jasanoff, *The Fifth Branch: Science Advisers as Policymakers* (Cambridge, Mass., 1990).

work—to smooth out the messy spaces between classes and to create the appearance of sharp divisions, or bright lines as lawyers call them.[22] In the process of classification, problematic hybrids and hard-to-fit entities or communities may be erased, either through forcible elimination or through administrative and symbolic moves, such as selective mapping or listing, that take the unclassifiable things out of the ruler's visual space. Thus, unproductive citizens may be cleared out of slums and city streets,[23] wildernesses replaced by planned forests, last names substituted for patronymics, and medieval streets overlaid with the familiar grid pattern of the surveyable and policeable modern city.[24] The political theorist James Scott refers to these simplifications as a process of creating legibility, a concept lying somewhere between Weberian administrative efficiency and Foucauldian governmentality. Modern statecraft, Scott argues, has consisted in the main of taking "exceptionally complex, illegible, and local social practices" and creating "a standard grid whereby it could be centrally recorded and monitored."[25]

While Scott and, to some extent, Foucault, stress the role of the state and its docile experts in making knowledge and order, others have asked (as indeed Foucault did in connection with the "mentality" component of governmentality) how subjects buy into the imperial projects of which they are part. James Morris's splendid popular account of the British Empire at what he calls the moment of its climax in 1897, the diamond jubilee of Queen Victoria,[26] provides one illustration on an imperial scale of the argument advanced by Benedict Anderson in his influential treatment of nationhood. A nation, Anderson suggested, is best regarded as "an imagined political community—and imagined as both inherently limited and sovereign."[27] Characterizing what holds a nation, or, in Morris's case, an empire together then becomes a task for history and ethnography, for the definition orients our attention to the practices through which the state and its minions train the collective imagination of a national or imperial community. Morris's imperial moment called forth an unprecedented outpouring of celebration and circulation of people, goods, vessels, language, profits, and plants that criss-crossed the empire on which, famously, the sun never set. But what of the work that was needed to produce such a worldwide convergence? To see this, we need more disciplined histories.

Anderson, his own imagination challenged by the unlikely agglomerate of the

[22] On the processes of boundary work in the sciences, see Thomas F. Gieryn, *Cultural Boundaries of Science: Credibility on the Line* (Chicago, 1999). On boundary work within government agencies, see Jasanoff, *The Fifth Branch* (cit. n. 21), 14, 234–6.

[23] Damian Collins and Nicholas Bromley, "Private Needs and Public Space: Politics, Poverty, and Anti-Panhandling By-Laws in Canadian Cities," in *New Perspectives on the Public-Private Divide,* ed. Law Commission of Canada (Vancouver, 2003), 40–67. Under India's prime minister Indira Gandhi, in close association with her son Sanjay Gandhi, the slogan *garibi hatao* (eradicate poverty) became equated with a program of forcible slum clearance—in other words, eradicating not poverty but the visibly poor.

[24] Laid out on modern lines in the 1950s by the French-Swiss architect Le Corbusier, at the behest of Prime Minister Jawaharlal Nehru, the city of Chandigarh, the capital of Punjab and Haryana, accommodates a degree of traffic surveillance that I have not encountered in other Indian cities. Just over a hundred years before Chandigarh was inaugurated, Baron Georges-Eugène Haussmann substantially rebuilt Paris for Napoleon III, razing many old districts and replacing winding streets with broad boulevards so that the state could better control potential revolutionaries.

[25] James C. Scott, *Seeing Like State: How Certain Schemes to Improve the Human Condition Have Failed* (New Haven, 1998), 2.

[26] James Morris, *Pax Britannica: The Climax of an Empire* (London, 1979).

[27] Anderson, *Imagined Communities* (cit. n. 6), 6.

Indonesian nation-state, stressed the role of structuring élites, in particular the unifying work of the print media and, in an elaboration of his original argument, also of the state-sponsored census, map, and museum. Whereas Anderson looks primarily to the public spaces and instruments of national identity-making, Ann Stoler, the feminist anthropologist, provides a Foucauldian account of the intrusions into private life undertaken by colonial regimes for the sake of creating and maintaining relations of dominance. In the Dutch East Indian colonies, she argues, carefully constructed rules governing sexual relations among Europeans and between whites and natives preserved necessary demarcations between the governors and the governed. For both Anderson and Stoler, making empires is an active, creative, and dynamic process of ordering, centering on producing and, especially for Stoler, reproducing a vision of the thing being made.

The emergence of the European Union (EU) as an autonomous political force in the late twentieth century illustrates one more modality of imperial construction, based on constitutional principles and the rule of law, and designed to further the free flow of goods and services in an open market. The EU's tightening integration through successive treaties, the admission of ten new member states in May 2004, and the signing of a constitution in Rome on October 29 of the same year marked the production of a new kind of empire, but one founded on the democratic consent of its citizens.[28] Declining turnout in EU parliamentary elections, widespread popular disenchantment with Brussels, and the stinging rejection of the EU constitution in French and Dutch referenda in 2005 all indicate that constitutionalism on such a scale carries huge risks of alienation along with the promise of enhanced economic and political integration. What matters for our discussion, however, is the very availability of a constitutional process, with all of its positive connotations for democracy, in creating the EU's supranational authority; even the fact of electoral rejection may be seen, at one level, as validating the idea of a common European project. We will return below to the implications of the constitutional model of imperialism for governing biotechnology globally.

BIOLOGY IN THE SERVICE OF EMPIRE

As if echoing the explosion of historical and political writing about empire, there has been an explosion of writing on the uses of science in the cause of imperial expansion, with the scientific management of nature commanding center stage. Colonial historians have observed that the human and biological sciences came into their own to serve imperial needs from the eighteenth century onward, in much the same way that Scott's twentieth-century planning states used engineering and social sciences to achieve legibility. Anthropology, botany, ecology, geography, linguistics, and even early forensic sciences have deep colonial roots: to rule effectively, occupying governments had to map their territories, classify populations into identifiable groups, and catalog flora, fauna, languages, and cultural practices.[29]

Making things grow, often under unfavorable natural conditions in nonnative habi-

[28] The ten new members met the so-called Copenhagen criteria, according to which they had to "be a stable democracy, respecting human rights, the rule of law, and the protection of minorities; have a functioning market economy; and adopt the common rules, standards and policies that make up the body of EU law." See http://europa.eu.int/comm/enlargement/enlargement.htm (accessed Nov. 2004).

[29] On colonial histories of the human and natural sciences, see Bernard S. Cohn, *Colonialism and Its Forms of Knowledge* (Princeton, 1996); Matthew H. Edney, *Mapping an Empire: The Geographic*

tats, gave a push to imperial ecology, conservation biology, and agricultural science.[30] Sometimes the motives were crassly extractive and exploitative, as in the harvesting of wild rubber in King Leopold II's Belgian Congo, where violence and force were the notorious instruments of colonial rule.[31] Elsewhere, colonists heedlessly harvested tropical timber or took commercially useful plants such as cinchona (from which quinine is derived) or breadfruit for cultivation in new territories.[32] Sometimes otherwise well-intentioned migrations had disastrous results. For instance, rabbits transported to Australia for hunting became an uncontrollable pest, as Morris colorfully records.[33] Yet more altruistic motives also prevailed. Richard Grove traces the roots of western environmentalism to early modern European encounters with tropical islands.[34] As self-contained and containable spaces, these islands appealed to voyagers' Edenic and Romantic sensibilities, as well as to their protective instincts. Lush islands brought to life idyllic conceptions of the gardens of paradise; at the same time, in those bounded preserves, travelers could easily observe the destructive effects of resource depletion and environmental degradation. The island of Mauritius, in Grove's account, became the site of some of the world's earliest systematic efforts at nature conservation and scientific forest management. These practices, in turn, provided practical models for conservation efforts in India and elsewhere from the 1830s onward.[35]

Colonial enterprise also laid the basis for western ideologies of development. Along with concerns for the moral and religious education of the strangers they went to live among, the rulers of empires exhibited a compelling desire to improve the new territories under their command. British engineers laid roads and railways, built irrigation systems, and left indelible architectural imprints throughout India. Just as pervasive was Britain's (and in other regions, France's) engagement with botany and agriculture. Already in the early nineteenth century, a coalition of professional scientists and administrators had converted the Royal Botanic Gardens at Kew into a publicly run center of knowledge for the productive management of nature.[36] Problems of sugar cane cultivation in the West Indies led to the formation of the Imperial Department of Agriculture at the end of the nineteenth century. A source of scientific expertise for West Indian sugar cane growers, the department also became, under the leadership of Joseph Chamberlain, the Liberal secretary of state for the colonies, a breeding ground for early discourses of development.[37] Like enlightened estate managers back home,

Construction of British India, 1765–1843 (Chicago, 1997); Kavita Philip, *Civilizing Natures: Race, Resources, and Modernity in Colonial South India* (New Brunswick, N.J., 2004). On the colonial origins of fingerprinting, see Simon A. Cole, *Suspect Identities* (Cambridge, Mass., 2001), 60–96.

[30] John MacKenzie, ed., *Imperialism and the Natural World* (Manchester, 1990); S. Ravi Rajan, ed., *Imperialism, Ecology, and Politics: Perspectives on the Ecological Legacy of Imperialism* (Delhi, 1996); Peder Anker, *Imperial Ecology: Environmental Order in the British Empire, 1895–1945* (Cambridge, Mass., 2001).

[31] Adam Hochschild, *King Leopold's Ghost* (New York, 1999).

[32] See, e.g., Kavita Philip, "Imperial Science Rescues a Tree: Global Botanic Networks, Local Knowledge, and the Transcontinental Transplantation of Cinchona," *Environment and History* 1 (1995): 173–200; Richard Drayton, *Nature's Government: Science, Imperial Britain, and the Improvement of the World* (New Haven, 2000), 206–11.

[33] Morris, *Pax Britannica* (cit. n. 26), 77–8.

[34] Richard H. Grove, *Green Imperialism: Colonial Expansion, Tropical Island Edens, and the Origins of Environmentalism, 1600–1860* (Cambridge, 1995).

[35] Ibid., 9–10, 168–263.

[36] Drayton, *Nature's Government* (cit. n. 32).

[37] William K. Storey, "Plants, Power, and Development: Founding the Imperial Department of Agriculture for the West Indies, 1880–1914," in Jasanoff, *States of Knowledge* (cit. n. 11), 109–30.

those entrusted with the welfare of colonial "properties" felt a need to ameliorate the conditions of life for the local poor. Promoting development abroad, they also thought, would transform the colonies into more advantageous trading partners, thereby producing useful returns for domestic constituencies. Improving agricultural production was a favored route to achieving these goals, although access to metropolitan knowledge remained stratified, with native farmers, in many cases, continuing to cultivate their lands without the benefits of modern science.[38]

The first half of the twentieth century cast the imperial project of biology in a darker light as the improvers' attention turned toward standardization for control, and broadened to include humans in addition to plants and animals. The enthusiasm of progressive social reformers for eugenics at the turn of the century led to decades of discrimination in the United States, including the exclusionary Immigration Act of 1924, numerous state sterilization laws, and *Buck v. Bell,* the infamous 1927 Supreme Court decision upholding the sterilization of a Virginia woman, Carrie Buck, on the ground that "[t]hree generations of imbeciles are enough."[39] The eugenicists' concern for selective breeding and race purity was carried to pathological extremes in the Nazi period, when millions of humans deemed undesirable by German race theorists—Jews, gays, Gypsies—were uprooted and eliminated throughout the Third Reich. For the sociologist Zygmunt Bauman, these atrocities were the natural descendants of the same enlightenment ideals that had led Frederick the Great of Prussia to exclaim, "It annoys me to see how much trouble is taken to cultivate pineapples, bananas and other exotic plants in this rough climate, when so little care is given to the human race."[40] The modern "gardening state," Bauman argues, turned Frederick's metaphor into crude reality by ruthlessly weeding out everything that its planners saw as standing in the way of reason, order, and progress.

In spite of these midcentury turmoils and disruptions, the alliance between biology and power has only grown more intimate and pervasive in subsequent decades. Foucault saw biopower and biopolitics as essential technologies with which modern states must control their populations—by assuming responsibility for the health, safety, and stability of citizens' collective lives.[41] Central to the exercise of biopower, then, is the state's ability to characterize human bodies and behavior in ways that rationalize and, in democratic societies, publicly justify that state's policies. Increasingly, the state asserts itself under the umbrella of epidemiology: as the master diagnostician of ills that threaten groups of people in society. The polarizing debates on gay marriage before and during the 2004 U.S. presidential campaign may be seen in this light as part of a more general discourse on sexuality and the family, with competing political factions claiming citizens' allegiance by defining what counts as deviance in sexual behavior and family mores. In the culturally heterogeneous United States, as in Stoler's East Indian colonies, the rules of sexual conduct serve as powerful instruments for building social cohesion, by decreeing who falls inside and who outside the accepted forms of domestic order.

Today as before, moreover, biopower extends into all of life on the planet, not only the lives of humans but also the natural worlds with which humans live in close sym-

[38] William K. Storey, *Science and Power in Colonial Mauritius* (Rochester, N.Y., 1997).

[39] Justice Oliver Wendell Holmes Jr., an enthusiast for eugenics, wrote the majority opinion in *Buck v. Bell,* 274 US 200 (1927).

[40] Zygmunt Bauman, *Modernity and Ambivalence* (Ithaca, 1991), 27.

[41] Michel Foucault, *The History of Sexuality,* vol. 1, *An Introduction* (New York, 1978).

biosis. Sick and failing plants, no less than sick and failing people, fall within the biopolitical imagination of the neoliberal state and its corporate partners, whose innovative capacity is as essential to underwriting state action as is the capacity of expert professionals to define and apply the technical criteria of governmentality.[42] Governing *bodies,* after all, proceeds not only through exclusion, or weeding out, but also through therapeutic processes of making whole and bringing the previously sick back into the community of viable beings. The ordering state is most powerful when it is at the same time, demonstrably, a healing state, and such a state engages science for therapeutic, as well as diagnostic, ends. Let us return, then, to agricultural biotechnology as a field of contemporary biopower that continues the historical partnership of the life sciences with the state and, in so doing, intersects with each of the modes of empire-building described above.

PLANTS FOR THE PLANET: THE EMPIRES OF BIOTECHNOLOGY

Apart from occasional radical social misfits such as the so-called Unabomber, Theodore Kaczynski,[43] few any longer question the vital role of science and technology in human development. Even opponents of particular technological projects—large dams,[44] for example, or genetically modified (GM) foods[45]—rarely dismiss technology outright; rather they favor smaller, more transparent, or more locally governable technological systems. The question that preoccupies students of science and technology, then, is not whether, but how, to integrate innovation into people's lives so as to make a positive difference. Years of research in the social psychology of risk perception[46] and public understanding of science[47] have established that popular fear or rejection of new technology often rests, at bottom, on an uneasiness about the ways in which technology is managed or, more accurately, governed. What do these observations imply for an industry with global ambitions, like agricultural biotechnology? How, more specifically, does biotechnology contribute to ways of political world-making beyond the nation-state, and what implications do the engagements between biotechnology and global politics have for democratic governance?

In reaching for answers, it is useful to think of biotechnology operating politically in several different registers. It is, of course, most plainly a material technology: it makes new instruments for warding off harm and disorder, such as plants that resist insects, weeds, or drought, and it redesigns pieces of nature, such as genes, to perform new tasks in new environments. In this respect, biotechnology is, concurrently, a

[42] For an account of the changing social contract among science, state, and industry with respect to the life sciences, see Sheila Jasanoff, *Designs on Nature: Science and Democracy in Europe and the United States* (Princeton, 2005).

[43] Theodore Kaczynski, a mathematician educated at Harvard and the University of Michigan, conducted a single-handed letter-bombing campaign against representatives of various industries from his cabin in Montana between 1978 and 1996. These attacks killed three people and injured many others. He was caught when his brother recognized as his work a long letter he had sent to the *New York Times.* See Kaczynski, *The Unabomber Manifesto: Industrial Society and Its Future* (Berkeley, 1995).

[44] Sanjeev Khagram, *Dams and Development: Transnational Struggles for Water and Power* (Ithaca, N.Y., 2004).

[45] On transatlantic divisions over genetically modified crops and food, see Thomas Bernauer, *Genes, Trade, and Regulation: The Seeds of Conflict in Food Biotechnology* (Princeton, 2003).

[46] See, e.g., Paul Slovic, *The Perception of Risk* (London, 2000).

[47] Brian Wynne, "Public Understanding of Science," in *The Handbook of Science and Technology Studies,* ed. Sheila Jasanoff, James C Petersen, Trevor Pinch, and G. E. Markle (Thousand Oaks, Calif., 1995), 361–88.

metaphysical device; it brings new entities into the world and through that process reorders our sense of rightness in both nature and society.[48] At the same time, biotechnology is a discourse: to some, of progress and improvement, beneficence and utility; to others, of risk, invasiveness, and domination from afar. Proponents of agricultural biotechnology tell particular stories about a world in which plant genetic modification is possible, and these stories carry political and cultural weight. Lastly, biotechnology is an institution of governance; it shapes forms of social life by influencing how people choose to, or are able to, live with the products of bio-industry. Each of these registers, as we see below, has been activated in the global politics of biotechnology.

The Resisting Multitude

In May 2004, a scientific journal reported that German researchers were keeping secret the locations of some thirty sites planted with GM corn for fear that anti-GM activists would destroy the crops, as they previously had elsewhere in Germany.[49] Failure to disclose these locations was contrary to the EU Directive 2001/18, which requires GM crop sites to be publicly registered. Noncompliance with European law in traditionally law-abiding Germany may have been newsworthy, but the threat to GM crops was anything but novel. From the late 1990s onward, attacks on field trial sites began evolving into a form of international protest that seemed to epitomize Hardt and Negri's thesis about an emerging, assertive, global multitude: in Britain, hundreds of demonstrators dressed in decontamination suits uprooted GM plants in test fields in 1999; in India and Brazil, farmers' unions organized similar protests; in France, José Bové, the charismatic head of the radical Confédération Paysanne (Peasant Confederation), became a folk hero by orchestrating the destruction of thousands of GM plants, as well as a partially built McDonald's outlet, in 1999. His subsequent trial, fine, and terms of imprisonment left him and his supporters undaunted, indeed ready to resume battle as much as five years after their initial transgressions.

Field trial sites were not the only theater of protest against GM agriculture. Antiglobalization activists early identified biotechnology as a symbol of the environmental, economic, and cultural homogenization they wished to resist. Demonstrations against Monsanto and GM corn (or maize), together with evocations of risks to nontarget species such as the monarch butterfly, were part of the repertoire of street protest during the Third Ministerial Conference of the World Trade Organization (WTO) in Seattle in 1999. In this and similar episodes, representatives of a loosely networked global citizenry asserted their right to debate technological futures in terms other than those conventionally used by nation-states and their expert advisers: the formal discourses of law, molecular biology, economics, risk assessment, and bioethics. At stake was who had power to determine how much global harmonization there should be and which scientific, technological, and economic innovations should be allowed to diffuse throughout the world. Those opting for more local, bottom-up visions won a salient victory when Monsanto decided, under rising public pressure, to withdraw its plans to develop sterile seed technology, through use of the so-called

[48] For an elaboration of this argument, see Sheila Jasanoff, "In the Democracies of DNA: Ontological Uncertainty and Political Order in Three States," *New Genetics and Society* 24(3) (2005): 139–55.

[49] Ned Stafford, "Uproar over German GM Corn," *The Scientist,* May 17, 2004, http://www.the-scientist.com/article/display/22179/ (accessed Jan. 2006).

Terminator gene;[50] later, citing a drop in global demand, the company also announced that it would put on hold its plans to market genetically modified Roundup Ready wheat.[51]

Ideology and Enforcement

Not everyone saw the antiglobalization movement as the promising vanguard of planetary resistance against an outmoded, corporate-dominated, neoliberal world order. Using the classical ordering machinery of science and the law, proponents of agricultural biotechnology sought to promote their visions of social and technological progress, stifling opposition and dissent.

At the February 2000 annual meeting of the American Association for the Advancement of Science, Senator Christopher "Kit" Bond, Republican from Missouri, Monsanto's home state, was openly dismissive of the Seattle protest. He represented it as a struggle between scientific expertise and the misguided, if exuberant, ignorance of youth: "The scientific debate is not being controlled by Ph.D.s but apparently by young people with a proclivity for street theater. . . . It's coming to the point that scientists are going to have to get dressed up as corncobs to get the attention of the media."[52] At the same meeting, Madeleine Albright, President Clinton's secretary of state, also cast the conflict as one between reason and unreason. "But science," she said, "does not support the 'Frankenfood' fears of some, particularly outside the United States, that biotech foods or other products will harm human health."[53] Both speakers, from different political parties, enlisted science as their ally in defending biotechnology against its critics. This invocation of scientific authority in support of technological innovation is a marker of America's commitment to a particular ideology of technoscientific progress.[54]

A look across the ocean at contemporaneous UK debates on biotechnology helps bring into relief the ideological dimensions of the American position. The term "Frankenfood" was widely used in the British tabloid press to reflect and, some said, reinforce public anxieties. But concerns were not restricted to the media and the ignorant public. The British scientific community had all along expressed greater uncertainty about the safety of GM crops than its American counterpart, particularly with respect to the environmental consequences of commercial use.[55] These doubts led British experts to reject the official U.S. position that the process of genetic

[50] The Terminator gene would have disabled grain seeds from sprouting in consecutive years. Farmers who had routinely planted seed stored from the previous year's harvest would then have been forced to buy new seed each year. The coalition that forced Monsanto to abandon this technology, at least for a time, included both indigenous organizations and the influential Rockefeller Foundation. Jasanoff, "In a Constitutional Moment" (cit. n. 12), 171.

[51] Roundup is a popular weed killer marketed by Monsanto, and Roundup Ready plants are genetically modified to withstand the use of that product. Many observers thought Monsanto's decision was motivated by opposition to GM crops in Europe and Japan. See "GM Wheat Put on Hold," *NewScientist.com* news service, May 11, 2004, http://www.newscientist.com/article.ns?id=dn4977/.

[52] Senator Christopher Bond, Annual Meeting, American Association for the Advancement of Science, Washington, D.C., Feb. 23, 2000.

[53] Secretary of State Madeleine Albright, Annual Meeting, American Association for the Advancement of Science, Washington, D.C., Feb. 21, 2000.

[54] Jasanoff, *Designs on Nature* (cit. n. 42), chap. 4.

[55] Ibid., chap. 2.

modification carries no special risks; all that matters for regulatory purposes is the end product. Scientific and public opinion in Britain united behind a more cautious approach, demanding more experimentation—for example, through farm-scale trials[56]—before authorizing the commercialization of GM crops. As doubts intensified, Tony Blair's government decided on a highly unusual three-pronged review of the science, economics, and public acceptability of these products to reevaluate the case for their introduction.[57] The immediate outcome of this process was a decision to approve the commercialization of only one variety of GM corn, at least to start. Thus, while American neoliberalism treated biotechnology as just another stream of products, adequately controlled by the market except for assessments of their safety to human health and the environment, Britain's more cautious and communitarian political culture granted the public some say in deciding which products they wanted to allow into the market.

Whereas consultative procedures such as Britain's GM debate and referenda in countries such as Denmark and Switzerland sought to defuse public opposition, elsewhere legal sanctions were employed to beat down what biotechnology promoters saw as unacceptable acts of intransigence. Thus, demonstrators such as José Bové who destroyed GM crops were prosecuted for damaging property in several countries. At the international level, the United States brought a case against the EU at the WTO for imposing an allegedly illegal moratorium on the importation of GM crops and foods. Foundational to the U.S. case was the argument that there were no good scientific reasons for keeping these products off the European market, and that the moratorium therefore amounted to illegal protectionism.[58]

Intellectual property law, too, has been invoked in safeguarding the investments made by multinationals such as Monsanto in GM crops. Particularly interesting were the prosecutions brought against farmers in the United States and Canada who were found to be growing GM crops patented by Monsanto without a license. In the best known of these cases, a seventy-three-year-old Saskatchewan farmer named Percy Schmeiser was sued for growing genetically modified Roundup Ready canola, which he claimed had blown into his fields from neighboring farms. A 5-4 decision of the Supreme Court of Canada upheld Monsanto's patent infringement claim, saying that Schmeiser's unlicensed use of seed containing Monsanto's patented gene was sufficient to constitute infringement.[59] In a Solomonic turn, though, the Court awarded no damages to Monsanto, on the ground that Schmeiser had not benefited economically from his unlawful act; equally, Schmeiser was not required to pay Monsanto's court costs. The case warned GM crop producers that, under Canadian law, they would have a difficult time collecting damages for patent infringement; at the same time, they could be subject to potentially unlimited liability if their seeds accidentally contaminated, and thus damaged, the products of certified GM-free organic farms.

[56] Agriculture and Environment Biotechnology Commission, *Crops on Trial,* Sept. 2001.

[57] The most unprecedented feature of this process was a nationwide public consultation known as *GM Nation?* See http://www.gmnation.org.uk.

[58] For details of the case, as well as an argument against the U.S. positions on science and risk assessment, see David Winickoff, Sheila Jasanoff, Lawrence Busch et al., "Adjudicating the GM Food Wars: Science, Risk, and Democracy in World Trade Law," *Yale Journal of International Law* 30 (2005): 81–123.

[59] *Monsanto Canada Inc. v. Schmeiser,* [2004] 1 S.C.R. 902, 2004 SCC 34.

Legibility

Advertisements for agricultural biotechnology frequently show fields of grain laid out in neat parallel lines, illustrating both the fertility and the increased control that genetic modification can allegedly deliver. One could hardly find more compelling images of the "legibility" described by Scott. Intrusive weeds, barren patches, unruly growth have all been eliminated in favor of healthy, predictable, quantifiable yields—achieved through the precision of genetic control. However, just as the midcentury grand planners' dreams of legibility were achieved at a cost, so legibility in modern GM agriculture demands unseen labors of standardization, and consequent elimination of ambiguity, to achieve its surface regularity. Four dimensions of standardization are worth noting: ontologies, epistemologies, socio-ecologies, and forms of life. All four maintain traditional relations of power between center and periphery, and all can be illustrated through the case of "golden rice," the poster crop for a new generation of nutrient-enriched GM crops to feed the developing world.[60] The name was given to a strain of rice bioengineered to produce beta-carotene, which colors the grain a pale gold; when ingested, it converts to vitamin A in the body and protects consumers against vitamin deficiency leading to possible blindness.

For the products of GM agriculture to locate themselves securely in global markets, there has to be broad agreement on what these entities actually *are*. This ontological question may seem straightforward at first—proponents of golden rice, for instance, claim that it is nothing more than a more nutritious plant variety—but food crops straddle too many categorical boundaries for their identity in the political domain to be anything but hybrid. There are, to begin with, regulatory classifications. Should a crop engineered to produce ingredients of medicinal value be considered a food or a drug? Even if such issues can be settled by formal administrative definitions, the North-South debate surrounding GM crops shows how difficult it is to achieve ontological closure around a commodity that is at once a natural kind (a plant with specific genes and traits) and a social kind (a product of particular economic and political orderings, and a potential reorganizer of society).[61]

How one should know the properties of GM crops is similarly open to question. U.S. authorities have insisted that the only proper basis on which to evaluate the impacts of these novel entities is through science-based risk assessment. Yet, as the dispute between the United States and Europe at the WTO graphically illustrates, vast disagreements persist about the epistemological status of risk assessment. Is it a "science" at all, in the sense of being a well-demarcated, uncontroversial, paradigmatic (in a Kuhnian sense) method of representing the world; or is it instead a patently political and culturally constructed instrument for managing the uncertainties that

[60] Sheila Jasanoff, "Let Them Eat Cake: GM Foods and the Democratic Imagination," in *Science and Citizens,* ed. Melissa Leach, Ian Scoones, and Brian Wynne (London, 2005), 183–98.

[61] Such ontological hybridity is taken as part of the order of things in the work of many science studies scholars. See, in particular, Michel Callon, "Some Elements of a Sociology of Translation: Domestication of the Scallops and the Fishermen of St. Brieuc Bay," in *Power, Action, and Belief: A New Sociology of Knowledge?* ed. John Law (London, 1986), 196–233; Bruno Latour, *We Have Never Been Modern* (Cambridge, Mass., 1993). Hybrids complicate the clean separation that philosophers such as Ian Hacking have sought to draw between natural ("indifferent") and social ("interactive") kinds. Hacking, *The Social Construction of What?* (cit. n. 20).

inevitably accompany large projects of reconfiguring nature or society?[62] To accept producers' contention that crops such as golden rice are "safe," one has to buy the former, not the latter, characterization. If, however, risk assessment is an expression of political culture by other means, then one should not be surprised if that form of analysis does not travel friction-free across political and cultural boundaries.[63]

GM crops are developed in the laboratory, usually in science-rich Western nations, tested in the field, and transported thence for commercial propagation in both naturally and socially variable environments. Monsanto, in this respect, is like the Kew Gardens of the nineteenth century: a metropolitan "center of calculation"[64] from which standardized products flow out to take root in the world's economic and political peripheries. Key to sustaining this mode of production is the assumption that socio-ecologies are as standard as the crops grown within them—put differently, that social and ecological circumstances at the periphery are not so radically different from those at the metropolitan center as to defeat the project of global technology transfer. Yet accidents occurring even within the boundaries of single nation-states show that transfers from the laboratory to the field can bring unpleasant surprises. For example, in one costly U.S. episode, ProdiGene, a GM corn variety containing an insulin precursor, trypsin,[65] was planted in an unmarked field in rural Iowa. The manufacturer agreed with the U.S. Department of Agriculture, which approved the field trials, that the field would be quarantined the following year so as to remove any volunteer plants.[66] In fact, the fields were not properly isolated and an undetermined quantity of the GM crop was harvested along with about 500,000 bushels of soybeans during the following season. Similar failures resulting from unforeseen couplings of technology, environment, and human behavior are all the more probable when transfers occur across disparate cultures of farming and of hazard control.

Expanding on this point, it has become clear that complex technological systems are forms of life, uniting human and nonhuman components in a common purposive framework, as much as they are targeted attempts to improve upon aspects of human life by physical or biological means. Thus, transportation systems do not only move people about from place to place. They remake social structures and self-understandings. A car culture, for example, gives rise to different visions, and valuations, of time, distance, autonomy, community, environmental quality, and the cost of life than a culture dependent chiefly on bicycles or public transportation does. Similarly, industrial agriculture is organized and managed on different principles from small family farms; the two systems of production rest on different economic, social, and technological infrastructures, and their impacts on human solidarity and on the environment are correspondingly divergent. Conventional risk assessment methods take little or no account of the social and ethical ramifications of technological systems, including the threats they pose to long-settled patterns of living. This blindness to technology's disruption of established forms of life, underwritten by the allegedly scientific power of

[62] Winickoff et al., "Adjudicating the GM Food Wars" (cit. n. 58)

[63] See Jasanoff, *Designs on Nature* (cit. n. 42), on the relationship of risk assessment to political culture.

[64] Latour, "Drawing Things Together" (cit. n. 11), 19–68.

[65] Bill Hord, "The Road Back: Prodigene and Other Biotech Companies Are Moving Ahead in an Environment of Increasing Fear of Crop Contamination," *Omaha World Herald,* Jan. 19, 2003, 1(d).

[66] "Volunteer" plants are those that emerge spontaneously, usually from a previous season's growth, in places where they were not intentionally planted.

risk assessment, has fueled much of the criticism of agricultural biotechnology in the global South.[67]

Identity and Community

Empires, no less than nation-states, engender and depend on feelings of belonging. Devices for producing imperial imagined communities have included, besides the grand, polarizing, ideological discourses of the cold war, mundane practices such as performing national celebrations,[68] teaching a common language, training administrative and judicial élites, and building infrastructures for commerce and communication. Science and technology, we have seen, have long served as agents of imperial governmentality, helping to produce the mission consciousness and the associated forms of knowledge and skill that serve as instruments for extending power. Modern biotechnology, similarly, provides a discourse of development that continues colonial traditions, although the agents, recipients, and specific mechanisms of the development project have been partially reconfigured in modern times.

The discovery of Africa as a site for biotechnological development, through the propagation of crops such as golden rice, offers perhaps the clearest illustration. In the rhetoric of development specialists, and the scientific and industrial institutions that serve them, Africa is represented through tropes of crisis and charity that render the continent's condition as dire and the offers of scientific and technological solutions as salvationary.[69] In one instructive example, Gordon Conway, former president of the Rockefeller Foundation, and a colleague wrote an article in the prestigious journal *Science* on biotechnology's capacity to help Africans. Though presented as scientific, the article merged the empiricist register of science with a narrative register that was little short of missionary. At the center of the discussion was a fictional African housewife, "Mrs. Namurunda," who the authors said was not a real person but "a composite of situations existing in Africa."[70] The story begins with Mrs. Namurunda, a farmer and single mother, eking out a hard-scrabble existence on fields infested with every form of insect blight, under adverse conditions of drought and soil degradation. It ends with scientific biotechnology solving her problems, enabling her to turn a profit and secure a brighter, better educated, more enlightened future for her children.

This script follows Foucault's delineation of biopower with uncanny precision. An entire continent becomes a medicalized body, requiring urgent therapeutic intervention, both as a collective and for its individual members. The fictional person of Mrs. Namurunda, unveiled in the pages of one of the world's leading scientific journals, becomes a symbol for Africa's "composite" ailments. Advanced societies' power to

[67] See, particularly, the arguments on this topic by the well-known Indian author and activist Vandana Shiva, *Monocultures of the Mind: Perspectives on Biodiversity and Biotechnology* (London, 1993); *Biopiracy: The Plunder of Nature and Knowledge* (Toronto, 1997); *Yoked to Death: Globalisation and Corporate Control of Agriculture* (New Delhi, 2001).

[68] Morris, for example, describes Victoria's jubilee celebrations in London as a crystallizing moment for the British Empire in 1897. *Pax Britannica* (cit. n. 26), 21–34. See also the account of the Imperial Assemblage of 1877 in Delhi by Bernard S. Cohn, "Representing Authority in Victorian India," in *The Invention of Tradition,* ed. Eric Hobsbawm and Terence Ranger (Cambridge, 1983), 165–209.

[69] Jasanoff, "Let Them Eat Cake" (cit. n. 60), 190–4.

[70] Gordon Conway and Gary Toenniessen, "Science for African Food Security," *Science* 299, no. 21 (2003): 1187–8.

develop and deliver the requisite treatments offers them the right, indeed the obligation, to engage in a new *mission civilisatrice*—built on a biomedical ethic of cure rather than, as in earlier times, a religious model of grace. But, this time, eschewing the forceful, state-led constellations of power that undergirded colonial rule, the neoliberal state works through a lightly regulated global industry and a largely self-regulating scientific community. *Their* expansion into new territories carries the promise of better jobs and higher incomes back in the home country, thereby allowing the economically more powerful state to justify itself where votes are counted, in its own national community of citizens. The sick and incapacitated recipient, however, has little or no say in either the diagnosis or the treatment of the alleged pathology.

The Constitutional Turn

We turn now to the fifth modality of empire-making identified above—the constitutional approach, which relies for its robustness on the formal consent of citizens. The European Union at the turn of the twenty-first century represents perhaps the most ambitious working out of this approach. With twenty-five member states as of May 2004, the EU brought within a single constitutional regime one of the most linguistically and culturally heterogeneous political assemblages ever created. In contrast to the institutionally inchoate, emergent empire discerned by Hardt and Negri, the EU is very much an orthodox space of governance, circumscribed by law and accountable to its members and (as illustrated by the French and Dutch "no" votes on the EU Constitution) to the particularities of their domestic politics.[71] On its Web pages, the public face it presents to the electronically plugged-in world, the EU takes considerable pains to explain itself: why it exists, how it was formed, its past achievements, and its hopes for the future. At one level, the talk is highly Weberian, a matter of official institutions and considered policies, justified in terms of an overall mission of peace, safety, solidarity, and a European model of society.[72] At this discursive level, *Europe* very much exists; the question is only how to realize, through concerted, practical action, its already formed sense of collective identity.

At another level, however, Europe's identity is still very much in the making, and its constitutional union is but a cover for working out varying conceptions of what it means to be European; domains in which European-ness remains an open question, subject to multiple interpretations, include the development and deployment of the life sciences to advance communal interests in the EU. Looking at European engagements with biotechnology, both in Brussels and in the member states, one gets some sense of the issues in this debate, as well as some of the ways in which Europe has approached the problem of coordinating differences among its members without erasing them. The European example offers, in this respect, an intriguing alternative to the totalizing, disciplining vision of global biopower.

To be sure, European policy for biotechnology has followed to some extent famil-

[71] This system of distributed accountability has resulted in a union whose members have not equally bought into all aspects of the EU vision. Thus, Sweden, Denmark, and Britain have not adopted the single currency (euro); Ireland and Britain are not parties to the Schengen agreement on frontier controls; and Britain thus far has not adopted the Community Charter of Fundamental Social Rights for Workers.

[72] See *Why the European Union?* http://europa.eu.int/abc/12lessons/index1_en.htm (accessed Nov. 2004).

iar modernist impulses toward standardization and central control. Brussels has sought for decades to foster technological innovation and create new jobs, partly for the sake of continued European economic growth and partly in response to perceived threats from U.S., and now Chinese and Indian, innovation. Older discourses on international competitiveness[73] have been joined of late to new worries about labor mobility within Europe, the out-sourcing of jobs to developing countries, and concomitant pressures to lower regulatory and ethical barriers to the free flow of scientists within the European research area. Since 1990, the EU has issued directives on research with genetically modified organisms (GMOs), release of GMOs into the environment, labeling of foods containing GMOs, and patenting of the products of biotechnology. In its efforts to counter popular resistance, the EU has also sponsored research on the public understanding of science—constituting in the process a citizenry whose needs the European state can characterize and cater to with aggressive programs of science and risk communication.[74]

These centralizing initiatives from Brussels, however, have run up against resistance from members states and their polities, showing that—at least in Europe—the prerogative of imagining technological futures no longer rests with governments alone but must be shared with increasingly knowledgeable publics. Those publics, moreover, approach the promises of biotechnology with significantly different ethical sensibilities toward nature and different attitudes toward uncertainty and responsibility from the industries wishing to commercialize the new technologies.[75] While public perceptions converge in important respects across Europe, the means through which people express their concerns and seek reassurance remain different, conditioned by national political culture and traditions. Thus, the nationwide public debate on GM crops held in Britain had no exact parallels anywhere else in the EU; other states conducted their own consultative exercises, in the form of citizen juries, consensus conferences, and referenda. The results, too, have varied, with member states disagreeing about how to establish the adequacy of data bearing on risk, as well as in the actions they have taken with respect to specific GM crops.

In sum, European experience with the governance of biotechnology indicates that, in an empire built on constitutional principles, there may be broad agreement in public attitudes toward technology and on the rulers' willingness to take account of public views and values while actively pursuing the agenda of technological development. At the same time, democratic consultation pursued with genuine respect for diversity may produce locally specific accommodations that bear little resemblance to the global legibility sought by some twenty-first-century multinational corporations, or striven for in vain by Scott's over-ambitious twentieth-century planning states.

CONCLUSION

Imperial projects, as many are arguing today, did not end with the end of colonialism but may be resurfacing in new guises with the passage of time. Since early modernity, these projects have benefited from the enterprises of science and technology, and the

[73] Herbert Gottweis, *Governing Molecules: The Discursive Politics of Genetic Engineering in Europe and the United States* (Cambridge, Mass., 1998).
[74] Jasanoff, *Designs on Nature* (cit. n. 42), chap. 3.
[75] Claire Marris, Brian Wynne, Peter Simmons, and Sue Weldon, *Public Perceptions of Agricultural Biotechnologies in Europe,* http://www.pabe.net (accessed Nov. 2004).

biological sciences in particular have been caught up for centuries in the spread of imperial forms of governance. It is no surprise, then, to find contemporary biotechnology enrolled in various modalities of empire-making, whether through bottom-up resistance, top-down ideological imposition, administrative standardization, or consensual constitutionalism. In particular, as shown above, the capacity to engineer the genetic characteristics of plants has blended seamlessly with state and corporate projects of managing human populations so as to legitimate the exercise of power. Both nation-states and, in an era of neoliberalism, the multinational corporations that states are in league with have displayed their readiness to deploy agricultural biotechnology in advancing their interests on a global scale.

Struggles over the governance of biotechnology complicate any easy, linear narrative of progress. Instead, the nexus of globalization and technological innovation emerges on closer inspection as a politically contested site, where opposing conceptions of how human societies should live, and what other life forms should sustain them, remain very much at play. The example of European integration around biotechnology strongly suggests that there is considerable cross-cultural variation in the lines that human societies, even closely similar ones, choose to draw between nature and culture and the extent to which they are willing to tolerate line-crossings between those two domains. Given a chance to express themselves democratically, moreover, stable societies often opt to retain old boundaries and forms of life, preferring gradual, internally motivated change to imported, alien visions of progress, no matter how glittering the offerings presented to them.

These observations should not be taken as closing the door on the global promises of agricultural biotechnology, which may be considerable, even if not immediately on the horizon. The genie of genetic manipulation is with us in any case: there are not many precedents for turning the clock back on what human inquiry has revealed of the workings of the natural world, although highly developed techniques have occasionally been lost or gone into long periods of recession. Nor *should* we seek refuge in regress from innovation. The challenge, rather, is to constitute in tandem with global advances in technology the institutional capacity that will permit citizens to participate meaningfully in debating the implications of the new technologies. This essay speaks for more enlightened uses of our knowledge and capacity, preferably employed within constitutionally governed systems—keeping in mind that enlightenment flows not only from ingenious ways of tinkering with the material world but also as much, or more, from reflecting on how we should deploy for the good our profoundly human ingenuity.

NOTES ON CONTRIBUTORS

Itty Abraham is the author of *The Making of the Indian Atomic Bomb* (London, 1998) and several articles on nuclear politics in India. He is currently a Fellow at the East-West Center, Washington, D.C., and former Program Director at the Social Science Research Council.

Kai-Henrik Barth is a Visiting Assistant Professor in the Security Studies Program at Georgetown University's School of Foreign Service, where he teaches classes on technology and security. He has published on the role of scientific experts in nuclear test ban negotiations and the cold war transformation of seismology. He is currently working on a book, *Experts in International Affairs: Scientists and the Making of the Comprehensive Test Ban Treaty,* and a new project on nuclear scientists and engineers as drivers of nuclear proliferation.

Alexis De Greiff, a theoretical physicist by training, received his Ph.D. in history of science from London University. He is Associate Professor and Vice-Rector of Universidad Nacional, Colombia, where he also directs the interdisciplinary research group on Social Studies of Science. He has published on discourses and practices of South-North scientific exchange in the twentieth century, particularly in the framework of the United Nations system.

Ronald E. Doel teaches history of science at Oregon State University. Coeditor with Thomas Söderqvist of *The Historiography of Recent Science, Medicine, and Technology: Writing Recent Science* (Oxford, 2006), and coauthor (with Nikolai Krementsov and Dieter Hoffmann) of "National States and International Science: A Comparative History of International Scientific Congresses in Hitler's Germany, Stalin's Russia, and Cold War United States," (*Osiris* 20, 2005), he is currently writing on the rise of the modern environmental sciences. He received the Carter Award for Outstanding and Inspirational Teaching (OSU College of Science) in 2005.

Paul N. Edwards is Associate Professor of Information at the University of Michigan. He has written extensively on the history, politics, and culture of information technology and its uses. He is author of *The Closed World: Computers and the Politics of Discourse in Cold War America* (Cambridge, Mass., 1996) and coeditor (with Clark Miller) of *Changing the Atmosphere: Expert Knowledge and Environmental Governance* (Cambridge, Mass., 2001). His next book is *The World in a Machine: Computer Models, Data Networks, and Global Atmospheric Politics* (MIT Press, forthcoming 2006).

Jean-Paul Gaudillière is a historian of science and medicine. He works as Senior Researcher at the Centre de Recherches Medecine, Sciences, Santé et Société in Paris. He has extensively published on the history of biological and medical research after 1945. His current research focuses on the industrialization of life in the twentieth century, with a peculiar angle on the history of biological drugs. Recent publications include *Inventer la biomédecine: La France, l'Amérique et la production des savoirs du vivant (1945–1965)* (Paris, 2002); and, as editor (with H.-J. Rheinberger), *The Mapping Cultures of 20th-Century Genetics* (London, 2003).

Jacob Darwin Hamblin is Assistant Professor of History at Clemson University. He is the author of *Oceanographers and the Cold War: Disciples of Marine Science* (Seattle, Wash., 2005) and has written several articles on the international dimensions of postwar science and technology. He received his Ph.D. from the University of California, Santa Barbara, and is a former Postdoctoral Fellow of the Centre Alexandre Koyré in Paris. His current book project, *Poison in the Well: Scientists and Radioactive Waste at Sea at the Dawn of the Nuclear Age,* will be published by Rutgers University Press.

Kristine C. Harper is Assistant Professor of History at the New Mexico Institute of Mining and Technology. She spent the 2004–2005 academic year as a Postdoctoral Fellow at the Dibner Institute for the History of Science and Technology. Her book, *Weather by the Numbers: The Genesis of Modern Meteorology,* will be published by the MIT Press. She continues to extend her research into the intersection of meteorology, technology, and the environment.

Gabrielle Hecht is Associate Professor of History at the University of Michigan. Her first book, *The Radiance of France: Nuclear Power and National Identity after World War II* (Cambridge, Mass., 1998), won prizes from the American Historical Association and the Society for the History of Technology. She is currently writing technopolitical histories of uranium, based on research in South Africa, Namibia, Gabon, Madagascar, Australia, France, Britain, and the United States.

Sheila Jasanoff is Pforzheimer Professor of Science and Technology Studies at Harvard University's John F. Kennedy School of Government. Her work centers on the comparative politics of science and technology in modern democracies, with a special focus on the construction of persuasive public arguments in legal and policy institutions. Her most recent book, *Designs on Nature,* was published by Princeton University Press in 2005. Her current research is on the globalization of environmental policy and politics.

Robert Kargon is the Willis K. Shepard Professor of the History of Science at the Johns Hopkins University. The author of *The Rise of Robert Millikan: Portrait of a Life in American Science,* his research interests have focused more recently on the borderlands between science and practice in the nineteenth and twentieth centuries. Kargon has coauthored a number of articles with S. W. Leslie growing out of their research on places of innovation such as Silicon Valley and Route 128. He is currently completing a study of science, technology, and urban design in the twentieth century.

John Krige is the Kranzberg Professor in the School of History, Technology and Society at the Georgia Institute of Technology. He has published extensively on the role of science and technology in postwar Europe. His next book, *American Hegemony and the Postwar Reconstruction of Science in Europe,* will be published by the MIT Press in 2006. He is currently extending this analysis to nuclear and space technology.

Stuart W. Leslie teaches the history of technology at The Johns Hopkins University. He has written about the cold war and American science and is currently studying the history of the laboratory, including laboratory architecture. With Scott Knowles, he is writing a "road book" on the American (de)industrialization, *We Can't Make It Here Anymore.*

Clark Miller is Assistant Professor of International Public Affairs and Science Studies and Senior Fellow in the Center for World Affairs and the Global Economy at the University of Wisconsin–Madison. He has published extensively on the global politics of science and the environment. He is the editor of *Changing the Atmosphere: Expert Knowledge and Environmental Governance* (with Paul Edwards, Cambridge, Mass., 2001). His current research is a comparative analysis of international expert institutions and global governance in the fields of public health, finance, and nuclear energy.

Index

ABC, 201
Abraham, Itty, 6, 10
Academy of Sciences (France), 222
access to medical testing, 258–59
Acheson-Lilienthal report, 146–47
Acheson-Lilienthal/Baruch plan, 50
activism, of scientists, 185–89, 204–6; and see collaboration; diplomacy
administrative erasure, and empire, 279
administrative rule, see governmentality
Adventures in Science, 149
Adventures of Hajji Baba of Isfahan, 125
Advisory Committee on Genetic Testing, 251
Advisory Committee on Technical Assistance (ACTA), 158
advocacy, patient, and cancer, 260–61
AEA, see Atomic Energy Authority
AEB, see Atomic Energy Board
AEC, see Atomic Energy Commission
Afghanistan, Soviet occupation of, 187–88
Africa, decolonization of, 8; as site for biotechnological development, 289
Afrikaners, culture of, 32
agencies, federal, and Department of State, 134, 148, 153–55; and UN, 154–55
Agency for International Development, 158
agribusiness and GMOs, 265; global, 267–68
agriculture, and Atoms for Peace, 162; and biotechnology, 252, 275, 283; and colonialism, 282; in India, 76–77, 81; and meteorology, 231; and Point Four Program, 158; and technical aid, 136
Ahlfors, Lars Valerian, 93, 96
Ahmadiyya Jamaat, 102
AIDS, 1; and biomedical practice, 260–61; in France, 265, 267
air traffic, and meteorology, 233
Air Weather Service, 243–44
airline navigation, and international alliances, 134
Akhromeev, Sergei, 194
Alaska, and nuclear testing, 70
al-Assad, Hafez, 101
Albright, Madeleine, 15, 17, 285
allegiances, and Atoms for Peace, 173
al-Qaddafi, Muammar, 101
ambivalence, and nuclear history, 10, 49–65; as term, 51, 55–56
American Academy of Arts and Sciences, 187
American Association for the Advancement of Science (AAAS), 15, 95, 97, 285
American Cancer Society, 261
American Chemical Society (ACS), 150
American empire, 18–19, 274; and ideology, 277
American Jewish Congress, 91
American Mathematics Society, 98
American Museum of Atomic Energy, 172
American Physical Society, 187
American Society for Engineering Education (ASEE), 114
Americanization, and GMOs, 264, 266
Amnesty International, 96, 184
Anderson, Benedict, 279–80
Anglo-American agreement, and bomb, 145
Antarctica, and IGY, 68, 246
anticolonialism, and nuclear power, 6; see also colonialism; decolonization
anticommunism, and South Africa, 32
antiglobalization and GMOs, 284
antiglobalization protests, and UN, 137
anti-Semitism, 90; and American universities, 96; and see Israel
antisubmarine warfare, 69, 83
apartheid, 8, and IAEA, 25–48; and inspectors, 46; and U.S. interests, 12
Apsara, swimming pool reactor (India), 57
Arab bloc, in UN, 89–93
Arab-Israeli War, 9
Arafat, Yasir, 89, 101
Archambeau, Charles, 196–97, 199, 202, 206
Argentina, and monitoring, 196; and nuclear program, 65; and U.S. reactor technology, 173–74
Arkin, William M., 191
Arms Control and Disarmament Agency, 188
arms control, 4, 145; international, 16, 134; and NGOs, 18; and scientists as advocates, 182–206
Aron, Raymond, 92
Arrow, Kenneth, 92
Artsimovich, Lev, 187, 194
Aryamehr University of Technology (AMUT), 11, 123–28
astronomy, across borders, 3
Atlantic coast, and radioactive waste dumping, 215
Atomic Development Authority, 147
Atomic Energy Act (U.S., 1946), 165, 169
Atomic Energy Act (India, 1948), 62
Atomic Energy Authority (AEA; Britain), and radioactive waste, 214, 217–19, 225–26
Atomic Energy Board (AEB; South Africa), 8, 37, 48
Atomic Energy Commission (AEC; U.S.), 3, 34, 61, 67, 70, 150; and Atoms for Peace, 165; and CIRUS, 57; and NRDC, 190; and radioactive waste, 209–28; and radioisotopes, 167–72; and reactors, 174; and scientific authority, 211; and scientific exchange, 146; and South Africa, 43–44; and UN, 162; and Union Carbide and Carbon Corporation, 175; see also IAEA
atomic energy establishment, and radioactive waste, 209–28
Atomic Energy Organization (Iran), 126
atomic energy, international control of, 143–47

atomic scientists' movement, 146
Atoms for Peace, 15, 27, 33, 60, 67, 159, 161–81; and Geneva conference, 175; relative size of program, 170
Australia, and computerized weather forecasting, 247; and nuclear program, 65; and origins of IAEA, 27; as uranium producer, 30
authority, scientific, struggle for, 214–15, 218–19, 222–224, 226–28; and WMO, 238
autonomy, and nuclear power, 181; see nation-state; sovereignty
autonomy, professional, 269; see independence
avian flu virus, 1

Bacher, Robert F., 170
bacteria, 1
Baïssis, Henri, 224
balance of power, postwar, 6
balance, global, and IAEA membership, 28
Bandung Conference (1955), 89
Barth, Kai-Henrik, 16, 19
Baruch Plan, 147
Baruch, Bernard, 147
bathyscaphe, 213
Bauman, Zygmunt, 282
Bayh-Dole Act, 257
Belgium, and computerized weather forecasting, 247; and IAEA, 27, 37–38; and Republic of Congo, 40; and U.S. reactor technology, 173–74
Bergen School, and meteorology, 234
Berger, Jonathan, 199, 206
Berkner Report, 133, 145, 159, 166–67
Berkner, Lloyd, 14–15, 145, 159, 166–67
Bernstein, Leonard, 98
Bers, Lipman, 93, 95–96
Bertocchi, Luciano, 106
beta-carotene, and golden rice, 287
Bethe, Hans, 92, 103
Bhabha, Homi K., 45, 56–57, 59, 113, 174
Bhakra Nangal hydroelectric dam (India), 63
Bharatiya Janata Party (BJP; India), 54
Bhutto, Zulfiqar Ali, 101–2
Bihar drought (India), 66, 76, 78, 80
Bikini atoll, 163
biological effects of atomic radiation (BEAR study), 212
biological internationalism, 251–72
biological weapons, 4; and see weapons
biologists, marine, and atomic energy agencies, 16–17
biomedical research, 21; and DNA, 255; history of, 253; and radioisotopes, 168
biopower, and biopolitics, 282
biotechnological model of research, 256
biotechnology, 21; defined, 276, 283–84; and empire, 273–92; in European Union, 290–91; global, 252
Birla Engineering College, 120
Birla Industries, 122
Birla Institute of Technology and Science (BITS), 11, 118–23, 129
Birla, G. D., 11, 118–21
Blair, Tony, 286
blindness, and GMOs, 287
Bocking, Stephen, 227
Bohr, Aage, 98–99, 107
Bohr, Niels, 169
bomb, and centrality of the nuclear, 159; and cold war, 162; and foreign policy, 67; hydrogen, 60; India and, 45, 47; and international cooperation, 144; and Japan, 143; Pakistani, 101; and role of science, 143–44; and scientific cooperation, 146; and Soviet Union, 29, 133, 145; and uranium safeguards, 45; U.S. loss of off Spain, 74; U.S. monopoly on, 133–34, 145; and see World War II
Bond, Christopher "Kit," 285
Book of Nomination Forms, 100
borders, national, and meteorology, 231
Born, Max, 57
botany, and empire, 281
Boumedienne, Houari, 101
boundaries, and empire, 279
boundary work, and empire, 279
Bové, José, 286
Bovine Spongiform Encephalopathy (BSE), 265
Bowles, Chester, 78–81
boycott, Arab of Israel, 91–92; defined, 87–88; of ICTP, effects of, 93–95, 108; analysis of, 99–102; and noncooperation, 86–109
Boyer, Paul, 143
brain drain, from India, 118, 129–30; and molecular biology, 253–55
Bravo, and bomb tests, 163
Brazil, and GMOs, 284; and international patent law, 258; and nuclear program, 65; and origins of IAEA, 27; and U.S. reactor technology, 173–74
BRCA2, 255–59, 268; and see cancer gene
breast cancer, see cancer genes
Brezhnev, Leonid, 186, 188
Britain, and computerized weather forecasting, 247; and Geneva conference, 174; and GMOs, 284–86; and IAEA, 37–39; and Indian technology, 58; and nuclear power, 6; and nuclear test bans, 188; and radioactive waste, 209–28; and reactors, 173; and safeguards, 44–45; and South Africa, 47–48; and uranium supply, 57; and VAP, 241
British Foreign Office, 219
Brodie, Bernard, 50
Bronk, Detlev, 215
Brookhaven National laboratory, 94
Brooks, Harvey, 97
Brown, Gordon, 11, 111–12, 120–21, 123, 125–28
Brune, James N., 199–200, 204, 206
Brussels Convention, 230
BSE, in France, 267
Buck v. Bell, 282
Buck, Carrie, 282
Budinich, Paolo, 87, 98, 102–4
Bugliarello, George, 123
buildup, weapons, 161, and see cold war
Bulletin of International Meteorological Observations Taken Simultaneously, 232
Bundy, McGeorge, 248
Bureau of Intelligence and Research, 77
bureaucracy, as empire, 278

Bush, George W. and Bush administration, and climate change, 250; and empire, 274; and Public Health Service, 252
Bush, Vannevar, 145–47
butterfly, monarch, 284
Buys Ballot, Christophorus, 232
By the Bomb's Early Light, 143
Byrnes, James F., 146

Cairo training center, 36–38, 40
California, and radioactive waste dumping, 215
Cambodia, weather modification in, 11
Canada, and Arab boycott, 91; and computerized weather forecasting, 247; and IAEA, 27, 31; and nuclear science, 56; and peaceful nuclear power, 61; and reactors, 173; and safeguards, 42–43
Canadian reactor design, 34
cancer genes, 251–72
cancer, and GMOs, 19
Cannes Film Festival, 222
capitalism, agricultural, 266; and empire, 278
Carnegie Corporation, 3
Carnegie Foundation, 199
Carr, Robert, 154
Carritt, Dayton, 214–16
Carson, Rachel, 72
Carter, Jimmy, and test site monitoring, 183, 205
Castells, Manuel, 229
Castle test series, 163
CBS, 149; and Geneva conference, 176
CCTA, see Commission for Technical Co-operation in Africa
CEA, see Commissariat à l'Énergie Atomique
center and periphery, 287
Center for International Studies (CENIS), 112, 114
Central African Republic, as source of uranium, 27, 33
Central Electronics Engineering Research Institute, 122
Central Intelligence Agency (CIA), 7, 15
Centre de Recherches et d'Études Océanographiques (France), 221
Centre National de la Recherche Scientifique (CNRS), 222, 259
Centres de Lutte Contre le Cancer (CLCC), 259
Chamberlain, Joseph, 281
Chamberlain, Owen, 92
Chandigarh, India, 63
Charney, Jule, 82
chemical agents, 1, 4
Chernobyl, 195–96, 200; in France, 267
Cherwell, Lord, 161
China Lake, CA, 78
China, and bomb, 75; and India, 75, 113; and nuclear science, 57; and nuclear test, 11; and WMO, 236–37; U.S. technical aid to, 155
Churchill, Winston, 161
CIA, and IIT Kanpur, 116
CIRUS reactor (India), 57
citizenry, global, 275, 284
civic-consumer regulation, 270–72; and see regulation
civilian nuclear power industry, 6–7, 164; and see Atoms for Peace
climate change, 1, 137, 246, 249–50
Clinch River Breeder Reactor Project, 190
clinical model of research, 256, 263
Clinton, Bill, and Clinton administration, 15, 251, 261, 285
Closed World, The, 242
cloud seeding, see weather modification
Cochran, Thomas, 182, 185–86, 190–93, 196–202, 204, 206
Cockcroft, John, 57–58, 165, 219, 224–25
Cohen, Avner, 52, 61
cold war, and biomedical research, 254; end of, 5; and IAEA, 25–48; and identity, 289; India and, 75; and influence in third world, 8; Johnson and, 83; and mathematicians, 96; and meteorology, 242–47; and military alliances, 134; and nuclear power, 6; and oceanography, 228; and radioactive waste, 210; and rivalry, 13–14, 16, 20, 133; and scientific community, 184; and scientific internationalism, 167; scientists and end of, 185; and South African identity, 32; and Soviet Union, 161; and U.S. support for science, 2; and UN agencies, 137
Cole, Sterling, 219
collaboration, of atomic energy establishments, 225–26; and biological internationalism, 252; in nuclear programs, 57
colonialism, and development, 281–82; and difference, 274; and imperial projects, 291–92; and see decolonization, postcolonialism
Colorado River, and Johnson, 71, 83
Combined Development Agency, and South Africa, 27
commerce and nuclear power, 33
commercialization of medical testing, 258–59
Commissariat à l'Énergie Atomique (CEA), and radioactive waste, 214, 220–24, 226
Commission du Génie Biologique (CGB), 265–66
Commission for Technical Co-operation in Africa (CCTA), 38–40
Commission on Synoptic Meteorology, 238
Committee for the Exploitation of the Sea, 221
Committee of Soviet Scientists against the Nuclear Threat (CSS), 193–94
Committee on Atmospheric Sciences, 82
Committee on International Security and Arms Control (CISAC), 187–88, 193
committees, and federal agencies, 154–55
Commonwealth, British, and origins of IAEA, 27
Commonwealth, South African withdrawal from, 40
communication, mass, and empire, 278
communism, and Atoms for Peace, 163–64; India and, 75; and science, 178; and security, 169; and technology, 61; and U.S. aid, 155–56; and see cold war
communities, imagined, 279, 289–90
competition, and biological internationalism, 252; and see cold war
Comprehensive Test Ban Treaty (CTBT), 54, 188–89, 202
computers, capacity in 1960s, 83; and cold war,

computers (*cont.*)
243; and empire, 275; and IIT Kanpur, 117–18; and meteorology, 247–48
Conant, James, 145
Confédération Paysanne, 266
Conference on the Human Environment, UN, 230
Conference on the Law of the Sea (UN, 1958, 1960), 68–69
Congo, Belgian, 280
Congo, Republic of, 40
Congress Monthly, 92
Congress Party (India), 84
Conseil Européen pour la Recherche Nucléaire (CERN), 95, 98–99, 103, 254
Constituent Assembly of India, 58
constitutional mode of empire, 280, 290–91
consumer-based access, 258–59
containment, as cold war strategy, 242–43
control, discourse of nuclear, 49–65; state and testing, 262
Convention Relating to the Regulation of Aerial Navigation, 244–45
Conway, Gordon, 289
cooperation, scientific, and U.S. foreign policy, 133–60; and see collaboration
co-optation vs. mutual orientation, 246
corn, genetically modified, 265, 284, 286; and see GMOs
CORONA spy program, 245
corporate partners, and state, 283
Council for Scientific and Industrial Research (South Africa), 37
Cousteau, Jacques-Yves, 16, 222–24, 226
credibility, scientific, 179–80, 220; and nuclearity, 27
Cressman, George, 237
cultural traditions, and science, 129
culture, and social organization, 288
Czechoslovakia, and IAEA, 27, 31

Dahl, Norman, 115, 118, 121
Dalal, Ardeshir, 113
Dalitz, Richard, 100
Dallek, Robert, 84
dams, and technological criticism, 283
Davis, Watson, 176
de Gaulle, Charles, 62, 223, 254
de Greiff, Alexis, 9
de Klerk, F. W., 48
declassification of nuclear knowledge, 165–66
decolonization, and IAEA, 25–48; and national weather services, 239–41; and scientific cooperation, 148; and UN, 89; see also Africa
defense, see security
definition, and empire, 278; and hybridity, 287
degree of nuclearity, 7–9; and global hierarchies, 29; and source materials, 27–30
Delbrück, Max, 166
Deming, Olcott, 150, 156–57, and UNESCO, 151
democracy, and empire, 280; in India, and crops, 82; and technology, 276; and U.S. aid, 157
demonstration, nuclear explosion as, 52
Denmark, and GMOs, 286; and nuclear science, 56
Department of Agriculture, 138, 153–54; and GMOs, 288
Department of Defense (DoD), 3; and NRDC, 198–99; and physics, 179; and Department of State, 203; and weather modification, 78–80, 83
Department of Energy (DoE; U.S.), 7
Department of State, 66; and Berkner Report, 166–67; and DCR, 140–42; and federal agencies, 148–49, 153–55; and foreign policy, 151; and ICSCC, 136–60; and India, 75, 79, 81; and institutionalization of experts, 134; and Interdepartmental Committee on Cooperation with the American Republics, 140–43; and Marshall Plan, 156; and MIT, 112; and NRDC monitoring, 198, 203; and radioisotopes, 171; and Smith-Mundt Act, 157; and UNESCO, 90
dependence, and nuclear power, 181
depoliticization of nuclear negotiations, in South Africa, 31
Depression, and meteorology, 235; and Roosevelt, 138
desalination, nuclear-powered, 73
developing world, see third world
development, and nuclearity, 33–39
DeWind, Adrian, 192, 197–98
diplomacy, and science, 66–85; and weather modification, 80
Diplomatic November Revolution, 89–93, 99
discourse, biotechnology as, 284; of development, 281; imperial, in Africa, 289; and see empire
discovery, nuclear-related, 56
Division of Cultural Relations (DCR) of Department of State, 140–42
divisions vs. departments, and AMUT, 123–24
DNA, and biotechnology, 276; and see biomedical research
Dobrynin, Anatoly, 200, 202
Doel, Ronald E., 10–11, 246
doing, ways of, 269
dominance, American in molecular biology, 253–55; and GMOs, 266; relations of, 280
Doty, Paul, 187
Downing, Thomas N., 215
Draper Laboratory (MIT), 127
Drell, Sidney, D., 97
Drew, Thomas, 119–22
du Plessis, W. C., 42–43
Dulles, John Foster, 163
Dunster, H. J., 217–19, 224
Dupree, A. Hunter, 84

Earth Summit, 252
East-West Center, 72
ecological impacts of radioactive waste, 221–22
economic development, and Point Four Program, 158; and see third world
economic power, U.S. and biotechnological research, 252
education, cooperative, in India, 118–23
Edwards, Paul N., 14, 20

Eells, J., 98
Egypt, and IAEA, 8, 46–47; and nuclearity, 36, 40, 47; and South Africa, 27
Egyptian Atomic Energy Establishments, 37
Eisenhower, Dwight D. and administration of, 4, 15, 66–67, 70, 72, 144; and Atoms for Peace, 7, 27, 159, 161–81; and reactor, 175; and test site monitoring, 183, 205
Eklund, Sigvard, 40, 103
ElBaradei, Mohamed, 160
electronics, and IIT Kanpur, 117
el-Sadat, Anwar, 101
Elusive Transformation, 18
Elzinga, Aant, 108
émigré scientists, 96
empire, and biotechnology, 273–92; European Union as, 20, 280; and GMOs, 20; modes of governing, 277; as social technology, 273–92; U.S. as, 18–19
Energy Laboratory (MIT), 125–26
engineering, at BITS, 122; and Indian culture, 115–16, 119; and see MIT
England, and nuclear science, 56–57
Ensminger, Douglas, 119–20
entrepreneurialism, scientific, and nation-state, 252–53
environmental activism, 264
environmental effects of technological experiments, 70
Environmental Protection Agency (EPA), 264
environmentalism, history of, 281; and Kennedy-Johnson administrations, 72–75
epidemiology, 282
epistemic communities, 184, 206
erasure, and empire, 279
Ericson, T. E. O., 95
Ethical, Legal, and Social Implications (ELSI) Research Program, 261
Ethiopia, and IAEA, 40
eugenics, as biological standardization, 282–83
Euratom, 40
Europe, and GMOs, 287; and ICTP boycott, 97; and international patent law, 258; postwar, 155–56; and radioisotopes, 169–71; reconstruction of, 4; and see European Union
European Cooperation Administration, 156
European Patent Office, 263
European Space Agency, 254
European Union, as conglomerate, 17; as empire, 280, 290–91; and GMOs, 284, 286
Evangelista, Matthew, 185, 206
Evernden, Jack, 196–97
evil empire, and Reagan/Bush administrations, 274
Executive Committee on Economic Foreign Policy, 154
Expanded Program of Technical Assistance for the Economic Development of Underdeveloped Countries (EPTA), and meteorology, 241
experts and expertise, and Acheson-Lilienthal report, 147; and BSE, 265; and developing nations, 36; and empire, 278; and foreign policy, 139–40, 143, 159; and international affairs, 185; and ICSCC, 144; and meteorology, 241–42; and satellites, 248; and state regulation of medical testing, 262; and test monitoring, 206; for nuclear explosion, 53; physicists as, 144; and public debate on GMOs, 266–68; role of, 3, 10, 17, 135; and technical credibility, 27–28; in U.S. foreign policy, 151, 153–54; and WHO, 152
Explorer VII, 245

faculty, and IIT Kanpur, 116
Fage, Louis, 221–22, 224
Fahmy, Ismael, 38–39
failures, technological, 288
Faisal, King, 101
Falicov, Leo, 94–95, 98
fallout, nuclear, 68, 70, 212; and weather, 243–44
farming, and GMOs, 19, 268–69
FDA, and genetic tests, 261; and medical services, 258
fear, public, 16–17, 149, 172, 285–86; and GMOs, 265–67; and management of technology, 283; and radioactive waste, 216–18, 224–25
federal agencies, and Department of State, 148–49, 159
federal employees, and foreign policy, 139; see expertise
Fédération Nationale des Syndicats d'Exploitants Agricoles, 266
Federation of American Scientists (FAS), 187, 192, 194, 196, 205
Federation of Atomic Physicists, 146
feminist movement, and breast cancer research, 19
Fermi, Enrico, 174
Fermi, Laura, 174
fetishizing the bomb, 10
Fialka, John J., 191
Field, M. J., 98
Fischer, David, 46–47
fisheries, Latin American, 138; and technical aid, 136
fishing, 69
"fishing expeditions" as IAEA interference, 32
fission, nuclear, 56, 59; as world historical event, 64
fissionable materials, and Atoms for Peace, 162, 164; controls on, 42–45
Five Continent Peace initiative, 196–97
Fleagle, Robert G., 82
Flerov, Georgii, 59
Flournoy, Michèle, 200
FNRS III, 213
food aid, U.S. to India, 75–77
Food and Agriculture Organization (FAO; UN), 89, 134, 150, 153
Food and Drug Administration (FDA), and GMOs, 251
food production, see GMOs; weather modification
food safety, see GMOs
Ford Foundation, 199; and ICTP, 102; and IITs, 111–13, 115, 118–20, 122
Ford Fund, and peaceful atom, 176
forecasting, computerization of, 243
foreign aid, U.S., 113; see technical assistance

foreign policy (U.S.), and Johnson, 73; post World War II, 148; and scientific cooperation, 133–60
Forland, Astrid, 41
Forman, Paul, 88, 180
Forti, A., 103
Foucault, Michel, 278–79, 280, 282, 289
Foundation of Economic Trends, 264
Fourth International Cancer Research Congress, 168
France, and boycott of UNESCO, 106; and cancer genes, 252; and Geneva conference, 174; and GMOs, 19, 264–66, 284; and IAEA, 37–38; and Indian technology, 58; and medical testing, 258–60, 262; and meteorology, 235; and NATO, 254; and nuclear power, 6, 57, 62; and opacity, 52; and origins of IAEA, 27; and radioactive waste, 209–28; and reactors, 173; and South African uranium, 34–35; and UN, 91; and VAP, 241
Frankel, Benjamin, 52
Frankenfood, 285
Frederick the Great, 282
Freeman, Orville, 77
freeze movement, 206
French reactor design, 34
Friends of the Earth, 266
Fuchs, Klaus, 178
Fuchs, Wolfgang, 93, 96, 98
funding, government, 2–3, 14; for IIT Kanpur, 115–16; and oil in Iran, 125; and scientific authority, 218; and see patronage
Furnestin, Jean, 220–21, 224

Gabon, and IAEA, 40
Gabon, as source of uranium, 27, 33
Gaffney, Frank J. Jr., 199
Gallucci, Robert L., 7, 10, 12
Gandhi, Indira, 79–81, 84
Gandhi, Mahatma, 60, 119
Gandhi, Rajiv, 54
gap, technological, and U.S., 73
gardening state, and eugenics, 282–83
Garwin, Richard, 194
Gaudillière, Jean-Paul, 16–17, 19
gay marriage, and biopower, 282
gays, and eugenics, 282
General Electric, and civilian nuclear power, 164, 181
genetic testing, and NBCC, 260–61; risks of, 262
genetically modified crops, see genetically modified organisms
genetically modified organisms (GMOs), 1, 16, 251–72, 264–68, 283; as American, 266; in European Union, 291; and patronage, 16–17; and public fear, 285–86; see also corn; golden rice; Roundup Ready
Geneva conference (1955), 161, 165–66, 174–80; and security, 174
Geneva, and WMO, 240
genocide, and bomb, 163
Gerasimov, Gennady I., 201
German academics, politicization of, 96
Germany, and computerized weather forecasting, 247; and GMOs, 284; and IIT Kharagpur, 114; and Indian technology, 58; and nuclear science, 56–57; and Pearl Harbor, 140; World War II defeat of, 134
Ghana, and IAEA, 39–40
Gibbons, Michael, 255
Giddens, Anthony, 239
glasnost, 197, 203
globalization contestation, 5
Global Atmospheric Research Program, 240
global health, and national autonomy, 18
global information infrastructure (GII), 229–30, 240; and Kyoto Protocol, 249–50
Global Transformations: Politics, Economics and Culture, 229
global warming, see climate change
globalism, and agribusiness, 269; and meteorology, 229–50
globalization of biotechnology market, 257; and nation-state, 17–20; and regulation of GMOs, 251–72
Gokhberg, Mikhail, 200, 204
gold mines, South African, 28
golden rice, 287, 289
Goldschmidt, Bertrand, 224
Good Neighbor Policy, 138
Gorbachev, Mikhail, and arms control, 16; and glasnost, 197, 203; and nuclear testing, 183–86, 188–89, 192, 194–98, 200, 202–3, 206; and perestroika, 47, 202
government, de facto, UN and, 137
governmentality, 278
Great Depression, see Depression
Great Society, 83
Greece, and monitoring, 196; and U.S. reactor technology, 173–74
Greece-Turkey Aid Act, 155–56
green biotechnology, and empire, 275
Green Revolution, 84, 276
Greenpeace International, 184, 265–67
grid, urban, and empire, 279
Gromyko, Andrei A., 50
gross national product (GNP), 2
Grove, Richard, 281
growth hormones, 252, 268
guerillas, North Vietnamese, 11
Gypsies, and eugenics, 282

Haas, Peter M., 184
Hacking, Ian, 278
Hahn, Otto, 56
Halfman, Robert, 116–17, 121
Halperin, Danny, 91
Hamblin, Jacob Darwin, 16, 17
Hammond, Paul, 76, 84
Handbook of Meteorology, 234
Harari, Haim, 94–96, 98, 100–102
Hardt, Michael, 276–77, 284, 290
Harper, Kristine C., 10–11
Harrison, J. M., 100, 103–4
Hays, Samuel P., 72
Heald, Henry, 115

health, and Point Four Program, 158; public, and international alliances, 134; and see cancer genes; fear; GMOs
Hecht, Gabrielle, 6, 8, 9, 62
Hela, Ilmo, 227
Hesselberg, Theodor, 235
Hewlett, Richard G., 178
Hewson, Martin, 229, 231, 239
hierarchies, of nuclearity, 29, 34; and power, 274
Hill, A. V., 113
Hiroshima, 49, 59, 148, 170
history, Iranian, and AMUT, 124
Ho Chi Minh trail, 11
Hofstadter, Robert, 92
Holl, Jack M., 178
Holland, and nuclear science, 57
Holloway, David, 58–59
Horn, David, 94
Hornig, Donald, 71, 73–75, 79, 82
Houston Post, 216
Hoyle, Fred, 93
hubris, technological, 137
Hull, Cordell, 142
Human Genome Project, 252, 260–61
Human Rights Committee (NAS), 95
human rights, and NGOs, 18
hunger, and Pugwash conference, 187
Hurd Deep, 217, 225
hurricane steering, see weather modification
Huxley, Julian, 92
hybridity, and GMOs, 287; and empire, 274–75, 279

IAEA, see International Atomic Energy Authority
IBM, and India, 118
ICTP, see International Centre for Theoretical Physics
identity, and cold war, 289; collective, and European Union, 290; and community, 289–90; and empire, 276; and globalization, 18; of IITs, 118; and nuclearity, 6, 54–55; and science education, 124; of South Africa, 30, 32; and see communities, imagined
IGY, see International Geophysical Year
Illinois, University of, 114, 123
image, scientific, and public relations, 253–54
IMO, see International Meteorological Organization
imperial constructions, 276–80
imperialism, industrial, 173
Incorporated Research Institutions for Seismology (IRIS), 204
independence of scientists, 15, 17, 150
independence, national, and technology, 58; see also nation-state; sovereignty
India, agriculture in, 76, 81; and awareness of Manhattan Project, 58–59; and bomb, 45, 47; conservation in, 281; and France, 57; and Geneva conference, 174; and GMOs, 284; and IAEA, 27–28, 30–32, 34, 40; industrial development in, 76; and international patent law, 258; MIT schools in, 11, 110–23; and monitoring, 196; and nuclear program, 10, 52–55, 61, 63; and Republic of Congo, 40; and safeguards, 41–42; and South Africa, 27; and UN, 59–60; and weather modification, 66, 68, 73–83
Indian Atomic Energy Commission, 58
Indian culture and engineering, 115–16, 119
Indian Institute of Science, 113–14
Indian Institute of Technology Bombay, 115
Indian Institute of Technology Kanpur (IIT Kanpur), 113–18, 120–22; results, 117–18; status of, 114
Indian Ocean, 70
Indonesia, and community, 280
industry regulation, 269–72; and see regulation
industry, education and, in India, 118–23
industry, Indian, and IIT Kanpur, 76, 117
inequality, perpetuation of in nuclearity, 34
infectious diseases, 1
information and communication technologies (ICTs), and globalism, 229–50, 275
information technology, and NGOs, 18
informational globalism, and meteorology, 232, 236–37, 239–42, 245–47, 249
Inozemtsev, Nikolai, 193
inspection, and radioisotopes, 172
Inspectors' Document, and IAEA, 46
Institut Curie, 251, 263
Institut National de la Santé et de la Recherche Médicale (INSERM), 259
Institut Océanographique (Monaco), 221
Institute for Industrial Technology and Applied Science (Korea), 73
Institute of Marine Research (Finland), 227
Institute of Physics of the Earth, 200, 203
intellectual property, 252, 257; and cancer gene testing, 263; and globalization, 19; and multinational corporations, 286; vs. use, 258–63; and see patentability
intelligence, scientific, 178; and Geneva conference, 166–67; and meteorology, 244
inter-American treaties, 138–39
Interdepartment Committee on Scientific and Cultural Cooperation (ICSCC), 144, 148–49, 151–53; legacy of, 158–60; and State Department, 136–60
Interdepartmental Committee on Cooperation with the American Republics, 138–42
interdependence of states, 18
interests, vs. infrastructure, 230
Intergovernmental Panel of Climate Change (IPCC), 249–50
international affairs, scientists and engineers in, 4, 182–206; and see diplomacy; collaboration
International Atomic Energy Authority (IAEA), 7, 9, 16, 87, 103; and autonomy from UN, 31; composition of, 28; establishment of, 25–48, 159–60; and Geneva conference, 177, 180; and radioactive waste, 213–14, 218–20, 224–25, 227; regulations for radioactive waste, 210
International Breast Cancer Research Consortium, 263
International Centre for Theoretical Physics (ICTP), 9, 86–109; funding of, 103–6; and

International Centre for Theoretical Physics (*cont.*) negotiations with UNESCO, 102–7; status of, 102
International Civil Aviation Organization, 134
International Commission for Aeronatuical Meteorology (CIMAé), 233
International Commission for Air Navigation (ICAN), 233–35
International Commission for the Scientific Exploration of the Mediterranean, 222
International Committee for Intellectual Cooperation, 135
International Co-operative Alliance (ICA), and IITs, 114
International Council of Scientific Union (ICSU), and UNESCO, 99–100
International Forum of Scientists for Stopping Nuclear Tests, 197
International Geophysical Cooperation year, 246
International Geophysical Year (IGY), 4, 68–69, 135–36, 248; and radioactive waste, 213, 220; and infrastructural globalism, 245–46; and national weather services, 247; and standardization, 240
International Health Conference, 151–52
International Labour Organization (ILO), 90–91
International Meteorological Committee, 233
International Meteorological Organization (IMO), 230–35; and CIMAé, 233; and politics, 234; and *Réseau Mondial,* 233; and standardization, 235
International Monetary Fund (IMF), 1, 134, 229–30; and brain drain, 130
International Telecommunications Union, 230
International Union of Pure and Applied Physics (IUPAP), 99–100
international world order, post cold war, 6
internationalism, scientific, and expertise, 133–60; and ICTP, 86–109; as policy instrument, 133–60; as political stance, 88
internet, see World Wide Web
Iran, and AMUT, 123–28; and ICTP, 103; and MIT school, 11
Iraq, and IAEA, 40
Isaacs, John, 210, 215–16
Isfahan University of Technology, and AMUT, 11, 128–30; and see Iran
isotopes, and ecological impacts, 216, 221–22; and radioactive waste, 214
Israel resolutions, and ICTP, 87, 89–92, 96–97, 103–4, 108
Israel, and atomic energy, 61; and meteorology, 241, 247; and nuclear science, 57; and nuclearity, 36; and opacity, 52–53; and South African uranium, 34–35
Israeli Economic Warfare Authority, 91
Italy, and ICTP, 87, 104–6; and nuclear science, 56–57
Ivory Coast, and IAEA, 40

Jackson, C. D., 162
Jackson, Henry M. "Scoop," 69
Japan, and computerized weather forecasting, 247; and Human Genome Project, 252; and intellectual property rights, 257; and oceanography, 213; and Pearl Harbor, 140; World War II defeat of, 134, 149
Jasanoff, Sheila, 19–20
Jerusalem, excavations in, 90
Jet Propulsion Lab, 57
Jews, and eugenics, 282
Jha, Lakshmi Kant, 80
Johnson, Lyndon B., and sciences, 10–11, 66–85; and see weather modification
Joint Committee on Atomic Energy (U.S.), 179
Joint Numerical Weather Prediction (JNWP) unit, 243
Joint U.S.-India Precipitation Experiment, 78, and see weather modification
Joliot-Curie lab (Paris), 57
Joly, Pierre-Benoît, 268
Jooste, G. P., 42
Jordan, Robert, 90
Joseph, Arnold, 214, 216–17
Josephson, Paul, 176

Kaczynski, Theodore, 283
Kaddoura, A., 106
Kalahari Desert, nuclear testing in, 12
Kanpur Indo-American Program (KIAP), 115–18, 120
Kargon, Robert, 10–11
Kashmir, war over, 116
Kastler, Alfred, 99, 103, 107
Kazakhstan, and nuclear program, 16, 65; NRDC monitoring in, 189, 199–203
Kelkar, P. K., 115, 121
Kelman, Herbert, 13
Kendrew, John, 253
Kennedy, John F., and administration of, 11, 66, 70–72, 74, 79, 113, 245, 248
Kennedy, Ted, 194
Ketchum, Bostwick, 221
Kew Gardens, 281, 288
Khan, A.Q., 12
Khomeini, Ayatollah, 11, 128
Khrushchev, Nikita, 186
Kidder, Ray, 191
Killian, James, 119–20
Kissinger, Henry, 90, 187
knowing, ways of, 269
knowledge production, mode of, 255–56
knowledge, genetic, uses of, 260; see also predisposition testing
knowledges, local, 2
Kohn, Walter, 93, 95–96, 98
Kokoshin, Andrei, 183, 192
Korea, and nuclear weapons, 50; U.S. assistance to, 73; war in, 59, 161
Krige, John, 6–7, 15
Kubbig, Bernd W., 187
Kurchatov Institute of Atomic Energy, 193
Kuwait, and MIT, 126
Kyoto Protocol, 249–50

labeling of genetically modified food, 267–68
laboratory, international, and radioactive waste, 224–27

Lang, Norton, 93, 98
language, and empire, 274; and identity, 289; and public fears, 285; science as, 152, 178, 186, 206, 232
Laos, weather modification in, 11, 78, 83
Latin America, and ICSCC, 158–60; and nuclearity, 36; and scientific internationalism, 137–43; and U.S. technical aid, 136, 155, 157; and World War II, 137–38
Laurence, William L., 176
Lauritsen, Charles C., 170
Law of the Sea, UN Conference on, 210, 213
Lawrence Livermore National Laboratory, 191
Lawrence, Ernest, 179
Le Corbusier, 63
Le Monde du Silence, 222
Le Monde, 223
League of Nations, 135, 137; and globalization, 229
legibility, 20, 279–80, 287–89
legitimacy, and CCTA, 39; and nuclear power, 36, 62–65; scientific, 179–80; and states, 56; and UN, 238–39, 241; and ways of knowing, 269; see also sovereignty
Leopold II, King, 280
Leslie, Stuart W., 10–11
Lewis, John, 57
Libya, and weather service, 240
Lilienthal, David, 61
Lilly, Eli, and Myriad Genetics, 256
Limited Test Ban Treaty (LTBT), 70, 186, 188
Little, Arthur D., 123
Los Alamos, 7
Lovett, Richard, 171
Low, W., 100
Lucas, George, 274
Lucky Dragon Five, 50, 163
Luedecke, Alvin, 225
Lunqvist, Stig, 97

MacArthur Foundation, 199
mad cow disease, see BSE
Madagascar, as source of uranium, 27
Maddox, John, 93
Magnuson, Warren G., 69
Manhattan Project, 193; international awareness of, 58–59; and radioisotope program, 168
manpower, as resource, 2
Marcos, Ferdinand, 73
marine policy (U.S.), 74
marine resources, and see oceanography, 1
market, and empire, 278, 280; financial, and international alliances, 134; and testing, 262
Marris, Claire, 268
Marshall Islands, 163
Marshall Plan, 134, 137, 148, 156–57, 255
Marshall, George, 156
Marxism, in Africa, 12
Massachusetts Institute of Technology (MIT), and ICTP, 94; military research at, 117; and MIT idea, 110–12; and overseas technical institutes, 11, 110–30; and radioisotopes, 169
mathematics, and ICTP, 103
Mathews, Jessica T., 18–19
Mauritius, 281
M'Bow, Amadou Mahtar, 90–91, 103–6, 108
McCone, John A., 215
McDonald's, 284
McDougall, Walter, 245
McNamara, Robert, 78, 81
Medaris, John B., 178
Médecin, Jean, 222
Medhurst, Martin J., 173
media, and empire, 278
medicine, and Atoms for Peace, 162; and biotechnology, 252; and GMOs, 287; and radioactive isotopes, 172, 175; and see cancer genes
Mediterranean, and radioactive waste dumping, 219–24
mentality, and government, 278–79
Menuhin, Yehudi, 98
meteorology, 21, 229–50; across borders, 3; aeronautical, 233; globalization of, 14; and India, 76; and international science, 152; in Johnson administration, 10, 66; and need for international data, 69; see weather modification
metropolis and periphery, 2, 273–92
Mexico, and monitoring, 196
Middle East vs. Africa in nuclearity, 37
military, and empire, 278; and meteorology, 231, 238, 242–47, 249; and support for science, 69, 117, 170, 211; and weather modification, 82
military-industrial complex, 72
military-industrial-academic complex, 13, 144
Miller, Clark A., 14, 236, 241
Millikan, Max, 11, 112, 114
Millionshchikov, Mikhail, 187
mining, seafloor, 68–69
Ministry of Information and Broadcasting (India), 63
missiles, and AMUT, 127; ballistic, 69, 73, 83, 134; guided, 2; Pershing II, 187
mission civilisatrice, 290
MIT, see Massachusetts Institute of Technology
Mitra, C. R., 122
modeling, numerical, 83
modernity, science as symbol of, 4, 9, 75, 170
modernization and civilization, 137
modification, environmental, 70; and see weather modification; GMOs
molecular biology, see biomedical research
molecularization of biology and medicine, 254
Monaco conference, 217–20, 224; and Cousteau, 222
monitoring of nuclear tests, 19, 182–206
monitoring, international, of atomic energy, 147
Monod, Jacques, 253
monopoly, Myriad Genetics and, 258; U.S. and bomb, 133–34, 145
Monsanto, 265, 267, 284, 286, 288
moratorium, on GMOs, 265, 268, 286; nuclear test, 183, 189, 196–97, 200–201
Morgenthau, Hans J., 75
Morocco, decolonization of, 27; and IAEA, 40
Morris, James, 279, 281
multinational corporations, and states, 292
Muny, A., 100
Murray, Thomas E., 164

Musée Océanographique de Monaco, 222–23, 227
Music Council (UNESCO), 98
Mutual Security Agency, 158
Myriad Genetics, 252, 255–56, 258–59, 262–63

Nagasaki, 49, 148
NAS, see National Academy of Sciences
NASA, and Johnson, 72; and meteorology, 243; and TIROS, 245
Nasdaq, and biotechnology, 257
Nasr, Seyyed Hossein, 123–25, 127–28
nation, defined, 279
National Academy of Sciences (NAS), 70, 82, 97; and radioactive waste, 210, 212
National Academy of Sciences Committee on Oceanography (NASCO), 214–15
National Breast Cancer Coalition (NBCC), 260–61
National Council of Jewish Women, 156–57
National Emergency Council, and foreign aid, 138
national informational infrastructures (NIIs), 229
National Institute of Oceanography (Britain), 218
National Institute of Oceanography (India), 122
National Institutes of Health (NIH), 3, 261, 264
National Physical Laboratory (India), 122
National Science Council, 102
National Science Foundation, 165
National Security Council (U.S.), 70, 187
national weather services, 231, 233–36, 239–40; and IGY, 247
nationalism, and nuclear power, 62
Nationalist Party (South Africa), 30, 32
nation-building in India and Iran, 110–30
nation-state, and biotechnology, 251–72, 283; and corporations, 292; and globalization, 17–20, 277; and GMOs, 288
NATO, and biomedical research, 254; and IAEA, 27; and radioisotopes, 169; nuclearization of, 173; and weather modification, 82
Natural Resources Defense Council (NRDC), 19, 182–206
Nature, 92
nature, control of 67–68, 280
Naval Ordnance Test Station, 11, 78
naval weather logs, 230–31
Nazis, and eugenics, 282
Near East, U.S. technical aid to, 155
Nebeker, Frederik, 231
Ne'eman, Yuval, 93–94, 96, 100
Negri, Antonio, 276–77, 284, 290
Nehru, Jawaharlal, 10, 57–58, 62–63, 114, 119
Nelkin, Dorothy, 227
Nersesov, Igor, 200, 204
Nevada Test Site (NTS), 16, 191, 198, 201, 203–4
New Deal, 71–72, 134
New Look military doctrine, 162, 173
new nations, cold war and, 173
New Scientist, 166
New York Times, 70–71, 90–92, 176–77, 198, 213, 215, 223
NGOs, see nongovernmental organizations
Niger, as source of uranium, 27
Nigeria, and IAEA, 40
Nitze, Paul, 66, 192
Nixon, Richard, 66, 188
Nobel Prize winners, 9, 87, 113, 160, 165, 186
noncooperation, 86–109
nongovernmental organizations (NGOs), 18–19; and GMOs, 266–67; and international patent law, 258; and IRIS, 204; as transnational actors, 184; and voluntary collaboration, 232–33; and weather, 232; and see NRDC
Non-Proliferation Test Ban Treaty, 54
nonproliferation, 20; and see Atoms for Peace
North Korea, and nuclearity, 10; and see Korea
Norton, Garrison, 152
Norway, 57, 65, 235
November Revolution, 94, 107
Nowotny, Helga, 255
nuclear ambiguity, see ambivalence
nuclear engineering, and IIT Kanpur, 117; and Iran, 126
Nuclear Non-Proliferation Treat, 160, 186
nuclear power, as symbol of modernity, see modernity
nuclear programs, centrality of, 5–9, 20; closed, 65; and see weapons
nuclear proliferation, and Iran, 126; post-World War II, 12;
nuclear test ban, and SAS/NRDC, 188–200
Nuclear Weapons Databooks, 191
Nuclear Weapons Freeze Campaign, 191, 206
nuclear winter, 194
nuclearity, defined, 26; and development, 33–39; and Geneva conference, 181; and IAEA, 25–48; manifestations of, 37–38; and postcolonialism, 10; and secrecy, 29
Nucleonics, 57
Nye, David, 60–61
Nye, Joseph S. Jr., 11

Oak Ridge National Laboratories, 168, 172, 175
observer status, and WMO, 237
oceans, and IGY, 245–46; and meteorology, 244; and see oceanography
oceanography, 21, 68–69; across borders, 3, 230; and atomic energy agencies, 16–17; and ICTP, 103; in Johnson administration, 10, 66; and patronage, 16; and radioactive waste, 209–28
Oceanographer, R/V, 74
Office of American Republic Affairs, and DCR, 141
Office of International Information and Cultural Affairs, 149
oil embargo (1973), 125
oil, and sea pollution, 213; and technical education in Kuwait, 126
oncogenetics, 259
opacity, defined, 52–53
OPEC, 125
operations research, 2
opinion, public, and radioactive waste, 211, 216–18, 222–23
Oppenheimer, J. Robert, 146, 171

opportunism, and scientific authority, 212, 216–17, 227–28, and see patronage
organic farms, and GMOs, 286
Organization for Economic Co-Operation and Development (OECD), 2, 264
orientalism, British, 125
orientation, mutual, 246
ozone depletion, 1, 246

Pacific coast, and radioactive waste dumping, 215
Pakistan, disputes with India, 75; nuclearity of, 12; and opacity, 52, 54; and Salam, 101; U.S. arms sales to East, 116; weather modification in, 66, 73–83
Palestine Liberation Organization (PLO), 9, 89–91
Panofsky, Wolfgang, 96, 107, 194
Paris Convention Relating to the Regulation of Aerial Navigation, 233
Parliamentarians for Global Action, 196
Parran, Thomas, 151–52
patents and patentability, of cancer gene testing, 263; definition of, 256–59, 262; and gene sequences, 256–58; of GMOs, 286, 291; of living entities, 252; and Myriad Genetics, 252
pathology, discourse of, 289–90
Pati, Jogesh, 94–95, 107
patient advocacy, 260
patient organizations, U.S., 269
patronage, 5, 120; and ICTP, 103; and oceanography, 218, 227; and military, 69, 117, 170, 211; state, 13–17
Patterson, Robert, 144
peaceful nuclear explosion (PNE), 52, 54, 188
Peaceful Nuclear Explosions Treaty (PNET), 188–89
peaceful use of nuclear power, 60–61; see also Atoms for Peace
Pearl Harbor, 140
Pearson, Lester, 60
peer pressure, 240
penicillin, 2, 254
Penney, William, 225
Pentagon, and weather modification, 79, 81, 85
Peres, Shimon, 61
perestroika, 47, 202
periphery, and metropolis, 2, 273–92
Perkovich, George, 47, 75
Perle, Richard, 198
Perrin, Francis, 222–24, 226
Persian culture, see Iran
Philippine Rehabilitation Act, 155–56
Philippines, U.S. assistance to, 73
Physicians for Social Responsibility, 187
Physics Today, 99, 107
physics, and national security, 144; rejection of cooperation in, 145
Pickering, Andrew, 107
Pickstone, John, 269
Pilani, and BITS, 121
Plowshares project, 60
plutonium, 1, 60
Point Four Program, 148, 158–59
Poland, and Indian technology, 58
polar front theory, 234
Politburo, 200
political physics, 194
politicipation of science, 86–109
politics of R&D, Johnson and, 71
pollution, 18, 68, 72; see also radioactive waste
Pontifical Academy of Sciences, 193
popularization, and Atoms for Peace, 165
Portugal, 89; and IAEA, 27, 37–38; and Republic of Congo, 40; and U.S. reactor technology, 173–74
postcolonialism, 5, 9–13, 58, 62, 274
power grid, 275
power, and global science organizations, 14, 250; see cold war; empire; prestige; and passim
prediction, and control discourse, 49–51, 64
predisposition testing, 261
President's Science Advisory Council (PSAC), 66, 72
prestige, as foreign policy tool, 67; and nuclear power, 35–36, 164; and scientific internationalism, 180; and technology, 65; and U.S. space program, 72; see also identity; legitimacy
Price, Melvyn, 179
Priestley, Keith, 201, 204
privatization of knowledge, 256
ProdiGene, 288
professional regulation, 269–72; and see regulation
profit motive, and medical testing, 259
Progressive Era, and federal agencies, 134
Project Chariot, 70
Project GROMET, 78–81, 83–84
Project Plowshare, 67
Project POPEYE, 78–79
Project STORMFURY, 78–79
proliferation, nuclear, reasons for, 50
propaganda, and Atoms for Peace, 165; and nuclear testing, 163; test bans as, 189
property rights, and genes, 256; and GMOs, 264, 286; see intellectual property
protest, 277, 284
Provisional International Civil Aviation Organization, 151
psychological warfare, and Atoms for Peace, 162
Public Health Service, 138, 153
public opinion, see opinion, public
public relations, and NRDC, 205; and radioactive waste, 226
public, and cancer research, 261; and GMO assessment, 266–67, 269; and regulation, 252; and research, 255; see also fear; opinion
Pugwash movement, 186
punishment, and empire, 278
Punjab, and weather modification, 80

research and development (R&D), 2, 3; Egyptian, 37, 39; U.S., 71, 73
rabbits in Australia, 281
Rabi, Isidor I. , 92, 165, 225
radar, 2, 144
radiation, protection from, 175
radicalism, student, 117, 124–25
radioactive waste, dumping, 16, 209–28

radioactivity, and organisms, 227
radioisotopes, 165, 168–74
rainforest resources, 252
Rainger, Ronald, 214
Rainier, Prince, 223
Raman, C. V., 113
Ramirez, Rodriguez, 237
RAND Corporation, 245
reactor, and Atoms for Peace, 173; construction of, 165; and Geneva conference, 166, 175; and nuclearity, 33–34; sale of, 7, 60; South African demonstration, 35; water, 164
Reagan, Ronald, and Reagan administration, 84; and empire, 274; and nuclear test monitoring, 183, 187, 189, 192–94, 197–98, 203, 205–6
real-time meteorological data, 230, 234–35, 248
reconnaissance satellites, 4
regional powers, India and China as, 75
Regional Training Centre for the Arab Countries (Cairo), 36–38, 40
regulating, ways of, 269–72
regulation and GMOs, 251–72, 287
Renou, Jean, 226
Réseau Mondial, 230–35, 247
restrictiveness, of Atoms for Peace, 163
Revelle, Roger, 212, 217
revolutionaires, Islamic (Iran), 127
revulsion, against nuclear weapons, 59–60
Reykjavik summit, 206
Rhodesia, 89
Richter, Burton, 94
Rifkin, Jeremy, 264
right to know, and cancer testing, 258–59
risk assessment, 266, 283, 287–89
rivalry, cold war, and scientific prestige, 133, 164, 177
Rockefeller Foundation, 3, 289
rogue nation, defined, 25; status of, 47–48
Rolef, Susan, 91
Romanovsky, Vsevolod, 221
Romantic sensibility, and environmentalism, 281
Roosevelt, Franklin D., 136, 143, 145; and Latin American policy, 138
Rosenfeld, Stephen, 198
Rostow, Eugene V., 188
Rostow, Walt, 11, 79, 81–82, 112
Rotblat, Joseph, 186
Roundup Ready canola, 286
Roundup Ready wheat, 285
Royal Botanic Gardens, 281, 288
Royal Swedish Air Force, and meteorology, 244
Rubinstein, Arthur, 98
Rusk, Dean, 77
Russia, and IAEA, 31

Sabato, Ernesto, 92
Sadovsky, Mikhail A., 200
safeguards, and AEC, 146–47, 160; and Atoms for Peace, 164; and nuclearity, 41; and reactor, 181; technopolitics of, 29
Sagan, Carl, 194
Sagan, Scott, 50
Sagdeev, Roald, 187–88, 193, 206
Salam, Abdus, and ICTP, 9, 87–109
Salomon, Jean-Jacques, 13
Sandia, 7
Sarkar, N. R., 113
SARS, 1
Sartre, Jean-Paul, 92
Sasaki, Tadayoshi, 213
satellites, and meteorology, 242, 244–45, 248
scapegoat, ICTP as, 102
Scheinman, Lawrence, 47
Scherr, S. Jacob, 198
Schmeiser, Percy, 286
scholarly interests, and national policy, conflation of, 51
Schrag, Philip G., 203
Schwartz, Thomas Alan, 83
Science and Foreign Relations, see Berkner Report
Science and Global Security, 194
Science Office (U.S. Department of State), 79
Science, 289
science, and foreign affairs, 3, 150, 209–28; militarization of, 13; see diplomacy; collaboration; and passim
Science, Technology, and American Foreign Policy, 18
scientific authority, in international arena, 5, 209–28
scientific internationalism, and Atoms for Peace, 161–81; at Geneva conference, 177–78; and isotopes, 171–72; and meteorology, 234; and security, 171–72;
Scott, James C., 20, 137, 279–80, 287, 291
Scripps Institute of Oceanography, 199, 203–4, 212, 215, 217
Seattle, and WTO protests, 284
Second Assessment Report, 249
secrecy, and atomic physics, 150; and bomb, 146; at Geneva conference, 166–67; and nuclearity, 29; and physics research, 144; and Soviet Union, 178–80; see also security
Secretary's Advisory Committee on Genetic Testing (SACGT), 261
security agreement, U.S. and Latin America, 138
security, and glasnost, 203; and transparency, 171–72
Security Council, 150
security, and atomic energy debates, 56, 63, 144, 165–66; and atomic weapons, 1, 145, 160; and geophysical sciences, 67, 69; and national autonomy, 13–15, 18, 20; and NGOs, 19; and prestige, 75; and radioisotopes, 169–71; and postwar science, 134, 148; and scientific internationalism, 135; and weather modification, 82; see also secrecy
seeds of GMOs, 267–68
seismic verification, see verification
seismology, 21, 230
Seitz, Fred, 179
Semipalatinsk test site, see Kazakhstan
Senate Foreign Relations Committee, and nuclear test ban treaties, 188–89
Senegal, and UN, 90
September 11 attacks, 275
sexual relations, 280, 282

shah of Iran, 11, 123–28
Sharif University of Technology, and AMUT, 128–30
Shaw, G. Howland, 141
Shevardnadze, Eduard, 185
Shippingport, and water reactor, 164
Shiraz Technical Institute, 126
Sidey, Hugh, 72
Silent Spring, 72
Silicon Valley, and IIT graduates, 129
Silone, Ignazio, 92
Sivin, Nathan, 127
Skolnikoff, Eugene, 18
Slater, A. H. K., 218
Sloan School of Management, and Iran, 126
Smith-Mundt Act, 157
Smithson, James, 149–50
Smithsonian Institution, 150
Smullin, Louis, 121
social capital, in international arena, 5
social organization, and culture, 288; and regulation, 269
Sole, Donald, 30–46, 48
source materials and nuclearity, 27–30, 44
South Africa, 65, 89–90; and dismantling of nuclear program, 48, 52; expulsion from IAEA, 26; and Marxism, 12; national identity of and nuclearity, 8, 25–48; and origins of IAEA, 27; and uranium, 30; and U.S. reactor technology, 173–74
South African Nuclear Energy Corporation (NECSA), 48
Southeast Asia Treaty Organization (SEATO), 101
sovereignty, Afrikaner, 42; and World Weather Watch, 249
Soviet Academy of Sciences (SAS), 182–206; and NGOs, 19; and see NRDC
Soviet engineering, and India, 116; and Iran, 123
Soviet foreign policy, and scientists, 195–96
Soviet reactor design, 34
Soviet Space Research Institute, 187
Soviet Union, and Atoms for Peace, 174; and awareness of Manhattan Project, 58–59; and Baruch Plan, 147; and bomb, 144–45; and Chinese program, 57; and cold war, 13–15, 161; collapse of, 6; and computerized weather forecasting, 247; and France, 254; and Geneva Conference, 166–67, 176–77; and ideology, 277; and IGY, 220; and IIT Kharagpur, 114; and Indian technology, 58; isolation of scientists in, 178–79; and Johnson, 73; and nuclear power, 6, 61, 159; and nuclear testing, 174; and nuclear science, 56; and radioactive waste, 213, 225; and reactors, 173; and South Africa, 27; and VAP, 241
Soviet-American Disarmament Study Group, 186
soy, genetically modified, 265
space program, U.S., 67, 70; and national defense, 72
Spain, and U.S. reactor technology, 173–74
Sputnik, 66–67, 69, 72, 178, 246, 254
St. Amand, Pierre, 78, 84
Stages of Economic Growth, 112
Stalin, 15, 178
standardization, 242; and biology, 282; and computers, 247–49; and empire, 278; and legibility, 286; and meteorology, 231–35; and WMO, 238, 240
Stanford Linear Accelerator Center (SLAC), 94–96, 98, 100, 102, 107
Star Wars, 274
Star Wars, see Strategic Defense Initiative
State Department, see Department of State
state, and biopower, 282; national security, 13–14; and regulation, 270; sovereign, empire as, 277
statistics, 2, 136, 138
status quo, and ICTP, 101
steilization, and standardization, 282
stem cell research, 252
Stern, Isaac, 98
Stettinius, Edward, 142–43
Stimson, 145
Stoler, Ann, 280, 282
Stone, Jeremy, 187, 192
Strassman, Fritz, 56
Strategic Defense initiative (SDI), 84, 189, 193–94
strategic value of sciences, 67
Stratton, Julius, 114, 120
Strauss, Lewis, 165, 169
Subcommittee on International Organizations (U.S.), and UN, 154–55
submarine, nuclear (U.S.), 164
Subramaniam, Chidambaram, 77
Sudan, decolonization of, 27
sugar cane, and empire, 281
Sunday Times (London), 219
superpower rivalry, 2, 5, 133, 185; and see cold war; rivalry
supranational conglomerates, 17
Supreme Court of Canada, 286
Supreme Court, U.S., 217, 282
surveillance, and empire, 278
Swarup, Govind, 59
Sweden, and computerized weather forecasting, 247; and monitoring, 196; and nuclear program, 57, 65; and VAP, 241
Swedish International Development Cooperation Agency, 102–3
Switzerland, and boycott of UNESCO, 106; and Geneva conference, 174; and GMOs, 286; and IMO, 232; and UN, 91
Sykes, Lynn, 196
symbol, nuclearity as, 10; and see modernity

Taiwan, and ICTP boycott, 98
Tanzania, and monitoring, 196
Tass, 213
Tata Iron and Steel Company, 113
technical assistance, U.S., 73, 133–60; and WMO, 240–41
Technical Commission for the Application of Meteorology to Aerial Navigation, 233
Technical Cooperation Administration, 158
Technological Decisions and Democracy, 227
Technological Institute of Textiles, 119
technological sublime, 61

technological systems, as forms of life, 288
technology transfer, global, 288
technology, conventional, and empire, 275
Teisserenc de Bort, Léon, 232–33
telegraph, and meteorology, 231–33
teletype, and meteorology, 234–35
Television and InfraRed Observation Satellite (TIROS), 245
Teller, Edward, 82, 84–85, 92, 178
Tennessee Valley Authority, 61
Terminator, 267, 284
testing, atomic, 49, 51, 54, 70, 163; medical, 258–59
Texas, Johnson in, 71
Thacker, M. S., 114
third world, 8; and ICTP, 9, 86–109; technology for, 155–58
thorium, availability of, 173
Thresher, 74
Threshold Test Ban Treaty (TTBT), 188–89, 201, 204–5
Time Incorporated, 162
Ting, Samuel, 94
transnational actors, defined, 184
transnational shipment of GMOs, 265
transparency of science programs, 15–16, 171; and Atoms for Peace, 164
treaty, nuclear test ban, 68, 70; and see testing
Tregub, Felix, 204
Trieste, see Italy
Truman, Harry and Truman administration, 134, 137, 143; and AEC, 168; and national science policy, 112; and Point Four Program, 148; and security, 170; and segregation, 217; and Smith-Mundt Act, 157–58
Truman-Attlee-King declaration, 146
trust, and international collaboration, 185, 205
truth as universal, 3
tuberculosis, 1
Tunisia, and IAEA, 40; decolonization of, 27
Turkey, and nuclearity, 36; and U.S. reactor technology, 173–74

United States, and boycott of UNESCO, 106; and computerized weather forecasting, 247; and Geneva conference, 174; and GMOs, 19; and IAEA, 27, 31; and ICTP, 87; and IMO, 232; and Indian technology, 58; and international patent law, 258; and radioactive waste, 209–28; and safeguards, 44–45; and South Africa, 47–48; and VAP, 241; and passim
U.S. Agency for International Development (USAID), 76, 82, 115, 117, 137
U.S. Army Signal Office, 232
U.S. Geological Survey (USGS), and nuclear tests, 191
U.S. News and World Report, 164
U.S. Weather Bureau, 243, 245
Ukraine, and nuclear program, 65
United Nations (UN), charter of, 156; and IAEA, 27, 86–109; at end of World War II, 143; reorganization of, 134; science at, 151–55; and passim
Unabomber, 283
Unarmed Forces: The Transnational Movement to End the Cold War, 185
UNESCO, 8–9; and Berkner Report, 167; and ICTP, 86–109; and IITs, 114; politicization of, 86–109; and role of science, 150; Statement on, 97; and U.S. policy, 155; see also UN, ICTP
Union Carbide and Carbon Corporation, and AEC, 175
United Arab Republic (UAR), and nuclearity, 36–40
United Kingdom, see Britain
United Nations Development Program (UNDP), 102
United Nations Scientific Committee on the Effects of Atomic Radiation (UNSCEAR), and radioactive waste, 212–13
uranium, and Atoms for Peace, 162; availability of, 6–8, 27, 29–30, 57, 173; enrichment of, 1; and IAEA, 7; market, 34; and nuclear industry, 33; and Republic of Congo, 40; and safeguards, 42–43, 45; and South Africa, 27, 32, 34, 45;
Urenco, 57
Uttar Pradesh, 78, 80

Van Hove, Leon, 98–99, 103, 108
Veksler, Vladimir, 166, 177–79
Velikhov, Yevgeny, 182–90, 192–97, 200, 202–3, 205–6
verification, seismic, 183, 188–90, 192, 196–98, 201–2, 204–5
Verwoerd, Hendrik, 40
veto, UN, 147
Victoria, Queen, 279
Vietnam War, and World Meteorological Congress, 237
Vietnam, weather modification in, 66, 73, 78, 83–84
violence, and colonialism, 281, 277
viruses, 1, 254
Vityaz, 213, 220
voluntarism, and globalism, 247
Voluntary Assistance Program (VAP), 240–42
von Hippel, Frank, 187, 192, 194, 196–97, 205–6
von Neumann, John, 243

Wade, Nicholas, 198
Wall Street Journal, 191
Wang, Jessica, 101
waste, nuclear, disposal of, 69–70, 209–28; and see oceanography
Water for Peace, 73–74
water policy, international, 73
Watergate, 188
waters, international, 68–69
wealth, distribution of, and empire, 275
weapons, 2, 6, 27, 179, 242–44, 275; and see bomb; military; missile; weather
Weart, Spencer, 170
Weather Bureau, 153
weather modification, 11, 66, 68, 70–71, 75–83, 84–85, 243
weather prediction, global, 20

weather weapon, see weather modification
Weber, Max, and standardization, 278–79, 290
Weisskopf, Victor, 98–99, 106–7, 193
Weizenbaum, Joseph, 126–27
Weizmann Institute, and ICTP, 94, 102
Welles, Sumner, 138, 141
Wentworth Institute of Technology, 126
West Germany, see Germany
West Indies, 281
Westinghouse, and civilian nuclear power, 164, 181
Westphalian internationalist model, and IMO, 237
wet space program, see oceanography
White Revolution, Iranian, 124
White, David, 121–22, 125
Whitehead, John C., 192–93, 198
WHO, 134, 150–53
Wicker, Tom, 198
Wiedemann, Howard, 77
Wiesner, Jerome, 125
Wigner, Eugene, 92
Wolf, André, 92
women's health movement, 260–61
women's rights, and NGOs, 18
World Bank, 1, 134
world data centers (meteorology), 246–47
World Health Organization (WHO), 33, 89
World Meteorological Congresses, 237–38, 240, 247
World Meteorological Convention, 152, 235–36
World Meteorological Organization (WMO), 74, 134, 153, 229–50; founding of, 235–242; as UN system, 238; and World Weather Watch, 239–42
World Trade Organization, 252; and GMOs, 286–87; and intellectual property, 257–58; protests against, 284
World War I, 88, 233
World War II, 2, 20, 49, 64, 88, 113, 133, 232–233, 238
World Weather Watch (WWW), 74, 229–50
World Wide Web, 230, 240
Wu, Tai Tsun, 100–101, 107

Xue Litai, 57

Yasser Arafat, 9
Yom Kippur/Ramadan War, 89, 91
Yukhnin, Nicolai, 204

Zaikov, Lev, 200
Zaire, as source of uranium, 27
Zarghamee, Mehdi, 123, 125, 128
Zenkevich, Lev, 213
Zhurkin, Vitalii, 192
Zia, president of Pakistan, 12
Ziman, John, 94, 97–99
Zinn, Walter, 179
Zionism, 90, 96
Zuckerman, Solly, 227
Zuoyue Wang, 72
Zwemer, Raymund L., 142–43, 149–51

SUGGESTIONS FOR CONTRIBUTORS TO OSIRIS

OSIRIS is devoted to thematic issues, conceived and compiled by guest editors who submit volume proposals for review by the OSIRIS Editorial Board in advance of the annual meeting of the History of Science Society in November. For information on proposal submission, please write to the Editor at Osiris@georgetown.edu.

1. Manuscripts should be submitted electronically in Rich Text Format using Times New Roman font, 12 point, and double-spaced throughout, including quotations and notes. Notes should be in the form of footnotes, also in 12 point and double-spaced. The manuscript style should follow *The Chicago Manual of Style*, 15th ed.

2. Bibliographic information should be given in the footnotes (not parenthetically in the text), numbered using Arabic numerals. The footnote number should appear as superscript. "Pp." and "p." are not used for page references.

a. References to books should include the author's full name; complete title of book in *italics*; place of publication; date of publication, including the original date when a reprint is being cited. *Example*:

[1] Mary Lindemann, *Medicine and Society in Early Modern Europe* (Cambridge, 1999), 119.

b. References to articles in periodicals or edited volumes should include the author's name, title of article in quotes, title of periodical or volume in *italics*; volume number in Arabic numerals; year in parentheses; page numbers of article; and, if required, number of the particular page cited. Journal titles are spelled out in full on the first citation and then abbreviated subsequently according to the journal abbreviations listed in *Isis Current Bibliography*. *Example*:

[2] Lynn K. Nyhart, "Civic and Economic Zoology in Nineteenth-Century Germany: The 'Living Communities' of Karl Möbius," *Isis* 89 (1999): 605–30, on 611.

c. Journal articles are given in full in the first reference. For succeeding citations, use an abbreviated version of the title with the author's last name. *Example*:

[3] Nyhart, "Civic and Economic Zoology" (cit. n. 2), 612.

3. Special characters and mathematical and scientific symbols should be entered electronically.

4. A small number of illustrations, including graphs and tables, may be used in each volume. Hard copies should accompany electronic images. Images must meet the specifications of The University of Chicago Press "Artwork General Guidelines" available from the Editor.

5. Manuscripts are submitted to OSIRIS with the understanding that upon publication copyright will be transferred to the History of Science Society. That understanding precludes consideration of material that has been previously published or submitted or accepted for publication elsewhere, in whole or in part. OSIRIS is a journal of first publication.

OSIRIS (SSN 0369-7827) is published once a year.

Subscriptions are $50.50 (hardcover) and $33.00 (paperback).

Address subscriptions, single issue orders, claims for missing issues, and advertising inquiries to *Osiris*, The University of Chicago Press, Journals Division, P.O. Box 37005, Chicago, Illinois 60637.

Postmaster: Send address changes to *Osiris*, The University of Chicago Press, Journals Division, P.O. Box 37005, Chicago, Illinois 60637.

OSIRIS is indexed in major scientific and historical indexing services, including *Biological Abstracts*, *Current Contexts*, *Historical Abstracts*, and *America: History and Life*.

Paperback edition, ISBN 0-226-45404-5

Osiris

A RESEARCH JOURNAL DEVOTED
TO THE HISTORY OF SCIENCE
AND ITS CULTURAL INFLUENCES

A PUBLICATION OF THE
HISTORY OF SCIENCE SOCIETY

EDITORIAL OFFICE
BMW CENTER FOR GERMAN & EUROPEAN STUDIES
SUITE 501 ICC
GEORGETOWN UNIVERSITY
WASHINGTON, D.C. 20057-1022 USA
osiris@georgetown.edu